T0190127

Statistical Distributions in Engineering

Engineers face numerous uncertainties in the design and development of products and processes. To deal with the uncertainties inherent in measured information, they make use of a variety of statistical techniques.

This book presents single-variable statistical distributions that are useful in engineering design and analysis. It lists significant properties of these distributions and describes methods for estimating parameters and their standard errors, constructing confidence intervals, testing hypotheses, and plotting data. Each distribution is worked through typical applications. Figures are used extensively to clarify concepts. Methods are illustrated by numerous fully worked examples in the form of Mathcad documents that readers can use as templates for their own data, eliminating the need for programming. Intended as both a text and reference, the book assumes an elementary knowledge of calculus and probability.

Graduate and advanced undergraduate students, as well as practicing engineers and scientists, will be able to use this book to solve practical problems connected with uncertainty assessment in a wide range of engineering contexts.

Dr. Bury is Professor of Mechanical Engineering at the University of British Columbia.

Statistical Distributions in Engineering

KARL BURY

CAMBRIDGE UNIVERSITY PRESS
Cambridge, New York, Melbourne, Madrid, Cape Town, Singapore, São Paulo

Cambridge University Press
The Edinburgh Building, Cambridge CB2 8RU, UK

Published in the United States of America by Cambridge University Press, New York

www.cambridge.org
Information on this title: www.cambridge.org/9780521632324

First published 1999

A catalogue record for this publication is available from the British Library

Library of Congress Cataloguing in Publication data

Bury, Karl V., 1935–
Statistical distributions in engineering / Karl Bury.
p. cm.
ISBN 0-521-63232-3 (hardcover). – ISBN 0-521-63506-3 (pbk.)
1. Engineering–Statistical methods. 2. Mathematical statistics.
I. Title.
TA340. B87 1998
620′.0072–dc21 98-22941
CIP

ISBN 978-0-521-63232-4 hardback
ISBN 978-0-521-63506-6 paperback

Transferred to digital printing 2008

Contents

Preface

This text presents several single-variable statistical distributions that engineers and scientists use to describe the uncertainty and variation inherent in measured information. It lists significant properties of these distributions and describes methods for estimating their parameters, constructing confidence intervals, and testing hypotheses. Each distribution is illustrated by working through typical applications including some of the special methods associated with them. The intention is to provide the professional with a ready source of information on a useful range of distribution models and the techniques of analysis specific to each.

The need to deal rationally with the uncertainties that enter engineering analysis and design appears now well recognized by the engineering profession. This need is driven, on the one hand, by the competitive pressure to optimize designs and, on the other hand, by market demand for reliable products. Hence, engineers design their products closer to the limits of the materials used, while improving product durability. The result of these opposing pressures is that the engineer needs to replace traditional "contingency factors" by careful uncertainty analysis.

What is perhaps less well understood by the professional is the need to choose a distribution model that closely represents the *entire range* of measured values. This need arises from the *skewness* typical of the frequency functions of engineering data, coupled with the usual focus of engineering decisions on the location of *distribution tails*. Hence, statistical analysis that is adequate for predicting *averages* of engineering variables is often not adequate for producing defensible conclusions near *data extremes*. The need for more sophisticated data analysis is further driven by the usual small size of engineering samples. This text presents useful modern methods of constructing defensible statistical conclusions to support engineering decisions.

In this connection, the view is taken that, to act responsibly, the professional must link every estimated quantity with a measure of its uncertainty. The text discusses and illustrates the various methods for doing so.

A significant development over the past few years is the computational environment that is now accessible with only a modest desktop computer. Complex calculations that were practically infeasible are now performed easily *without the need to program*. Accordingly, many of the methods presented in this text require a modern computational aid. Most of the commercially available aids can implement the methods of this text. However, this text presents all worked examples as Mathcad documents, which the reader may use as Mathcad templates for his or her own data. The reason for choosing Mathcad is its *transparency*: All equations, formulas, constants, etc. that enter a calculation must show on the screen. Thus, with text and graphics added, the printout of a completed Mathcad session is in essence an engineering report. It tells the reader not only the results but also precisely the model, and other information, used in the analysis.

This text assumes that the reader is familiar with elementary calculus and has taken the usual introductory "Probability and Statistics" course. Thus, the text does not develop the subject afresh. However, Part One on Statistical Background briefly surveys those concepts that directly support the chapters that follow. Each chapter in Parts Two and Three stands on its own and provides the information for working with the distribution discussed. Occasional references to Part One link to more general concepts. Thus, to use a particular distribution does not require the reader to work through the entire text, although one might find perusing Part One useful for recalling some basic statistics and broadening one's perspective. For readers who wish to dig deeper, a most readable general reference for the concepts surveyed in Part One is

M. G. Kendall and A. Stuart, *The Advanced Theory of Statistics*, Volume I (1967) and II (1969), Griffin, London.

Additional material and extensive source references to the distributions discussed in Parts Two and Three can be found in

N. L. Johnson and S. Kotz, *Discrete Distributions* (1969) and *Continuous Distributions* (1970), Volumes I and II, Houghton Mifflin, Boston.

Throughout the text the "engineer" is referred to by a masculine pronoun, which is intended in the generic sense only. No slight to the many accomplished female engineers and scientists is implied.

PART ONE

STATISTICAL BACKGROUND

CHAPTER ONE

Introduction

UNCERTAINTY IN THE ENGINEERING CONTEXT

1.1 Uncertainty

The engineering function is to design, produce, test, and service structures, devices, materials, and processes that meet a market need, reliably and at a competitive cost. Much of the work involved with that function deals with *mathematical models* of the engineer's designs and the physical phenomena encountered by them. These models efficiently aid the engineer in the design of his artifacts and in the prediction of their behavior.

Consider, for example, the mathematical model for the deflection of the unsupported end of a beam whose other end is rigidly fixed (such a beam is called a "cantilever"). For a beam of length L and cross-sectional moment of inertia I, the deflection D of the free end is

$$D = \frac{WL^3}{3EI},$$

where W is a point load on the free end of the beam and E is a material property called the *modulus of elasticity*. One design problem is to choose the shape and dimensions of the beam cross section so that the deflection is limited to a specified value D.

To use these models, the engineer requires *information* on the constants, parameters, and functional variables that enter the model. The seasoned engineer will be aware of the *uncertainties* that often come with these values. He will therefore know that his designs and predictions will also be associated with uncertainties.

In the above example, the engineer will need to know the magnitude of the load W. Suppose that his design is to support an overhead conveyor track for transporting manufactured parts to an assembly operation. He will notice that W varies from instant to instant. He therefore needs to choose a design value W_D so that his design meets the specified deflection limit D. Similarly, he may realize that the E-value *varies* among material test specimens. He therefore needs to choose a design value E_D so that his design is likely to perform adequately.

How an engineer deals with the uncertainties he faces depends on their relative magnitudes and their likely consequences. If he expects that only small variations of a parameter may occur around a large central value, or if the effect of the expected variation on the performance of the device is small, he may ignore the uncertainty altogether and assume a suitable constant value. If, however, the expected variation is large, or its effect may produce a failure with serious consequences, then the uncertainty is *significant*, and he must deal with it explicitly.

Continuing with the preceding example, the engineer may realize that the E-value for his material is likely to depart from its average value by no more than a few percent. He may therefore declare it to be constant for his purposes and work with its average value or, to be conservative, some low percentile value. However, because the loads W that are likely to occur may vary across an order of magnitude for different parts carried to assembly, he needs to be careful about his choice of design load W_D.

How the engineer accommodates significant uncertainty in his calculations depends on the situation. He may simply take the most detrimental value that he considers possible and multiply it by a "safety factor" to arrive at a deterministic design value. The result is typically a costly overdesign. Furthermore, it is not possible to assess the *risk of failure* for such a design when the extreme cannot be accurately predicted.

In the above example, it may be perfectly acceptable to take the weight of the heaviest part to be transported by the conveyer track and multiply it by the factor of, say, two to obtain the design load W_D. The increased cost for a few such cantilevers is unlikely to be significant. However, if the design is to be mass produced, the increased cost *will* be significant, so that an arbitrary safety factor may not be acceptable. Similarly, if the maximum load W for a range of design applications cannot be predicted, and only sample values of possible loads are available, the likelihood of adequate performance of the design cannot be assessed under this deterministic approach.

The modern climate of competitiveness demands the efficient use of materials, as well as more reliable designs. The former requirement pushes the design closer to the possibility of failure, whereas the latter demand obviously pushes in the opposite direction. To deal with this dilemma in a professional manner, the engineer must treat the significant uncertainties of his information base. The *rational* approach to dealing with such uncertainties is to construct a *mathematical model* that describes the uncertainty aspect of engineering information, similar to representing a physical phenomenon. The *measure* of this uncertainty aspect of engineering quantities is *probability*. This text deals specifically with some useful ways to model the uncertainty and variation associated with the quantitative information encountered in a wide variety of engineering contexts.

1.2 Sources of Uncertainty

The question might be asked at this point: "If engineering information is often uncertain, where does that uncertainty come from?" The *sources* of uncertainty can be broadly classified as follows.

1. data uncertainty,
2. statistical uncertainty,
3. event uncertainty, and
4. model uncertainty.

Data Uncertainty

The majority of the quantities the engineer measures or observes feature *inherent variability*. That is, the measured value is caused, or influenced, by many chance factors whose effects aggregate to produce a measured value. No matter how carefully one measures such a quantity, variability among measured values is an inherent reality, so that the actual value of a future measurement is uncertain.

Examples of inherently variable quantities are: the yield of a chemical batch process, the specific strength of an engineering material, the gust load on an aircraft wing, the time-to-failure of equipment, the cost of an engineering project, the duration of an assembly task, the throughput time of a production order, the propagation rate of a combustion flame front, and the number of stress reversals to the fatigue failure of a metal specimen.

Inherent variability is a source of uncertainty the engineer encounters commonly. However, even when the engineer measures a quantity that is inherently constant, such as the distance between two survey stations, he will find that his observations vary at the limit of precision of his measuring instruments. That is, the process of measurement itself often introduces uncertainty, regardless of whether the measured quantity is inherently constant or variable. This *measurement uncertainty* may show up as instrument *bias* where the true value differs from its measured value by some consistent amount. *Calibration* of the instrument by a standard of sufficient accuracy reduces this uncertainty. Measurement uncertainty also appears in the form of *random* differences between true and measured values, in much the same way that inherent variability occurs. Using instruments of sufficient *precision* and carefully designing the instrumentation system render this uncertainty insignificant compared to the precision required in the measurement.

Statistical Uncertainty

An important source of uncertainty is the limited amount of information that is typically available on a measurable quantity. That is, the quantity of interest may be measurable an unrestricted number of times, but time and budget constraints permit only few observations. Clearly, the more observations are on hand, the more information on that quantity is available. Thus, *limited* information implies uncertainty about the true nature of the quantity.

For example, if only five prototypes of a newly developed device are available for performance testing, the project engineer could only form a rather approximate impression of the design's performance. A characteristic such as average performance would then be highly uncertain. Other important characteristics, such as 5-percentile performance, could not even be expressed sensibly. If, however, one had fifty prototypes to test, one could draw conclusions on design performance with more assurance that they are close to true values.

An assessment of statistical uncertainty is made in the form of *standard errors*, or related measures, attached to the predictions of the quantity in question. Error statements, and the like, are important means of communicating the presence and magnitude of uncertainty to the decision maker who can then evaluate the *risk* associated with his decision. In this text the methods for constructing these uncertainty measures are presented and illustrated.

Event Uncertainty

In specifying his design, the engineer will have to guard against the effects of unfavorable events that may occur during the design's mission. That is, he is concerned with the *occurrence* of events. The events of design significance are usually of the kind that happen only rarely, so that typically there is little information available on their likely occurrence. The resulting design decision is often highly uncertain.

For example, the engineer would consider the possibility of a major meteorite impact in the design of a spacecraft structure. He would consider the chance of a high wind load occurring during the construction phase of a suspension bridge. In the design of a communications tower he would consider the possibility of a high load induced by a major earthquake.

Model Uncertainty

The mathematical models the engineer uses in his work typically represent only one, or a few, of the important features of the physical phenomenon in question. That is, a model represents a restricted version of reality. Furthermore, the model's description of a real problem is often an idealization. The model therefore deviates from reality. When the model's "lack of fit" to reality significantly affects the conclusions drawn from it, these conclusions are in error. To reduce this type of uncertainty, one needs to construct more realistic models.

For example, material *failure* mechanisms involve complex interactions of many contributing causes. Mathematical models of failure only consider one or two of the important factors and relate these to failure events by simple algebraic relations. Predictions of material behavior based on these models often differ widely from what is observed. In contrast, although many strain models are

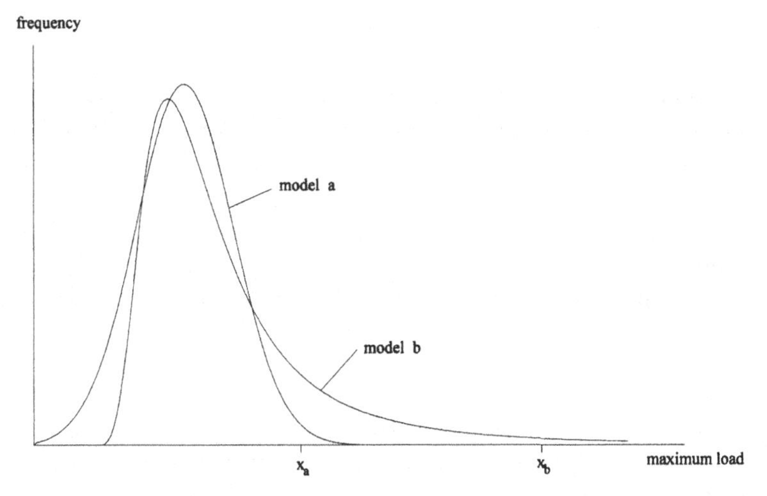

Figure 1.1. Two uncertainty models of a load variable.

linear idealizations of nonlinear reality, their predictions are quite accurate, at least for small strains.

The problem of model-fit, when describing physical phenomena, also appears when a *probability model* is chosen to describe the uncertainty aspect of engineering information. Again, when the model does not accord with reality, the conclusions from the model are in error.

For example, suppose that an engineer has on hand a good set of observations on the maximum of a load to which his equipment design will be exposed during its mission. Suppose further that he needs to know the 99-percentile maximum load as an input to his design process. Figure 1.1 shows two probability models representing the maximum load variable. The two models have about the same average and spread, in line with the data, and both are acceptable as fair representations of the bulk of the data. However, the estimate x of the 99-percentile load differs by a factor of about *two* between the models. Thus, the question of which model provides a better description of this load phenomenon clearly needs further and careful attention.

This text alerts the practicing engineer to a variety of useful models available to him, and it describes the practical situations in which each model is appropriate.

1.3 Population versus Sample

It is useful to explore the notion of *statistical uncertainty* further to make the important distinction between a *population* of individuals and a *sample* of individuals from that population. When all individuals of a population are known (i.e., when all possible measurements on a variable are on hand), one has *complete information* on that population (i.e., the true nature of the measured phenomenon is known). In practice, one cannot usually know all individuals in a population. Instead, only a small selection of individuals (a sample) is known. The sample thus represents *incomplete information* on the population: It is all the quantitative information one has. However, one naturally wishes to form conclusions about the population, given the incomplete information of the sample. Conclusions, then, pertain to populations (i.e., a measurement phenomenon in general), whereas the available information base only comprises the knowledge residing in the sample (i.e., the limited set of actual observations on hand). Thus, sample information is fully known, but conclusions about the population are necessarily shrouded in uncertainty.

For example, the engineer may have a sample of twenty-five material test specimens, loaded to failure. There is nothing uncertain about the sample itself, barring experimental error. However, with respect to the strength property of this type of material, the sample information is incomplete. Nevertheless, given this incomplete information, the engineer wishes to draw conclusions on the strength of this material in general.

1.4 Statistics versus Probability

The specific measurements in a sample are the data that comprise the engineer's quantitative information base. Given that information, he needs to construct a probability model that describes the variability of the population as accurately as possible. The process of constructing such a model on the basis of data is termed "statistical inference." The main inferential procedures are point estimation, confidence interval estimation, and hypothesis testing (see Chapter 3). These procedures are *inductive*, as they proceed from the specific (sample) to the general (model), and belong to the subject of *statistics*. They answer questions about the population.

Continuing with the preceding example, the engineer may need to establish, on the basis of the twenty-five test results, whether this material is substantially stronger than a specified value. He would do this by means of a hypothesis test.

Once a population model is inferred (or assumed), it forms the basis for answering questions about a *sample*. The procedures used are essentially the rules of *probability*. These rules are *deductive* as they proceed from model to experimental data.

In the above example, suppose the engineer has inferred a probability model of material strength from the data. Given that model, he could deduce the probability that a further test specimen breaks below a specified strength value.

Figure 1.2 shows the distinctions involved. This text focuses on the *statistics* of inferring, from measurement data, the specifics of a distribution, chosen from a number of models that are prominent in engineering and the sciences.

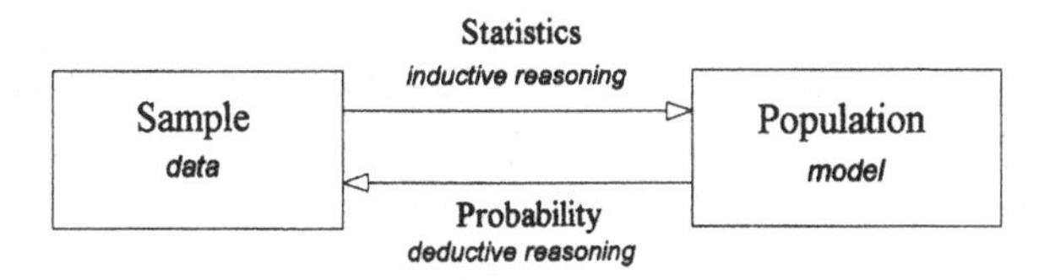

Figure 1.2. The relation between sample and population.

DATA

1.5 Sample Measures

An *experiment* in which a variable of interest X is measured or observed n times produces the data x_i that the engineer works with. Such an experiment may be set up in a laboratory specifically to produce the required data, or it may take place in the field and consist of gathering data as they happen.

For example, a fatigue experiment in a materials testing lab measures the number of stress reversals to failure for several specimens. These measured stress reversals are the *data*. Keeping records of operating times between failures of transmissions for a fleet of trucks produces information on the *life* of these transmissions. These times-to-failure are the data.

The collection of *raw* data in a *sample*,[1] denoted by $\{x_i\}_n$, is by itself not particularly informative. To bring out salient features of the sample, the data are processed to generate *descriptive statistics*, usually in the following sequence:

1. The data are *ordered* according to increasing magnitude, resulting in a rearranged set $\{x_{(i)}\}_n$. Thus, $x_{(1)}$ is the smallest observation in the sample, $x_{(2)}$ is the next-larger observation, and so on. These rearranged values are called "order statistics." The *range* of the data, given by $[x_{(n)} - x_{(1)}]$, tells how widely dispersed the data are.
2. Some useful *sample measures* are calculated, such as the
   - sample mean: $\bar{x} = \frac{1}{n} \sum_{i=1}^{n} x_i$,
   - sample mode: x_m = most frequent measurement,
   - sample median: $\tilde{x}$ = middle-most ordered measurement,
   - sample variance: $s^2 = \frac{1}{n-1} \sum_{i=1}^{n} (x_i - \bar{x})^2$,
   - sample standard deviation: $s = \sqrt{\text{sample variance}}$,
   - sample coefficient of variation: $scv = s/\bar{x}$.

   These measures summarize the sample information by describing (a) where the bulk of the data are located on the measurement axis (i.e., $\bar{x}$, x_m, $\tilde{x}$) and (b) how dispersed the sample is (i.e., s (in units of X); scv (dimensionless).

[1] The statistical procedures of this text assume that the sample is "random." This means that the data are drawn independently from a common population; see also Section 2.2.

3. If the sample size n is sufficiently large (at least 25), it is instructive to *group* the data into (usually equal) measurement intervals, so that a sample *frequency distribution* (called a *histogram*) can be displayed in a table or on a graph. That is, if the data range is split into k intervals Δx_j, the frequencies

   $$q_j = \text{the number of observations } x \text{ in } \Delta x_j$$

   can be obtained. The *relative* frequencies

   $$f_j = \frac{q_j}{n}$$

   describe in some detail the variability among the data $\{x_i\}_n$. The *cumulative relative* frequencies

   $$F_j = \sum_{i=1}^{j} f_i$$

   indicate the *distribution* of the likelihood of the values x_i over the data range. The graphical display of f_j provides an informative picture of the variation among the data x_i. See Example 1.1 for a Mathcad document where a sample of 29 measurements is processed as described above.

MODELS

1.6 Distribution Function

When one cannot know in advance the values of repeated measurements on a quantity of interest, it is practical to describe that quantity as a *random variable*, denoted by a capital letter such as X. This random variable refers to the *population*. The collection of all possible values in the population is called the *sample space* S. The *realizations* of X, that is, measurements on X, are denoted by lowercase letters such as x. A set of these realizations is the *sample*. Thus, a sample is a subset of values in the larger sample space S.

The *uncertainty* aspect of the random variable X is modeled by a statistical distribution $F(x; \theta)$. Here F is a mathematical function of the values x that the variable X can take in its sample space S. The model F is indexed by parameters θ. Thus, $F(x; \theta)$ comprises a *family* of distributions, indexed by the values θ can take in its *parameter space* Ω. See Chapters 4 and 9 for a little more detail on the nature of the distribution function F.

For example, suppose an engineer is interested in a random variable X that represents the yield of a chemical process under specified process conditions. Observations x on process yield are realizations, and a set of these constitutes a sample.

Suppose the engineer postulates that a Normal distribution (see Chapter 10) adequately models the variability of process yield. This *population model* has the mathematical form

$$F(x; \mu, \sigma) = \frac{1}{\sigma\sqrt{2\pi}} \int_{-\infty}^{x} \exp\left\{-\frac{1}{2}\left(\frac{t-\mu}{\sigma}\right)^2\right\} dt.$$

As it stands, this function is a *family* of distributions, since it is indexed by two parameters μ and σ. To model his specific variable X, the engineer would have to "pick" a particular *member* of this family by specifying values for μ and σ. The inference procedure of doing so is "point estimation." Given estimated values for μ and σ, the above expression becomes a mathematical function of x only. This function models the variability of the process yield X.

As *descriptive statistics* summarize the information available in a sample, so there are *properties* of the model $F(x; \theta)$ that succinctly describe it. To introduce some of these properties, we must distinguish between *discrete* and *continuous* random variables. The former term refers to variables X that can take on only discrete values x, that is, at most a finite, or countably infinite, number of different values. Such variables are often associated with *counting* processes so that x is integer valued; see also Chapter 4. The latter term, continuous, refers to variables X that can, at least in principle, take on any value x in a continuum, or interval, of points. See also Chapter 9 for additional discussion of continuous distribution functions.

For both types of random variables the function $F(x; \theta)$ is the basic probability model. It is termed the *cumulative distribution function* and is abbreviated as *cdf*. It is defined as

$$Pr(X \leq x) = F(x; \theta),$$

meaning that the probability of the random variable X realizing a value less than or equal to x is specified by the function F. A related definition, for a *discrete* random variable X, is the "probability mass function" $p(x; \theta)$, abbreviated as *pmf*:

$$Pr(X = x) = p(x; \theta).$$

For a *continuous* random variable X the equivalent definition is the *probability density function* $f(x; \theta)$, abbreviated as *pdf*:

$$Pr\left(x - \frac{dx}{2} \leq X \leq x + \frac{dx}{2}\right) = f(x; \theta)\, dx.$$

These functions are related to F by

$$F(x; \theta) = \sum_{t \leq x} p(t; \theta)$$

and

$$F(x;\theta) = \int_{-\infty}^{x} f(x;\theta)\,dx,$$

respectively.

1.7 Expectation

A basic mathematical operation for constructing useful properties of F is that of an *expectation*. The expectation of a function h of a random variable X is defined as

$$E\{h(X)\} = \sum_{all\ t} h(t)p(t;\theta)$$

for a discrete random variable. For a continuous random variable, that definition is

$$E\{h(X)\} = \int_{x\in S} h(x) f(x;\theta)\,dx, \tag{1.1}$$

provided the sum (or integral) is finite. Thus, the probability $p(x;\theta)$, or $f(x;\theta)\,dx$, is a *weighting factor* determining the expectation as the weighted average value of $h(X)$.

1.8 Moments

Analogous to the "moments of an area" in mechanics, the *moments* of the random variable X characterize the model $F(x;\theta)$. The rth moment of X *about the origin* is defined by the operation (1.1) for $h(X) = X^r$. This type of moment is denoted by the symbol μ'_r. For $r = 1$, the *expected value* of X results: $E\{X\} = \mu'_1$ gives the average value of all possible measurements on X. The *mean square* of X is obtained for $r = 2$: $E\{X^2\} = \mu'_2$.

Instead of moments about the origin, consider the moments of X *about the mean* $E\{X\}$. The rth moment about the mean is defined by the expectation (1.1) for $h(X) = [X - E\{X\}]^r$. This type of moment is denoted as μ_r. A particularly useful moment corresponds to $r = 2$ and is termed the *variance* of X:

$$\mathrm{Var}(X) = \mu_2 = E\{[X - E\{X\}]^2\} = E\{X^2\} - (E\{X\})^2. \tag{1.2}$$

This moment characterizes the spread of the model F. Its square root is the *standard deviation* $s(X)$.

For some models F, the moments about the origin can be obtained from a *moment generating function* $M(t)$, defined by the operation (1.1) for $h(X) = \exp\{tX\}$. Hence,

$$M(t) = \int_{x\in S} \exp\{tx\} f(x;\theta)\,dx, \tag{1.3}$$

when X is a continuous random variable. The rth moment μ'_r is then obtained as the rth derivative of $M(t)$ with respect to t, at $t = 0$. See Section 8.3 for an example of a moment generating function.

The moments about the origin and about the mean are related by

$$\mu_r = \sum_{j=0}^{r} \binom{r}{j} \mu'_{r-j}(-\mu'_1)^j, \tag{1.4}$$

where $\binom{r}{j} = \frac{r!}{j!(r-j)!}$ and $\mu_0 = \mu'_0 = 1$. In particular,

$$\mu_2 = \mu'_2 - (\mu'_1)^2,$$
$$\mu_3 = \mu'_3 - 3\mu'_2\mu'_1 + 2(\mu'_1)^3,$$
$$\mu_4 = \mu'_4 - 4\mu'_3\mu'_1 + 6\mu'_2(\mu'_1)^2 - 3(\mu'_1)^4.$$

1.9 Additional Measures

Besides the expected value $E\{X\}$, there are other measures of *location* of the model F that may be of importance to the decision maker:

a. The *quantile of order q*, denoted as x_q, is defined by the equation

$$F(x_q; \theta) = q \tag{1.5}$$

for continuous variables X. This quantity gives the value below which the proportion q of possible measurements on X are likely to fall. For example, the quantile of order 0.05 (also termed the *5-percentile*) is exceeded by 95% of all measurements. The quantile of order 0.5 is the *median* value below which half the measurements on X are likely to fall.

b. The *mode* value x_m is defined by the equation

$$\frac{d}{dx_m} f(x_m; \theta) = 0 \tag{1.6}$$

for continuous variables X. This measure locates the most likely value of measurements on X, that is, where the pdf is maximum.

The standard deviation $s(X) = \sqrt{\mathrm{Var}(X)}$ describes the *spread* of the model F, in the units of measurements x. An alternate description measures the spread *relative to the mean*. This dimensionless measure is of particular importance in engineering. It is termed the *coefficient of variation* and is defined as

$$cv(X) = \frac{s(X)}{E\{X\}} = \frac{\sqrt{\mu_2}}{\mu'_1}. \tag{1.7}$$

1.10 Shape Measures

In addition to properties of F that describe its location and spread, there are two dimensionless measures of *shape* that characterize F. The *first shape factor* γ_1, termed "skewness," is defined by

$$\gamma_1 = \frac{\mu_3}{(\mu_2)^{3/2}}. \tag{1.8}$$

This factor indicates the extent of *skew* exhibited by F: Positive values indicate that the right tail of the pdf is longer than the left tail, negative values indicate the reverse, and a zero value implies symmetry.

The *second shape factor* γ_2, related to what is called "kurtosis," is defined by

$$\gamma_2 = \frac{\mu_4}{(\mu_2)^2}. \tag{1.9}$$

The Normal pdf (see Chapter 10) has the value $\gamma_2 = 3$. For the distributions covered in this text, $\gamma_2 > 3$ implies that the pdf is more sharply peaked than the Normal pdf, whereas for $\gamma_2 < 3$ the pdf is more flat topped that the Normal pdf. Thus, these two factors indicate some aspects of the shape character of a distribution F.

From their definitions, the above properties of $F(x; \theta)$ are functions of the parameter θ. These functions can be determined for a given model F. Competing models can then be compared in terms of these properties. Model properties are listed for each family of distributions covered in this text.

1.11 Univariate Models

In its most basic form, the uncertainty of engineering information relates to a single quantity. The probability model $F(x; \theta)$ thus represents only one variable X and is termed a *univariate* model. The purpose of that model is to describe as accurately as possible the nature of the uncertainty associated with that single quantity. When several *statistically independent* quantities need to be considered, their uncertainties are conveniently analyzed separately, meaning that each variable is modeled by its own distribution, independently of the others. Even when it is known that some of these quantities are somewhat related, the engineer often assumes that they are independent, in order to simplify the analysis to one involving independent univariate models. The statistical distributions discussed in this text are all univariate models.

In some engineering work several quantities of interest are more strongly related. It is then usually of prime interest to investigate the relation among them, rather than the specifics of their individual uncertainty aspects.

For example, the time-to-failure of an engineered component may be affected by the vibration level as well as the extent of thermal cycling during the component's operation. What is, at least initially, of interest here is the degree to which the latter two factors influence component life, rather than the details of how the uncertainties of these factors are distributed.

The appropriate statistical models for this kind of work relate to *regression analysis, analysis of variance,* and *statistical design of experiments*. These models are not treated in this text.

Sometimes it is desired to describe the joint uncertainties of several variables. The model F would then be a *multivariate* one. Such models are somewhat specialized and more difficult to deal with. They go beyond the scope of this text and are not discussed further.

DECISIONS

1.12 Importance of Distribution Tails

Outside of engineering the *decision* quantities of interest usually relate to the *central region* of the distribution F. That is, one is primarily concerned with the average and the variance of a random variable X. Engineering decisions, in contrast, often focus on the *tails* of the distribution of X.

In a typical design situation, for example, the engineer needs to choose the *maximum* load to which he dimensions his design, as well as the *minimum* material strength value that he can tolerate among specimens of his design.

Thus, it is of considerable importance in engineering to model the extremes of measurement variables as accurately as possible. To do so, the engineer needs to have at his disposal suitably skewed distributions beyond the familiar bell-shaped Normal model. The distribution models presented in this text provide a good range of skewed distributions that adequately model most engineering random variables.

1.13 Uncertainty of Decisions

An engineering decision will lead to some kind of action from which flow economic, social, environmental, and other *consequences*. Since the information base available to the engineer is usually limited, his decisions, even after careful analysis, will feature *statistical* uncertainty (see Section 1.2).

For example, the warranty period of a product is chosen so that the majority of product specimens (99%, say) survives that period without requiring warranty repairs. There will still be the 1% of products that can be expected to return for (costly) repair. Note that the warranty decision focuses on the extreme lower tail of the product's life distribution. Whatever the estimated value of the warranty period may be, this decision value is inherently uncertain since the information base will be incomplete.

Thus, engineering decisions are associated with *risk*: A value x may occur beyond what the engineer anticipated, possibly resulting in an unfavorable event (e.g., failure) with costly consequences. To manage that risk, *the decision maker needs to know a quantitative measure of the uncertainty surrounding the estimated decision quantity.* These measures are a function of the amount of information on hand: The smaller the information base, the larger the statistical uncertainty.

It is a common feature of engineering samples that they are small. For example, a sample as small as eight is not uncommon! Because consequences of engineering decisions are often severe, it is particularly important with such small samples to pay close attention to estimating the statistical uncertainty of decision quantities. The common measures of statistical uncertainty are the *standard error* and a *statistical confidence interval.* The nature of these measures is described in Chapter 3, and computational details appropriate for the various models are presented in Parts Two and Three of this text.

EXAMPLE 1.1

A sample of 29 specimens of a newly designed fastener was tested for holding strength and the following values (in kg) were obtained:

205, 51, 376, 119, 113, 108, 211, 146, 401, 183, 163, 186, 353, 287,
616, 513, 175, 107, 50, 27, 113, 156, 150, 72, 85, 173, 92, 143, 46.

Display the descriptive statistics that summarize the features of this sample.

sample size $n := 29$ let $i := 0..n-1$

Input raw data:

$x_i :=$ ordered data: $d := \mathrm{sort}(x)$

x_i	d_i
205	
51	27
376	46
119	50
113	51
108	72
211	85
146	92
401	107
183	108
163	113
186	113
353	119
287	143

Sample measures:

range:	$r := d_{n-1} - d_0$	$r = 589$
mean value:	$xb := \frac{1}{n} \cdot \sum_i x_i$	$xb = 187$
median value:	$xm := d_{\frac{n-1}{2}}$	$xm = 150$
variance:	$s2 := \frac{1}{n-1} \cdot \sum_i (x_i - xb)^2$	$s2 = 20028$
standard deviation:	$s := \sqrt{s2}$	$s = 142$
coefficient of variation:	$cv := \frac{s}{xb}$	$cv = 0.76$

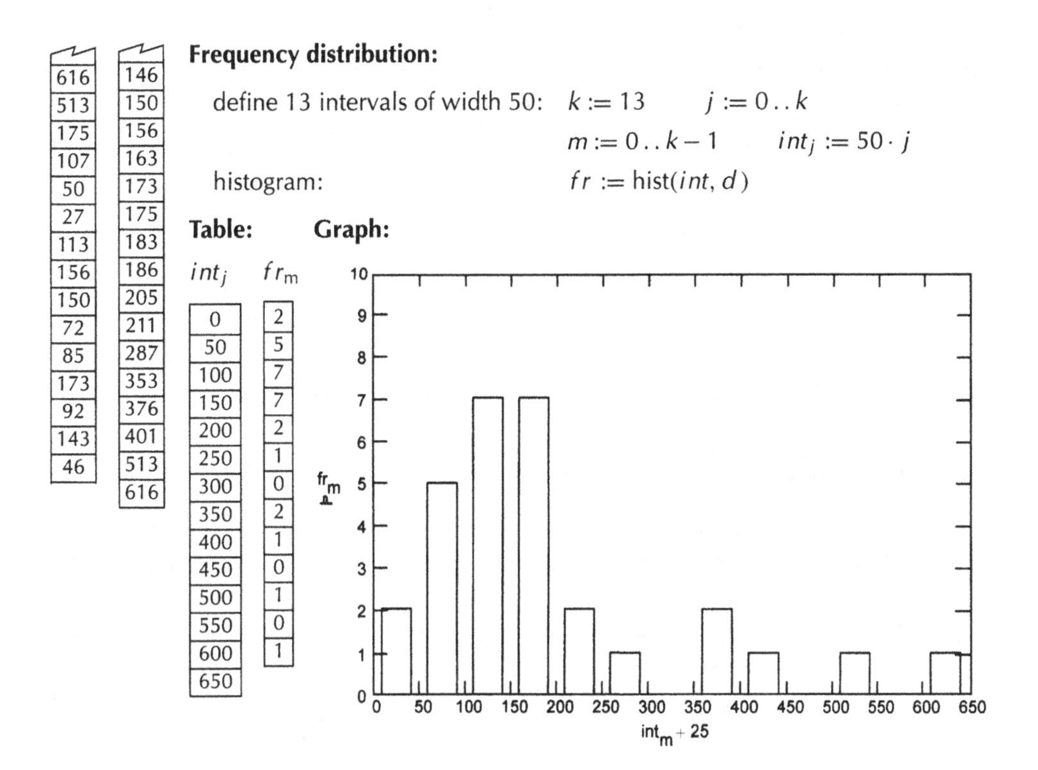

Frequency distribution:

define 13 intervals of width 50: $k := 13$ $\quad j := 0..k$

$m := 0..k-1$ $\quad int_j := 50 \cdot j$

histogram: $fr := \text{hist}(int, d)$

(data, cont.)	(data, cont.)
616	146
513	150
175	156
107	163
50	173
27	175
113	183
156	186
150	205
72	211
85	287
173	353
92	376
143	401
46	513
	616

Table:

int_j	fr_m
0	2
50	5
100	7
150	7
200	2
250	1
300	0
350	2
400	1
450	0
500	1
550	0
600	1
650	

Graph:

CHAPTER TWO

Statistics

SAMPLING DISTRIBUTIONS

2.1 Data Functions

Consider an experiment where n specimens of an electronic component are tested to failure. The time-to-failure (TTF) is represented by the random measurement variable X, and the n observations x_i are the data. Now consider a *function* of the *data*, such as the sample average $\bar{X}$, the first-order statistic $X_{(1)}$, etc. Any such data function is called a *statistic* T. The importance of some data functions is that they are the vehicle by which a population characteristic is inferred from experimental data (see Fig. 1.2 of the preceding chapter).

For example, a particular group of statistics, called *estimators*, serves to select a specific *member* of a chosen family of distributions by determining a *value* of the distribution parameter θ from a given sample. For instance, if the time-to-failure of an electronic component is to be modeled by an Exponential distribution (see Chapter 12), then the data average $\bar{x}$ estimates the Exponential parameter σ. Given this value, the Exponential model representing TTF is then completely specified.

A realization t of a statistic T is the result of evaluating the data function T for a specific sample $\{x_i\}_n$. It is important to understand that, before the experiment, the value t that a statistic T will attain is unknown. Furthermore, repeated experiments (with the same sample size) will result in different values t. Thus, T itself is a *random variable* with some probability distribution $f(t)$. Such a distribution of a statistic is termed a *sampling distribution*. As a mathematical function it is often different from the model $f(x; \theta)$ of the underlying random variable X.

For example, if the time-to-failure X of an electronic component is modeled by an Exponential distribution, then the *sampling distribution* of the *data average* $\bar{X}$ is Gamma (see Section 12.4).

2.2 Product Model

The most general statistic is the sample $\mathbf{X} = \{X_i\}_n$ itself. When the observations x_i are *independent* of each other and have the *common distribution* $f(x; \theta)$, the sample $\mathbf{X}$ is said to be a "simple *random* sample." The probability of observing a single value x_i in the interval $(x_i \pm \frac{dx}{2})$ is $f(x_i;\theta)\,dx$ (see Section 1.6), and because the observations are independent, the probability of obtaining the random sample $\mathbf{X}$ in the interval $\mathbf{dx}$ is the product of the probabilities of observations x_i:

$$Pr\left(\mathbf{x} - \frac{\mathbf{dx}}{2} \leq \mathbf{X} \leq \mathbf{x} + \frac{\mathbf{dx}}{2}\right) = \prod_{i=1}^{n} f(x_i;\theta)\,dx_i. \tag{2.1}$$

From this *product model* the exact sampling distributions of other statistics can, in principle, be derived. To demonstrate this derivation process, consider the following:

Continuing the preceding example, we have that the average $\bar{X}$ of n Exponential observations with scale parameter σ has a sampling distribution that is Gamma. This Gamma distribution has a scale parameter σ/n and a shape parameter n. There are several simple ways to arrive at this result, but for illustration let us start from the product model (2.1) for the case $n = 2$.

Substituting the Exponential pdf (12.3) into (2.1) gives the *joint* probability of the Exponential sample:

$$\frac{1}{\sigma}\exp\left\{-\frac{x_1}{\sigma}\right\}\frac{1}{\sigma}\exp\left\{-\frac{x_2}{\sigma}\right\}dx_1\,dx_2.$$

Let the average be $t = (x_1 + x_2)/2$, and transform the above probability from x_1 to t, with Jacobian 2. The result is the joint probability of t and x_2:

$$\frac{2}{\sigma^2}\exp\left\{-\frac{2t}{\sigma}\right\}dt\,dx_2.$$

The *marginal* probability of t is then

$$\frac{2}{\sigma^2}\exp\left\{-\frac{2t}{\sigma}\right\}\int_0^2 t dx_2\,dt = \left(\frac{2}{\sigma^2}\exp\left\{-\frac{2t}{\sigma}\right\}2t\right)dt.$$

The expression in parentheses is therefore the pdf of t and can be rewritten as

$$\frac{t^{2-1}}{(\sigma/2)^2\Gamma(2)}\exp\left\{-\frac{t}{\sigma/2}\right\}, \tag{2.2}$$

which is recognized as the Gamma pdf (13.1) with scale parameter $\sigma/2$ and shape parameter 2.

Thus, given the pdf $f(x;\theta)$ of a measurement variable X, and a statistic $T = T(\mathbf{X})$, the sampling model $f(t)$ could be computed from the product model (2.1). In many practical situations, however, this process is analytically intractable, and approximate sampling distributions are used instead.

The importance of sampling distributions is that they allow the assessment of the statistical uncertainty of inferences (see also Section 1.2). For example, confidence intervals on a *population* characteristic can be calculated from the sampling distribution of a relevant statistic. Evidently, if the sample size n is large, the confidence interval will be narrow; if n is small, the interval will be wide, reflecting the amount of sample information on hand.

Continuing with the preceding example, we have that a 90% equal-sided confidence interval on the *population* average $\bar{X} = \sigma$ of an Exponential process is obtained from the 5-percentile and 95-percentile of the sampling distribution (2.2), when the sample size is $n = 2$.

Without the known (exact or approximate) sampling distribution of a relevant statistic, the uncertainty of an inference connot be determined. Statistics without known sampling distributions should therefore be avoided. The exact sampling distributions of useful statistics are listed, when available, for the statistical models discussed in this text.

SUMMATION STATISTICS

2.3 Central Limit Theorem 1

A particularly useful statistic is the data average $\bar{X}$, which tells where the bulk of the data values are located. Some models (e.g., Poisson, Normal, Gamma) have the *reproductive property* that $\bar{X}$ is distributed like X, with known parameters. For example, the average of n observations from a Normal process is itself exactly Normally distributed with mean μ and variance σ^2/n (see Section 10.4). Thus, for these models the exact sampling distribution of $\bar{X}$ is known for any sample size n.

For most other models such an exact result is not available. However, the following *large-sample* result can be used to approximate the unknown exact sampling distribution of $\bar{X}$. Stated for a linear combination $T = \sum a_i X_i$, where the a_i are constants, this result is known as a "central limit theorem."

1. **If a random variable X is distributed with expected value μ'_1 and finite variance μ_2, then the summation statistic $T = \sum a_i X_i$ is asymptotically *Normally* distributed with expected value $\mu = \mu'_1 \sum a_i$ and variance $\sigma^2 = \mu_2 \sum a_i^2$.**

So, for example, if $a_i = 1/n$, then $T = \bar{X}$ and $\mu = \mu'_1, \sigma^2 = \mu_2/n$. That is, regardless of the nature of the measurement model $f(x; \theta)$, with increasing sample size the sampling distribution of $\bar{X}$ approaches the Normal model, which can therefore be used to obtain approximate confidence intervals on $\bar{X}$.

To get a sense of how the central limit theorem operates, compare the exact sampling distribution of $\bar{X}$ with its Normal approximation for an Exponential

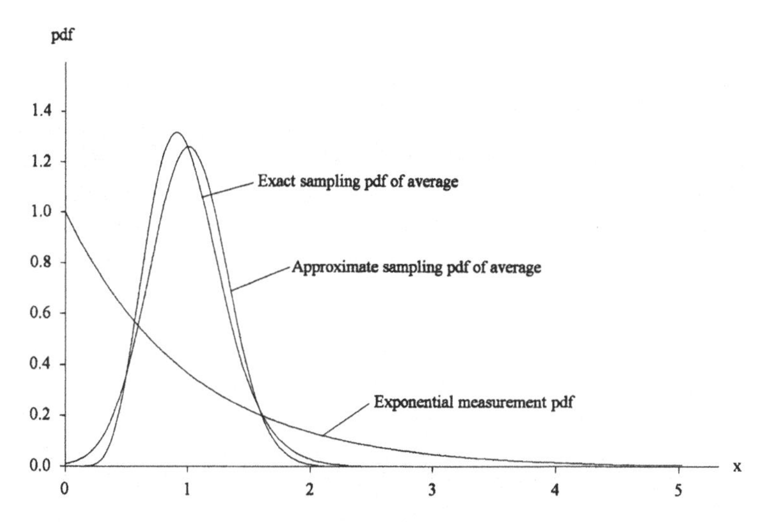

Figure 2.1. Sampling pdfs for the average of 10 Exponential observations.

sample. Figure 2.1 shows the Exponential pdf for $\sigma = 1$, as well as the exact Gamma pdf of $\overline{X}$ for $n = 10$ and its Normal approximation. The approximate sampling pdf is close to the exact pdf, even though X is highly non-Normal and the sample size n is rather small.

2.4 Central Limit Theorem 2

A more general central limit theorem applies to a linear combination of *different* variables Y_i:

2. **If n random variables Y_i are independent, with expected values $\mu_1'(i)$, finite variances $\mu_2(i)$, and finite third moments, then the summation statistic $T = \sum a_i Y_i$ is asymptotically *Normally* distributed with expected value $\mu = \sum a_i \mu_1'(i)$ and variance $\sigma^2 = \sum a_i^2 \mu_2(i)$.**

This result can be used to *motivate* the Normal distribution as a measurement model $f(x; \theta)$. That is, if it is reasonable to postulate that *many* independent underlying causes contribute *additively* to produce the measured variable X, then the Normal model is a good assumption for the distribution of X.

For example, the cost X of an engineered system is usually made up of many independent cost components Y_i, each possibly with a different distribution. It is then reasonable to postulate a Normal distribution for the system cost X, since cost components add up. If the averages $\mu_1'^{(i)}$ and variances $\mu_{2(i)}$ of the cost components are known, the parameters of the system cost distribution are determined from the above theorem.

ORDER STATISTICS

2.5 Exact Sampling Distributions

Of the many data functions T that are of interest in practical engineering work, the *order statistic* $T = x_{(i)}$ is one of the simplest and most useful, because it allows the decision maker to focus on a specific region of the distribution. In particular, the extreme $x_{(1)}$ or $x_{(n)}$ is important because it is often a required design input.

For example, if a device is to be designed to stand up to n repetitions of a random load X, then the maximum $x_{(n)}$ among these loads is the appropriate load input to the design process. As another example, if n identically distributed, *critical* components with random variable operating life X are assembled in a system, then the shortest-lived component $x_{(1)}$ determines the system life.

The *exact* sampling pdf of the ith order statistic is known:

$$f_i(x; \theta, n) = \frac{n!}{(i-1)!(n-i)!}[F(x;\theta)]^{i-1}[1 - F(x;\theta)]^{n-i} f(x;\theta), \tag{2.3}$$

where f and F are the pdf and cdf of the measurement variable X, respectively.

For the special case of $i = 1$, the pdf of the smallest observation $x_{(1)}$ in a sample of size n is

$$f_1(x; \theta, n) = n[1 - F(x;\theta)]^{n-1} f(x;\theta), \tag{2.4}$$

with a corresponding cdf of:

$$F_1(x; \theta, n) = 1 - [1 - F(x;\theta)]^n. \tag{2.5}$$

Similarly, the pdf of the largest observation $x_{(n)}$ in a sample of size n is

$$f_n(x; \theta, n) = n[F(x;\theta)]^{n-1} f(x;\theta), \tag{2.6}$$

with cdf:

$$F_n(x; \theta, n) = [F(x;\theta)]^n. \tag{2.7}$$

Thus, for a given measurement model $f(x;\theta)$, the exact sampling models of order statistics are defined. The resulting expressions are often analytically intractable but are easily calculated with a modern computational tool such as Mathcad. See Example 2.1 for an illustration.

2.6 Extreme Value Distributions

The Gumbel model (Chapter 15) and the Frechet model (Chapter 16) have the *reproductive property* that the sample extreme $X_{(n)}$ is distributed like X, with known parameters. Similarly, the Weibull distribution (Chapter 17) has the reproductive property that the sample extreme $X_{(1)}$ is Weibull-distributed with known parameters. Thus, the relevant exact sampling distribution, (2.7) or (2.5), is a member of the respective parent distribution $F(x;\theta)$. For these models the sampling distribution of the relevant sample extreme is therefore easy to work with.

For other models such simple results are not available. However, just as a central limit theorem prescribes the *asymptotic* sampling distribution of a summation statistic, there is a similar *asymptotic* result for the extreme order statistics $X_{(1)}$ and $X_{(n)}$. Specifically, their sampling distributions converge with increasing sample size n to one of three types of *extreme value* distributions, depending on the nature of the measurement variable X (see Section 15.2 for more detail). One model of each type is of particular interest in engineering: the type I model of maxima (Gumbel), the type II model of maxima (Frechet), and the type III model of minima (Weibull).

What is of importance here for engineering statistics is that these asymptotic results can be used to *motivate* one of the extreme value distributions as a *measurement model* $f(x;\theta)$. That is, if it can be argued that the measured variable X is the result of an underlying extreme value process, then the corresponding extreme value model is an appropriate measurement model for X.

For example, in the design of a flood-control structure, one of the important variables is the maximum of daily flow values during the year, for the watershed in question. This quantity is the *annual flood* X against which the structure has to be designed. Thus, X is inherently a maximum data extreme, and since there are many underlying daily data, an extreme value distribution of maxima (Gumbel or Frechet) can be expected to provide a reasonable measurement model for X.

As another example, consider the tensile strength of lumber for a specified wood species and sawn dimensions. Failure under tensile load is likely to occur at the location of the largest defect (usually a knot). If defects are randomly distributed throughout the volume of the test board, then the board can be thought of as a *chain* of many small volumes, each possibly containing defects. The *weakest* link in that chain is the one to precipitate board failure. Thus the measured board strength X equals the strength of the weakest element volume, and an extreme value distribution of minima (Weibull) can be expected to provide a reasonable model for the measured board strength X.

Because the size n of engineering samples is often small, it is usually difficult to discriminate several model candidates $f(x;\theta)$ on the basis of available data alone. However, an argument based on the physics underlying the measured phenomenon may lead to a specific distribution $f(x;\theta)$. Such an argument is a valuable addition to the engineer's information base. The physical reasons supporting the models discussed in this text are given in corresponding chapters, where appropriate.

SIMULATION

2.7 Background

The probabilistic behavior of large-scale or complex engineering systems is often very difficult to model analytically on the basis of known component behavior. The reason for this difficulty is typically the complexity of the system's structure. It follows that the sampling distributions of relevant system statistics are likewise difficult to obtain analytically.

For example, a simple, yet difficult, structure is a two-component *standby* system. The *main* component carries the system function, for example, a gas-turbine power unit in an unmanned pumping station on a gas-pipeline. A *standby* component, for example, a diesel power unit, takes over the system function when the main component fails, provided a failure-sensing/switchover unit functions properly. Although the life models of the components may be known, the system life model cannot be derived analytically. Hence, the sampling distribution of, say, the 5-percentile of system life is not available.

To determine the system's probabilistic behavior, system statistics need to be *generated* from known component characteristics and the known system structure. This is accomplished by *simulating* system specimens. That is, a random observation of each component behavior is simulated, and the system response is calculated according to the system structure, to simulate the response of one system specimen. Repeating this process many times simulates a *sample* of system specimens. This procedure is termed "Monte Carlo simulation." From this sample of system responses, relevant system statistics can be calculated and their sampling distributions can be estimated. Although this procedure can consume considerable computing time, it may be the only alternative available to the engineer.

Continuing with the preceding example, we simulate the time-to-failure (TTF) t_m of the main component. Then the switch is simulated: If it fails, t_m is a simulated observation of system TTF. If the switch functions, the TTF t_s of the standby component is simulated and the corresponding observation on system TTF is given by $(t_m + t_s)$. The result is the simulation of *one* system specimen. Repeating this process many times generates a *sample* of simulated system TTFs. From this sample a system statistic, such as the 5-percentile TTF, can be directly calculated, or the pdf of system TTF can be estimated. In addition, by repeating the Monte Carlo simulation itself a number of times, the sampling distribution of the 5-percentile can be obtained.

2.8 Statistical Basis of Simulation

A crucial part of simulation is to generate a sample of *component observations* $\{x_i\}_n$ so that it closely represents the specified component pdf $f(x; \theta)$. A powerful result is available to facilitate this "sampling from distributions," called the *probability integral transformation*:

Any cumulative distribution function, considered as a function of its random variable X, is itself a *Uniform* random variable on the interval $(0, 1)$:

$$F(X;\theta) = U. \tag{2.8}$$

(See Section 14.6 for details on the Uniform distribution.)

Since random observations u_i are easily simulated (these are the "pseudo-random numbers" supplied by a computer's random-number generator[1]), the corresponding simulated observations x_i are obtained by inverting expression (2.8). When $F(x;\theta)$ is of closed form, this inversion is accomplished analytically.

For example, the Weibull cdf (17.2) is

$$F(x;\sigma,\lambda) = 1 - \exp\left\{-\left(\frac{x}{\sigma}\right)^{\lambda}\right\}.$$

Equating to u and solving for x gives

$$x = \sigma\left[\ln\left(\frac{1}{1-u}\right)\right]^{\frac{1}{\lambda}}.$$

Thus, a random number u, substituted in the above expression, produces a simulated random Weibull observation x from $F(x;\sigma,\lambda)$.

When the cdf cannot be expressed in closed form, special techniques are available, or expression (2.8) can be inverted numerically. For each of the models discussed in this text, a simulation routine is suggested.

EXAMPLE 2.1

The uncertainty of a project cost variable X has been modeled by a Normal distribution with mean 6.8 and standard deviation 1.3 (in $10,000). Sixteen such projects are

[1] For engineering purposes of small-scale simulations, the strings of pseudo-random numbers generated by reputable computers can be taken as sufficiently random. Still, it is recommended to check the adequacy of these numbers by comparing at least the sample mean and variance with their theoretical values of $\frac{1}{2}$ and $\frac{1}{12}$, respectively. For large-scale simulation, possible sequential correlations in these numbers may seriously affect the reliability of simulated output. In that case it is wise to subject the computer-generated numbers to a "shuffling" procedure that effectively eliminates sequential correlation. See *Numerical Recipes* by William H. Press, et al., Cambridge University Press, 1986, p. 195, for one such algorithm. In general, for serious simulation work a good text on the subject should be consulted. See, for example, *A Guide to Simulation* by Paul Bratley, et al., Springer Verlag, New York, 1987.

currently underway, and the manager requires an estimate of the 80-percentile cost of the most costly current project.

ensemble size $n := 16$

Given: $\mu := 6.8 \qquad \sigma := 1.3$

The decision function is the *n*th-order statistic $x_{(n)}$

pdf of project cost: $f(x) := \frac{1}{\sigma \cdot \sqrt{2 \cdot \pi}} \cdot \exp\left[-0.5 \cdot \left(\frac{x-\mu}{\sigma}\right)^2\right]$

(see Chapter 10)

cdf of project cost: $F(x) := \int_0^x f(t)\,dt$

From (2.7) the cdf of $x_{(n)}$ is $\quad Fn(x) := F(x)^n$

The 80-percentile $x80$ of $x_{(n)}$ is that value x for which $Fn(x) = 0.8$

guess value for x: $\quad x := 10$

solution: $\quad x80 := \text{root}(Fn(x) - 0.8, x) \qquad x80 = 9.66$

(*Note:* Since no information is given on the sample that gave rise to the estimated model $f(x)$, the uncertainty associated with the estimate of $x80$ cannot be assessed.)

Display the cost pdfs:

From (2.6) the pdf of $x_{(n)}$ is $\quad fn(x) := n \cdot F(x)^{n-1} \cdot f(x)$

plot 100 points: $\quad i := 1..100$

$$t_i := 2 + \frac{i}{10}$$

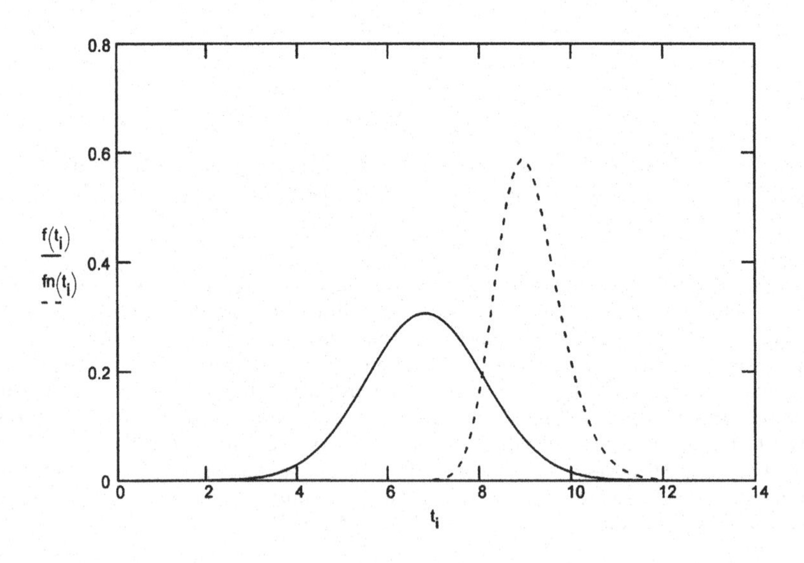

CHAPTER THREE

Inference

INFORMATION

3.1 Likelihood Function

Statistical inference is the process of drawing a conclusion on some probabilistic aspect of a *population*, on the basis of a *sample* of observations from that population (see Fig. 1.2). That is, the information contained in the sample is manipulated to infer a probability model that describes the uncertainty or variability of the observed phenomenon. In the context of this book, that model takes the form of a specific distribution function $F(x; \widehat{\theta})$, where F is a *chosen* family of distributions and $\widehat{\theta}$ is an *estimated* parameter value.

Given a chosen distribution family F, all sample information available for estimating the unknown parameter θ is encapsuled in the probability of occurrence of the *random* sample $\mathbf{X}$:

$$Pr\left(\mathbf{x} - \frac{d\mathbf{x}}{2} \leq \mathbf{X} \leq \mathbf{x} + \frac{d\mathbf{x}}{2}\right) = \prod_{i=1}^{n} f(x_i; \theta)\, dx_i; \tag{3.1}$$

see also Section 2.2. The right side of (3.1) is a fixed value if the observations x_i are given and θ is a constant. The *value* of θ is, however, not known and needs to be estimated from the given sample $\mathbf{x}$. The information base for estimation is the product-term of (3.1). That product is called the *likelihood* of the given sample $\mathbf{x}$. Considered as a function of the unknown parameter θ, it is the *likelihood function* (LF) of the sample $\mathbf{x}$:

$$L(\theta) = \prod_{i=1}^{n} f(x_i; \theta). \tag{3.2}$$

The amount of information *available* in a given sample $\mathbf{x}$ for inferring an estimate $\widehat{\theta}$ is defined by the following expectation:

$$I_\theta = -E\left\{\frac{\partial^2 \ln L(\theta)}{\partial \theta^2}\right\}. \tag{3.3}$$

The value of this *expected information* varies linearly with the sample size n: The larger n is, the more information on θ is available.

For example, the one-parameter Exponential pdf is defined in Chapter 12 as

$$f(x;\sigma) = \frac{1}{\sigma}\exp\left\{-\frac{x}{\sigma}\right\},$$

so that the LF of an Exponential sample **x** is

$$L(\sigma) = \frac{1}{\sigma^n}\exp\left\{-\frac{1}{\sigma}\sum_i = 1^n x_i\right\}.$$

The second derivative of the log of $L(\sigma)$ is therefore

$$\frac{\partial^2 \ln L(\sigma)}{\partial\sigma^2} = \frac{n}{\sigma^2} - \frac{2}{\sigma^3}\sum_{i=1}^{n} x_i.$$

Now, the expectation of $\sum_{i=1}^{n} x_i$ is $nE\{X\}$, and from (12.5), at $\mu = 0$, the expectation of X is $E\{X\} = \sigma$. Hence, the expected information is

$$I_\sigma = +\frac{n}{\sigma^2},$$

which provides a measure of the amount of information available in an Exponential sample of size n for estimating the Exponential parameter σ.

When θ is multivalued (i.e., the measurement model $f(x;\theta)$ is indexed by several parameters), the expected information (3.3) generalizes to the *expected information matrix*

$$[I_{i,j}(\theta)] = \left[-E\left\{\frac{\partial^2 \ln L(\theta)}{\partial\theta_i\partial\theta_j}\right\}\right], \tag{3.4}$$

which can be expressed in closed form for some distribution models f. See expression (14.22) for an example of an expected information matrix.

3.2 Minimum Variance Bounds

Since an estimator of θ is always some data function T, it is associated with a sampling distribution, (see Section 2.1). That is, repeated samples of size n produce different estimates $\widehat{\theta}$, and the sampling distribution of T models this variation. The larger the sample, the smaller is the spread of the sampling distribution.

For example, in Chapter 12 the data function $\bar{x}$ is recommended as the estimator of the parameter σ for the one-parameter Exponential model. The sampling distribution of this data function is Gamma with scale parameter σ/n and shape parameter n (see Section 12.4). Figure 3.1 shows that sampling pdf for several sample sizes n, indicating how increased sample information reduces the uncertainty surrounding the estimate $\widehat{\sigma}$.

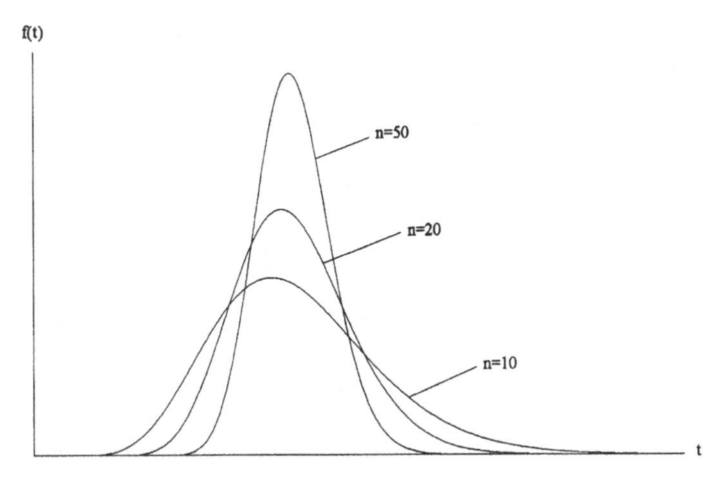

Figure 3.1. The sampling pdf of the data average for n Exponential observations.

Similarly, among competing estimator functions, the more information an estimator utilizes, the smaller will be the spread of its sampling distribution.

The relation between the amount of information *utilized* by an estimating function T and the amount of information I_θ *available* in the sample can be expressed by the sampling variance of T as

$$\mathrm{Var}(T) \geq \frac{1}{I_\theta}, \tag{3.5}$$

so that $1/I_\theta$ is the smallest sampling variance any estimating function of θ can attain. That is, an estimating function T can at best achieve equality in (3.5), thereby utilizing all the available information I_θ. Such data functions are called *minimum-variance-bound* (*MVB*) estimators with variance

$$MVB_\theta = \frac{1}{I_\theta}. \tag{3.6}$$

Because engineering data are costly to obtain, it is important to utilize as much as possible of the available information. Hence, MVB estimators are highly desirable. In this text, the MVB estimators of distribution parameters θ are presented, when available.

When the parameter θ is multivalued (i.e., the measurement distribution is indexed by several parameters), distinct data functions T_i estimate the parameter components θ_i, and the relation (3.5) generalizes to the *covariance matrix*

$$[\mathrm{Cov}(T_i, T_j)] \geq [I_{i,j}(\theta)]^{-1}, \tag{3.7}$$

where $[I_{i,j}(\theta)]$ is given by (3.4). The right side of (3.7) is termed the *covariance matrix of MVBs*:

$$[V] = [\mathrm{Cov}(\theta_i, \theta_j)] = [I_{i,j}(\theta)]^{-1}. \tag{3.8}$$

This matrix can be expressed in closed form for some distribution models f. See expression (13.14) for an example.

3.3 Standard Error of Parameter Functions

The square root of the sampling variance of the estimator T is termed the *standard error* of the corresponding estimate $\widehat{\theta}$:

$$se(\widehat{\theta}) = \sqrt{\text{Var}(T)}. \tag{3.9}$$

It is the standard deviation of the estimating function T and tells the decision maker the degree of statistical uncertainty associated with the estimate $\widehat{\theta}$. *It is recommended that any parameter estimate $\widehat{\theta}$ be quoted with its standard error $se(\widehat{\theta})$*:

$$\widehat{\theta} \pm se(\widehat{\theta}). \tag{3.10}$$

This statement not only summarizes the utilized sample information as a specific estimate $\widehat{\theta}$, but it also indicates by the standard error the relative amount of information (or lack of it!) that produced that estimate. When the standard error is small, relative to the estimate, the decision maker would be more comfortable with the estimate than if the standard error is relatively large. This relative amount of information can be pictured as the spread of the sampling pdf of T; see for example Fig. 3.1.

The quantity of interest to the decision maker is often a *function* $g(\theta)$ of several parameters. As for a parameter estimate $\widehat{\theta}$, *an estimate $\widehat{g}$ should be quoted with its standard error*. The sampling variance of the function g can be obtained approximately from the *error propagation formula*:

$$\text{Var}(g) \doteq \sum_i \sum_j \left(\frac{\partial g}{\partial T_i}\right)\left(\frac{\partial g}{\partial T_j}\right)\text{Cov}(T_i, T_j), \tag{3.11}$$

where the terms $\text{Cov}(T_i, T_j)$ are the elements of the estimators' covariance matrix (3.7). When the estimators are MVB, the covariance terms are the elements of matrix (3.8). The square root of (3.11) gives the *standard error* of $g(\theta)$:

$$se(g) = \sqrt{\text{Var}(g)}. \tag{3.12}$$

For example, an engineer may need to focus on the 95-percentile of a Gumbel load process. From (15.22) that decision quantity is a function of the two parameters μ and σ:

$$x_{0.95}(\mu, \sigma) = \mu - \left[\ln\left(\ln\frac{1}{0.95}\right)\right]\sigma = \mu - c\sigma.$$

To find the standard error, we need the derivatives of $x_{0.95}$ with respect to μ and σ. These are 1 and $-c$, respectively. Hence, the standard error of $x_{0.95}$ is

$$se(x_{0.95}) \doteq \sqrt{\text{Var}(T_\mu) + c^2\text{Var}(T_\sigma) - 2c\text{Cov}(T_\mu, T_\sigma)},$$

where the Ts are the estimators of the parameters.

Note that formula (3.11) approximates Var(g) reasonably well when the function g is not too severely nonlinear and the variance/covariance terms are small. In engineering practice, the resulting standard error $se(g)$ usually provides an acceptable measure of the uncertainty in g.

3.4 Standard Error of Measured Functions

In the preceding section the error propagation formula was used to approximate the standard error of a parameter function in terms of the standard errors of parameter estimates. Sometimes the decision quantity of concern to the engineer is a function not of distribution parameters but of m measured variables X_k.

Consider, for example, the efficiency η of a hydroelectric turbine as a decision function. The efficiency is

$$\eta = a\frac{P}{HQ},$$

where a is a constant. The power output P, the hydraulic head H, and the flow rate Q can be measured repeatedly at some fixed operating condition, and η needs to be estimated and its uncertainty assessed.

Suppose that n sets of simultaneous observations on m variables X_k are on hand. The covariance matrix $[V]$ of these variables can then be calculated directly as follows, and the approximate standard error of the decision quantity is obtained from (3.12).

Let $[D]$ be the $n \times m$ data matrix from which the $1 \times m$ vector $[A]$ of average values $\bar{x}_k$ can be calculated. If the vector $[A]$ is subtracted from each of the n data rows of $[D]$, the *mean-adjusted data matrix* $[M]$ is obtained. The $m \times m$ covariance matrix is then given by

$$[V] = \frac{1}{n-1}[M]^T[M], \tag{3.13}$$

where $[M]^T$ is the transpose of $[M]$.

It is also useful to calculate the *correlation matrix* for the variables X_k to see how strongly they are correlated. Let $[S]$ be the *standardized data matrix*, obtained by dividing each element of $[M]$ by the square root of the corresponding diagonal element of $[V]$:

$$S_{i,k} = \frac{M_{i,k}}{\sqrt{V_{k,k}}}. \tag{3.14}$$

The $m \times m$ correlation matrix is then

$$[C] = \frac{1}{n-1}[S]^T[S]. \tag{3.15}$$

See Example 3.1 for an illustration of computing $[V]$ and $[C]$ for measured variables, as well as estimating a decision function and its standard error.

POINT ESTIMATORS

3.5 Optimal Estimators

When making a measurement, the engineer would like his measuring system to operate without *bias*. That is, the system should deliver *accurate* measurements that are, on average, *equal* to the quantity measured. In a similar vein, when estimating a parameter θ by a data function T, one would want the values t, produced by T from different samples, to be equal to the unknown parameter θ, on average. That is, one wants

$$E\{T\} = \theta, \tag{3.16}$$

regardless of the sample size n. Statistics T with this property are called *unbiased* and are clearly preferred to those that are biased.

Similarly, the engineer would like his measurement system to be *precise* by producing measurements with the least possible amount of *random variation*. A similar requirement of an estimating function T is that its sampling distribution shows the least amount of spread for any sample size n. That is, $\mathrm{Var}(T)$ should be the minimum attainable variance among all relevant data functions. Statistics T that meet this requirement are termed *minimum variance* (MV) estimators. Those that meet both of the above requirements are *optimal* and are called *minimum variance unbiased* (MVU). It is clearly desirable to work with optimal estimator functions, particularly when the MV reaches the lower bound MVB_θ.

3.6 Maximum Likelihood Method

A universal method for determining large-sample optimal estimators is that of *maximum likelihood* (ML). This method is based on the intuitive notion that the *point estimate* $\widehat{\theta}$ of an unknown parameter θ should be chosen so that the likelihood (3.2) of a *given* sample $\mathbf{x}$ is maximized, at which point the slope of the function is zero:

$$\left.\frac{\partial L(\theta)}{\partial \theta}\right|_{\widehat{\theta}} = 0, \tag{3.17}$$

or equivalently,

$$\left.\frac{\partial \ln L(\theta)}{\partial \theta}\right|_{\widehat{\theta}} = 0. \tag{3.18}$$

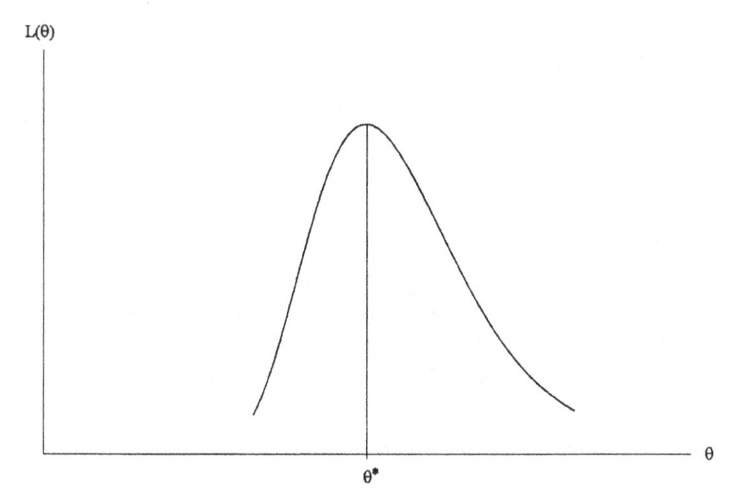

Figure 3.2. The likelihood function of a sample.

The latter is called the *ML equation*. It is usually more convenient to solve than (3.17). Figure 3.2 shows the typical appearance of a given sample's likelihood function. When the parameter θ is multivalued, expression (3.18) turns into a *system* of simultaneous estimating equations.

For some distributions (3.18) gives a closed-form solution for the ML *estimator* $\widehat{\theta}(\mathbf{X})$. See, for example, the solutions (11.28) and (11.29) for the Log-Normal model. The *point estimate* $\widehat{\theta}(\mathbf{x})$ is obtained by *evaluating* that estimator for the given data set **x**. For other distributions, (3.18) must be solved for the specific point estimate $\widehat{\theta}(\mathbf{x})$, for each given data set **x**. See the estimating equations (17.40) and (17.41) of the Weibull model for an illustration. Although it may appear tedious to solve a system of equations for every new data set, a modern computational tool like Mathcad simplifies this calculation to the point where it is no more difficult than evaluating a closed-form estimator. Furthermore, there are several benefits of using ML estimators, as described in the next section.

3.7 Properties of Maximum Likelihood Estimators

It can be shown that, as the sample size n increases, ML estimators have the following *large-sample* properties:

1. They are unbiased.
2. They achieve the minimum-variance-bound.
3. Their sampling distributions are Normal.

Thus, for large n, ML estimators are optimal and their statistical uncertainty can always be assessed. That is, the sampling variance of a ML estimator is asymptotically given by $MVB_\theta = 1/I_\theta$ of (3.6) so that the approximate standard error (3.9) of the ML estimate $\widehat{\theta}$ is defined. Moreover, the Normal sampling distribution with mean value θ and variance $1/I_\theta$ can be used to establish approximate confidence intervals on the parameter θ (see the next section).

For small sample sizes n the above properties become looser approximations to the true behavior of these estimators. In particular, the true sampling distribution becomes *skewed* for small n, and the estimators become somewhat *biased*, although it is sometimes possible to determine a function $b(n)$ that corrects that bias. Although the standard error $se(\widehat{\theta})$ is sufficiently accurate even for small n, confidence intervals on θ need to be viewed with caution when n is small. Nevertheless, when the exact sampling distribution of a ML estimator $\widehat{\theta}$ is not known, its Normal approximation is useful for indicating the statistical uncertainty of the estimate $\widehat{\theta}$, even when n is small.

An additional important property of ML estimators is that of *invariance under transformation*. That is, when a *function* g of parameters θ is the decision quantity of interest, the ML estimate $\widehat{g}$ of that function is obtained simply by substituting the ML parameter estimates into that function:

$$\widehat{g} = g(\widehat{\theta}). \tag{3.19}$$

The above large-sample properties of $\widehat{\theta}$ transfer to the function-estimate $\widehat{g}$. In particular, the sampling distribution of $\widehat{g}$ is approximately Normal, with mean g and variance approximated by the *error propagation formula* (3.11). The terms $\text{Cov}(T_i, T_j)$ in (3.11) are obtained from matrix (3.8). See Example 15.2 for an illustration of computing the standard error of a decision function (a Gumbel 95-percentile), where the covariance matrix (3.8) of minimum variance bounds is used.

When the *expected* information matrix (3.4) cannot be calculated, so that the covariance matrix (3.8) cannot be obtained, one can equivalently invert the *local* information matrix

$$[I_{i,j}(\theta)] = \left[-\frac{\partial^2 \ln L(\theta)}{\partial\theta_i \partial\theta_j} \right] \Bigg|_{\widehat{\theta}}, \tag{3.20}$$

which can always be calculated. That is, the ML parameter estimates are substituted directly into the elements of the above matrix, without the need to perform the expectation operation. The *local* matrix (3.20) measures the information content of the specific sample $\mathbf{x}$ on hand, whereas the *expected* matrix (3.4) gives the *average* information content of samples $\mathbf{X}$. Many statisticians prefer to use the local information matrix, even when the expected matrix can be obtained. For the models discussed in this text, both versions are presented, when they differ. See Example 14.4 for an illustration of computing the standard errors of parameter estimates, using the local information matrix (14.23) for a four-parameter Beta variable.

Sometimes, data samples in engineering are *censored*. That is, not all observations in the sample are actually known. See Section 12.2 for an introduction to censoring. The problem of estimating the parameters θ from a censored sample is generally more difficult than for a complete (noncensored) sample. However, the likelihood function of censored samples is easily expressed, and the ML method proceeds with only minor computational alterations. Compare, for example, the ML equations (17.40) and (17.41) for a *complete* Weibull sample with the equations (17.54) and (17.55) for a *censored* Weibull sample. The ML method of estimation is often the only feasible option for censored samples.

Because the assessment of the uncertainty associated with estimated decision functions $g(\theta)$ is of crucial importance in engineering, estimators for which sampling distributions are not at all available should *not* be used under any circumstances, when a sample of observations is available. Because of their asymptotically optimal properties, and the universality of their application, *ML estimators* are the recommended choice for engineering statistics.

This text uses the ML estimation method. ML estimators, or estimating equations, are given and illustrated for the distribution models discussed. For models that are often applied to phenomena for which censored samples are common, corresponding results are presented for censoring.

CONFIDENCE INTERVALS

3.8 Constructing a Confidence Interval

Instead of focussing on a *single best estimate* $\widehat{\theta}$ of a parameter θ and indicating its statistical uncertainty by its standard error $se(\widehat{\theta})$, the given decision situation may instead require a plausible *range of values* $(\widehat{\theta}_l, \widehat{\theta}_u)$ that brackets the unknown θ with some known *statistical assurance.* The required interval is developed along the following line of reasoning.

Before an experiment is conducted, the estimator T of the parameter θ is a *random data function* with a sampling pdf $f(t)$. From this pdf an interval (t_1, t_2) can be obtained such that the following probability has a specified value $(1-\alpha)$:

$$Pr(t_1 \leq T \leq t_2) = 1 - \alpha. \tag{3.21}$$

If one imposes the requirement that the probability tails outside the above event are of equal size $\alpha/2$, then the interval (t_1, t_2) is unique. The interval endpoints t_1 and t_2 are the $\alpha/2$- and the $(1 - \alpha/2)$-quantiles of the sampling pdf $f(t)$. Figure 3.3 illustrates such an interval.

For example, suppose one were to sample from a Normal process with *known* variance σ_k^2. The sampling distribution of the statistic $\overline{X}$ for estimating the unknown Normal mean μ is also Normal, with parameters μ and $\sigma_k/\sqrt{n}$ (see Section 10.4). The interval endpoints covering the statistic $T = \overline{X}$ with probability $(1 - \alpha)$ are therefore obtained as the Normal quantiles

$$\left(t_1 = \bar{x}_1 = \mu + z_{\frac{\alpha}{2}} \frac{\sigma_k}{\sqrt{n}}, t_2 = \bar{x}_2 = \mu + z_{1-\frac{\alpha}{2}} \frac{\sigma_k}{\sqrt{n}}\right),$$

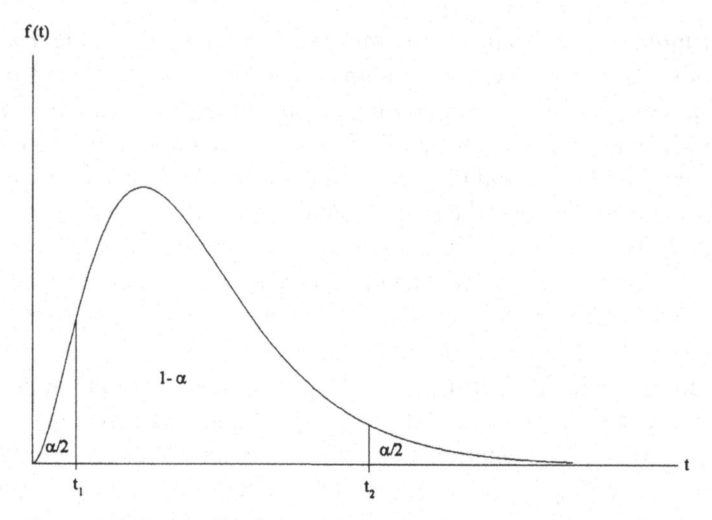

Figure 3.3. An equal-sided $(1-\alpha)$ interval (t_1, t_2) from a sampling pdf $f(t)$.

according to (10.11), where z_q is the standard Normal q-quantile and $z_{1-\alpha/2} = -z_{\alpha/2}$. The probability (3.21) then reads

$$Pr\left(\mu - z_{1-\frac{\alpha}{2}} \frac{\sigma_k}{\sqrt{n}} \leq \bar{X} \leq \mu + z_{1-\frac{\alpha}{2}} \frac{\sigma_k}{\sqrt{n}}\right) = 1 - \alpha.$$

The interval endpoints t_1 and t_2 are expressions involving the unknown parameter θ. However, the *random event* $(t_1 \leq T \leq t_2)$ can be re-expressed as a *random interval* $\{l_1(T) \leq \theta \leq l_2(T)\}$ that covers the fixed parameter θ, with the same probability $(1-\alpha)$:

$$Pr\big(l_1(T) \leq \theta \leq l_2(T)\big) = 1 - \alpha. \tag{3.22}$$

Continuing with the preceding example and re-arranging the random event as a random interval, gives the functions l_1 and l_2 as:

$$l_1(\bar{X}) = \bar{X} - z_{1-\frac{\alpha}{2}} \frac{\sigma_k}{\sqrt{n}} \quad \text{and} \quad l_2(\bar{X}) = \bar{X} + z_{1-\frac{\alpha}{2}} \frac{\sigma_k}{\sqrt{n}},$$

with probability

$$Pr\left(\bar{X} - z_{1-\frac{\alpha}{2}} \frac{\sigma_k}{\sqrt{n}} \leq \mu \leq \bar{X} + z_{1-\frac{\alpha}{2}} \frac{\sigma_k}{\sqrt{n}}\right) = 1 - \alpha.$$

Once a sample $\mathbf{x}$ is on hand, the random interval $\{l_1(T), l_2(T)\}$ becomes fixed at $\{l_1(t), l_2(t)\}$. Expression (3.22) is then no longer a probability statement. The fixed interval $\{l_1(t), l_2(t)\}$ is called a *confidence interval*, and the value $(1-\alpha)$ is termed the *statistical confidence* in that interval. The interpretation of this interval is that, since repeated samples would each produce slightly different calculated intervals, the proportion of intervals containing the unknown value θ

approaches $(1 - \alpha)$, as the number of repetitions increases. Thus, $(1 - \alpha)$ is a measure of the statistical assurance that a specific interval estimate covers the unknown parameter θ. The value $(1 - \alpha)$ is specified by the decision maker. Clearly, the higher the required confidence level, the wider will be the corresponding interval (l_1, l_2).

3.9 Methods of Obtaining Confidence Intervals

To construct a confidence interval on a parameter θ, one needs to know the sampling distribution $f(t)$ of a suitable estimating statistic T. When the exact sampling distribution of T is known, exact confidence intervals are obtained as in the preceding section. This is possible for only some statistics and for a few of the models discussed in this text; these cases are presented in corresponding chapters. Usually, exact sampling distributions are not known and approximate methods must be used. This is particularly true for estimating functions of several parameters (e.g., a population percentile).

One approach is to rely on the asymptotic behavior of the ML estimator $\widehat{\theta}$. That is, for large sample sizes n the sampling distribution $f(\widehat{\theta})$ approaches the Normal model with mean value θ and variance MVB_θ. Approximate confidence intervals on θ can therefore be constructed directly from the Normal pdf (see Section 10.2) as

$$\{l_1, l_2\} = \theta \pm z_{\frac{\alpha}{2}} \sqrt{MVB_\theta}, \tag{3.23}$$

where the parameter θ is evaluated at its ML estimate $\widehat{\theta}$, and z_q is the *standard Normal quantitle of order q*. Also, the invariance property of ML estimators can be used to construct similar confidence intervals on a decision function $g(\theta)$.

For example, exact sampling distributions for estimators of Gamma parameters are not available. To construct approximate confidence intervals on these parameters, their ML estimates are first obtained from a sample of observations, with (13.32) and (13.33) as the estimating equations. The covariance matrix (13.14) is then evaluated. The diagonal terms of that matrix give the approximate sampling variances of these estimators. (See Example 13.1 for a worked illustration.) Approximate confidence intervals on the parameters are then obtained from the Normal sampling distribution with mean value equal to the estimated parameter and variance equal to the estimated *MVB*. (See Example 13.3 for a worked illustration.)

For large sample sizes these approximate intervals are accurate. For small to moderate sized samples these approximations are less accurate. An alternative approximate approach, which is more difficult to evaluate but produces more accurate results compared to the Normal approach, is available. This approach is based on the fact that the following ratio of likelihoods is asymptotically distributed as a *Chi-squared* variable (see Section 10.6) with $\nu = 1$ *degree of*

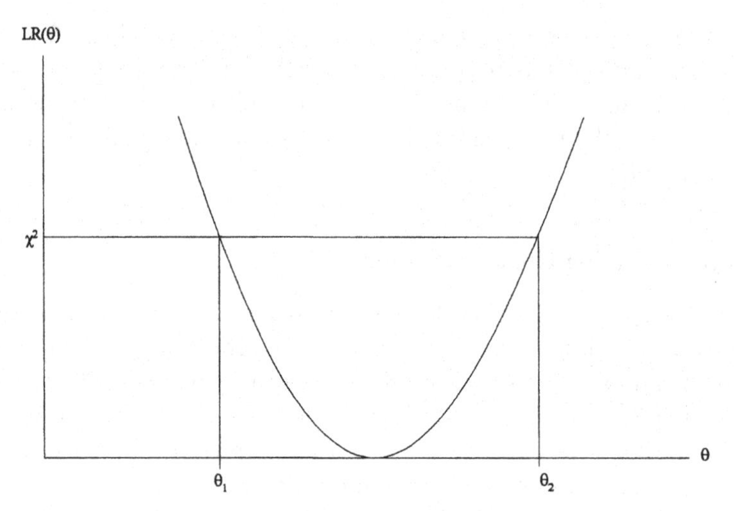

Figure 3.4. The likelihood ratio function for a sample.

freedom[1]:

$$LR(\theta) = \ln\left(\frac{L(\widehat{\theta})}{L(\theta)}\right)^2 = 2\ln[L(\widehat{\theta})] - 2\ln[L(\theta)], \tag{3.24}$$

where θ is a single unknown parameter. The likelihood ratio approaches its χ^2-distribution more rapidly with increasing sample size n than the ML estimator $\widehat{\theta}$ approaches its Normal distribution. The likelihood-ratio function therefore provides more accurate confidence intervals for small to moderate sample sizes. Figure 3.4 shows the typical appearance of the likelihood ratio function for a given sample. A $(1-\alpha)$-level confidence interval is thus obtained as the set of values of θ that satisfy

$$LR(\theta) \leq \chi^2_{1,(1-\alpha)}, \tag{3.25}$$

where $\chi^2_{1,(1-\alpha)}$ is the $(1-\alpha)$-quantile of the Chi-squared distribution with $\nu = 1$ degree of freedom. Refer to Section 13.6 for the Chi-squared distribution.

When the parameter θ is multivalued, all parameter components in $LR(\theta)$ are expressed in terms of the component in question by their ML equations, illustrated as follows:

Continuing the preceding example, we have that the likelihood function of a Gamma sample is given by (13.31), and so its logarithm is

$$-n\ln[\Gamma(\lambda)] - n\lambda\ln[\sigma] + n(\lambda-1)\ln[g] - \frac{n\bar{x}}{\sigma}.$$

[1] The term "degree of freedom" designates the number of unconstrained ways in which data values enter the corresponding statistic.

Here g and $\bar{x}$ are the geometric and arithmetic sample means, respectively (see Section 13.9). Substituting in the above expression for σ its ML equation (13.32), $\sigma = \bar{x}/\lambda$, gives the log-likelihood as a function of λ only:

$$LL(\lambda) = -n \ln[\Gamma(\lambda)] - n\lambda \ln[\bar{x}/\lambda] + n(\lambda - 1) \ln[g] - n\lambda.$$

The confidence interval endpoints l_1 and l_2 on the parameter λ are then obtained from the condition

$$LR(\lambda) = 2LL(\widehat{\lambda}) - 2LL(\lambda) = \chi^2_{1,(1-\alpha)}.$$

See Example 13.3 for a worked illustration.

The process of calculating such intervals is somewhat involved. However, a modern computational aid such as Mathcad reduces that process to a simple evaluation of an equation solver. The following chapters present and illustrate those cases for which useful results can be obtained by the likelihood-ratio method.

TESTS

3.10 Significance Tests

Sometimes the objective of a statistical investigation is not to find a single best estimate or a plausible range of values for a parameter θ. Rather, the objective is to determine if a *given* parameter value characterizes the measurement process. If sample information indicates that this given parameter value is reasonable, a certain action follows. If that value is not reasonable, a different action is taken. The statistical inference that leads to such a decision is called a *test*.

For example, suppose that the critical dimension of a mass-produced shaft is the diameter of an interference fit with a bearing, with nominal value x_d. As the edge of the cutting tool wears during production, the diameters X produced by the lathe increase over time, and at some point the engineer wants to know if the production average μ is still acceptable. If a sample of measured shaft diameters indicates that the proposition $\mu = x_d$ is still reasonable, he leaves the lathe alone. If, however, the sample indicates that this proposition is not supported by the data, he replaces the cutting tool.

The proposition stating the given parameter value is called the ***null hypothesis*** and is denoted as $H_0 : \theta = \theta_0$. This hypothesis usually represents the existing condition of the investigated phenomenon measured by X. The statistical test decides on H_0 by evaluating a suitable statistic T from a sample of observations on X. The difficulty here is obviously that the *value* t produced by a sample will include sampling variation, so that the decision on H_0 may be in error. In particular, even when H_0 is true, the decision may be to reject H_0, resulting in an error. This error is called *type I*, and its probability of occurrence α is controlled at an acceptably low level as follows.

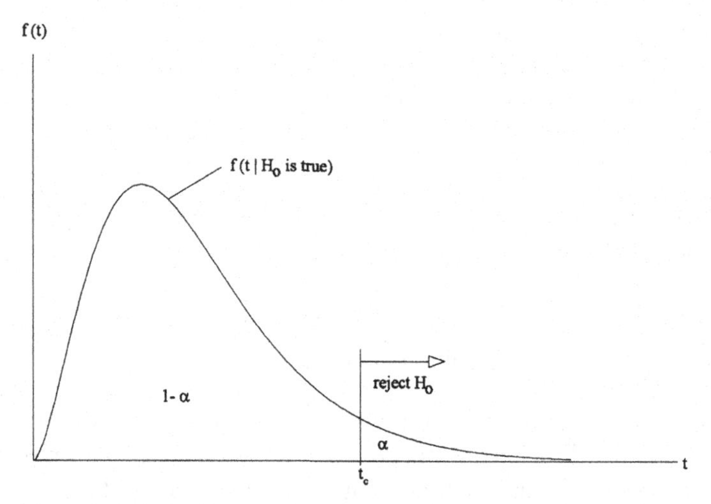

Figure 3.5. A significance test on a hypothesis H_0, at the significance level α.

The statistical variation of the data function T is modeled by its sampling distribution $f(t)$, given that H_0 is true. From that distribution a *critical value* t_c is chosen so that the type I error probability is kept at an acceptable *significance level* α. That is,

$$Pr(T \geq t_c \mid H_0 \text{ true}) = \alpha, \tag{3.26}$$

for the case where an *increase* in θ would be of practical significance. This critical value then establishes the decision rule: "if $t \geq t_c$, reject H_0 at the significance level α." However, if $t < t_c$, one concludes "there is no evidence to reject H_0." Such a test, involving a single hypothesis H_0 is called a *significance test*. Figure 3.5 pictures a typical sampling pdf of a test statistic, the rejection region $t \geq t_c$, and the significance level α, which is commonly chosen as 5%.

Continuing with the preceding example, we suppose that the machined diameter follows a Normal process, set at the mean value x_d. In Section 10.7 it is stated that the statistic

$$T = \frac{\bar{X} - \mu}{S/\sqrt{n}}$$

is *t-distributed* with $\nu = n - 1$ *degrees of freedom*. For an α-level test on $H_0 : \mu = x_d$, the critical value t_c of the above statistic is therefore the $(1 - \alpha)$-quantile $t_{1-\alpha}$ from that t-distribution, corresponding to the probability

$$Pr\left(T = \frac{\bar{X} - x_d}{S/\sqrt{n}} \geq t_{1-\alpha} = t_c\right) = \alpha.$$

A sample value $t = \frac{\bar{x} - x_d}{S/\sqrt{n}} \geq t_c$ would reject H_0, and the engineer would change the cutting tool.

3.11 Hypothesis Tests

Rather than focusing on a single hypothesis H_0, one may wish to discriminate between two or more hypotheses on the parameter θ. One would be the null hypothesis H_0, and another would be a specific *alternative hypothesis* $H_a : \theta = \theta_a$. In distinction to a significance test, this type of test is called a *hypothesis test.* Again, a suitable statistic T, evaluated from a given sample of observations, would lead to a decision: Either reject H_0, implying acceptance of H_a, or vice versa. As before, the former decision may be in error with probability α, the choice of which determines the critical test value t_c. However, the latter decision of accepting H_0, and therefore rejecting H_a, may also be in error when H_a is in fact true. This error is termed *type II*. Its probability of occurrence β is determined from the sampling distribution of T, given that H_a is true:

$$Pr(T \leq t_c \mid H_a \text{ true}) = \beta. \tag{3.27}$$

A good hypothesis test features a small value of β, typically around 10%.

With t_c determined by (3.26), the probability β is a function of the alternative hypothesis H_a. This function is called the *operating characteristic* of the test and needs to be evaluated for a proposed hypothesis test to check if the type II error probability is sufficiently small. The complementary probability $(1 - \beta)$ is called the *power of the test* to discriminate between the competing hypotheses. If β is too large, either the sample size n needs to be increased or the alternative H_a needs to be separated further from H_0. Figure 3.6 sketches the concepts involved for an adequate test.

For example, suppose that a research engineer at a materials manufacturer has come up with a process modification that promises to improve the mean strength μ_0 of a manufactured material. A test batch has been produced in the lab and n

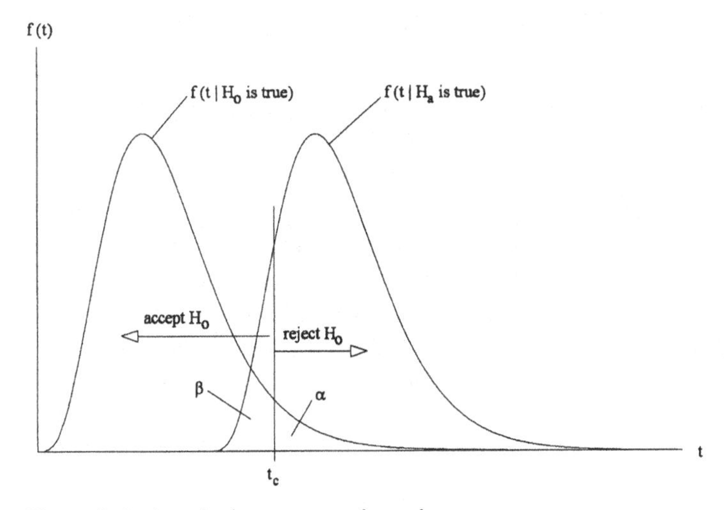

Figure 3.6. A typical test on two hypotheses.

test specimens have been tested for ultimate strength. The question of concern would be: "Has the modification improved the mean strength *significantly*?" The engineer would *choose* a new mean strength value μ_a that represents a sufficient improvement to warrant a costly process change. It is then required to decide, on the basis of the data, between the existing state $H_0 : \mu = \mu_0$ and the alternative state $H_a : \mu = \mu_a$. If the decision is H_0, the process is left alone. If the decision is H_a, the process is changed.

To establish a hypothesis test, one needs the sampling distribution of a suitable statistic T under the hypotheses H_0 *and* under H_a. Only a few practical cases are available for which the exact sampling distributions are known. These are mainly tests on the mean and variance of Normal, Log-Normal, and Exponential variables. Details on these tests are provided in the chapters that cover these models. For other distributions and more interesting parameter functions, the asymptotic Normal properties of ML estimators, or the likelihood ratio statistic, can be used to construct approximate hypothesis tests.

3.12 Test of Fit

When a family of distributions F has been chosen to model a random variable X, and its parameters have been estimated from data on X, it is important to check the conformity of the estimated model $F(x; \widehat{\theta})$ with the data. That is, a *significance test* on the null hypothesis $H_0 : F_0(x; \widehat{\theta})$ needs to be performed, typically using the *same* data set from which the parameters were estimated.

There are several measures of discrepancy between the estimated model and the *sample* distribution, for which the sampling distribution is known, at least approximately. The one test statistic recommended in this text is the *Anderson–Darling* statistic

$$A = -n - \frac{1}{n}\sum_{i=1}^{n}(2i - 1)[\ln(w_i) + \ln(1 - w_{n-i+1})]. \tag{3.28}$$

Here w_i is the cdf $F(x; \theta)$, which is evaluated at the order statistic $x_{(i)}$ and the ML estimates $\widehat{\theta}$. The reason for choosing A is that it tends to be more powerful than competing statistics T in detecting discrepancies in the distribution tails. Furthermore, critical test values t_c are available[2] for the practical case when the parameters θ are estimated from the *same* data as are used for the test-of-fit. This statistic can in most cases be modified to account for the effect of the sample size n, so that only a small set of critical test values is required. In addition, the test statistic A is simple to evaluate with a modern computational tool like Mathcad. Details are provided for the models covered in this text, when results are available.

[2] See, for example, R. B. D'Agostino, M. A. Stephens, *Goodness-of-Fit Techniques*, Marcel Dekker Inc, New York 1986.

PROBABILITY PLOT

3.13 Constructing a Plot

A statistical test-of-fit provides a useful check on the adequacy of an estimated distribution $F_0(x;\widehat{\theta})$ to model a measured variable X. However, a single test measure, such as the statistic (3.28), is usually not capable of detecting a significant departure from the data in the model tails. In engineering the tail-fit is often of overriding importance. Hence, the engineer routinely checks the model's fit to the data graphically, in addition to doing a statistical significance test. This graphical technique is called *probability plotting.*

The idea is to plot the estimated *model* cdf $F_0(x_{(i)};\widehat{\theta})$ and the *sample* cdf $p(x_{(i)})$ together on a graph to see if the model *fits* the data. Since the human eye can more easily judge how a straight line, rather than some curved graph, fits data points, the scales of the graph are chosen to produce a linear plot of both the model and the data.

The definition of the sample cdf for the data plot is controversial, as many choices exist. However, since probability plotting should only be used for a visual check on the model fit, any one choice (called a *plotting position*) is about as good as any other. In this text the following plotting position is recommended:

$$p_i = \frac{i - 0.3}{n + 0.4}. \tag{3.29}$$

This value approximates the *median* of the distribution-free estimate of the cdf $F(x_{(i)})$ and produces good comparisons with a model estimated by maximum likelihood. The approximation is good to one digit in the third decimal, even for small n.

When the model cdf is of *closed form,* its linearization is often straightforward, resulting in a relation of the form

$$g_1(F_i) = g_2(\theta) + g_3(\theta)g_4\big(x_{(i)}\big). \tag{3.30}$$

Here F_i stands for the estimated model $F_0(x_{(i)};\widehat{\theta})$ and the functions g are suitable transformations. Plotting $g_1(F_i)$ versus $g_4(x_{(i)})$ gives a straight line for the estimated *model*, and the *data* plot $g_1(p_i)$ versus $g_4(x_{(i)})$ will follow this straight line approximately if F_0 models the data well. Figure 3.7 shows one probability plot with the y axis in the *natural* coordinates F_i and p_i and another plot with the y axis in the *linear* coordinates $g_1(F_i)$ and $g_1(p_i)$. For both cases, $g_4(x_{(i)}) = x_{(i)}$ is plotted on the x axis.

For example, the Weibull cdf (17.2) is $F(x;\sigma;\lambda) = 1 - \exp\{-(\frac{x}{\sigma})^\lambda\}$. Rearranging and taking logs twice gives

$$\ln[-\ln(1 - F)] = -\lambda\ln(\sigma) + \lambda\ln(x).$$

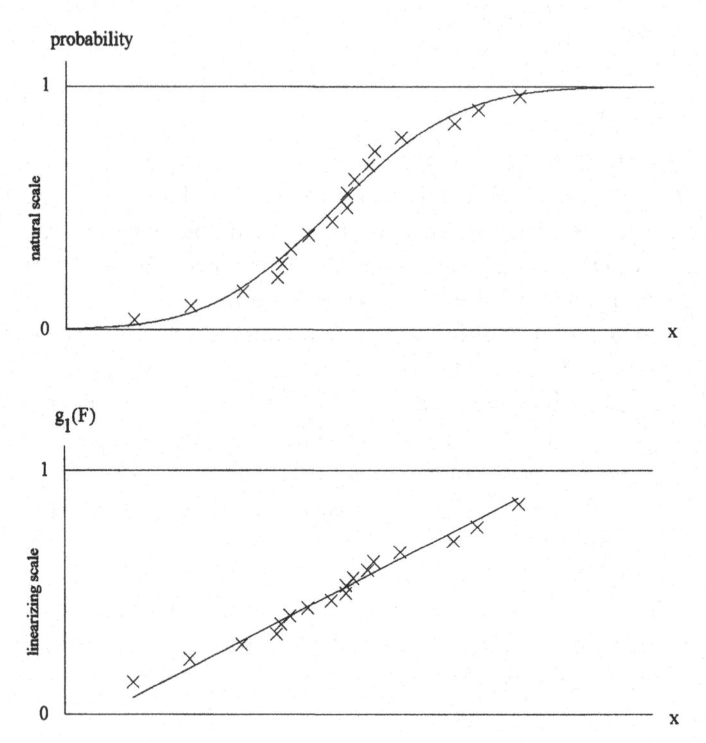

Figure 3.7. Two versions of a probability plot.

Hence $g_1(F) = \ln[-\ln(1 - F)]$ and $g_4(x) = \ln(x)$. Plotting the right side of the above expression with estimated parameters

$$\widehat{\lambda}\{\ln(x_{(i)}) - \ln(\widehat{\lambda})\}$$

versus $\ln(x_{(i)})$ produces a straight line for an estimated Weibull *model,* and plotting $g_1(p_i) = \ln[-\ln(1 - p_i)]$ versus $\ln(x_{(i)})$ linearizes Weibull *data.* See Example 17.4 for an illustration.

When the model F_0 is not of closed form, the plot can be linearized numerically. For location-scale models (see Section 9.2) or models that can be transformed to location scale, the linearization can be computed without knowing the estimates $\widehat{\theta}$. For an illustration, see Section 10.10 for the Normal model and Section 11.7 for the Log-Normal model. In other cases, the estimates $\widehat{\theta}$ must be on hand to compute this linearization. See Section 13.7 for the Gamma model as an illustration.

3.14 Interpreting a Plot

Linearization of a data plot does not require the estimates $\widehat{\theta}$, when the distribution has location-scale structure or is of closed form. For these models the data plot can be constructed *before* the model is estimated. This is sometimes done to explore several model candidates, before the estimation of the chosen candidate is undertaken. It is then tempting to estimate two parameters θ from the intercept $g_2(\theta)$ and the slope $g_3(\theta)$ of a straight line that is somehow fit to the data. Since the choice of plotting position is subjective, and the sampling distributions of the resulting estimators are not known, this temptation should be resisted, particularly because defensible ML estimates can be computed easily.

Although probability plotting does not yield quantitative inferences, relying as it does on the judgment of the viewer, valuable information is sometimes gleaned from the plot. Figure 3.8 shows several idealized possibilities of the plot's appearance. Plot (a) indicates that the measurements x came from two distinct populations. Plot (b) suggests that the true distribution has a longer upper tail (i.e., it is more skewed to the right) than the postulated model F_0. Plot (c) suggests that the true distribution is more skewed to the left than F_0. Plot (d) indicates the presence of a location parameter. In practice, of course, probability plots do not look as "clean" as in Fig. 3.8. That is, sampling variation will scatter the data points, particularly for small samples. Hence, one needs experience with these plots to build up a basis for judgment.

3.15 Plotting Censored Samples

In engineering, samples are sometimes *censored* (see Section 12.2 for an introduction to the circumstances of censoring). That is, not all observations of the sample are known. This means that the true rankings of some of the available observations are not known either, and so the correct plotting positions (3.29) cannot be determined past the first censored value.

To accommodate censoring information in the probability plot, the well-known Kaplan–Meier *distribution-free* estimate[3] of the *reliability function* $R(x) = 1 - F(x)$ can be used:

$$\widehat{R}(x) = \prod_{q=1}^{k} \frac{n-q}{n-q+1}. \tag{3.31}$$

Here n is the total sample size (including the censored items), q is the order of the *actual* observation $x_{(q)}$ in the *total* sample, and $x_{(k)} \leq x \leq x_{(k+1)}$. Thus, the order q includes the number of items censored before $x_{(q)}$ occurred. The *product limit* (3.31) is a maximum-likelihood estimate. Its standard error is approximately

$$se(\widehat{R}) \doteq \widehat{R}\sqrt{\sum_{q=1}^{k}\{(n-q)(n-q+1)\}^{-1}}. \tag{3.32}$$

[3] Kaplan, E. L., Meier, P., "Nonparametric Estimation from Incomplete Observations," *J. Am. Stat. Assoc.*, Vol. 53, pp. 457–481, 1958.

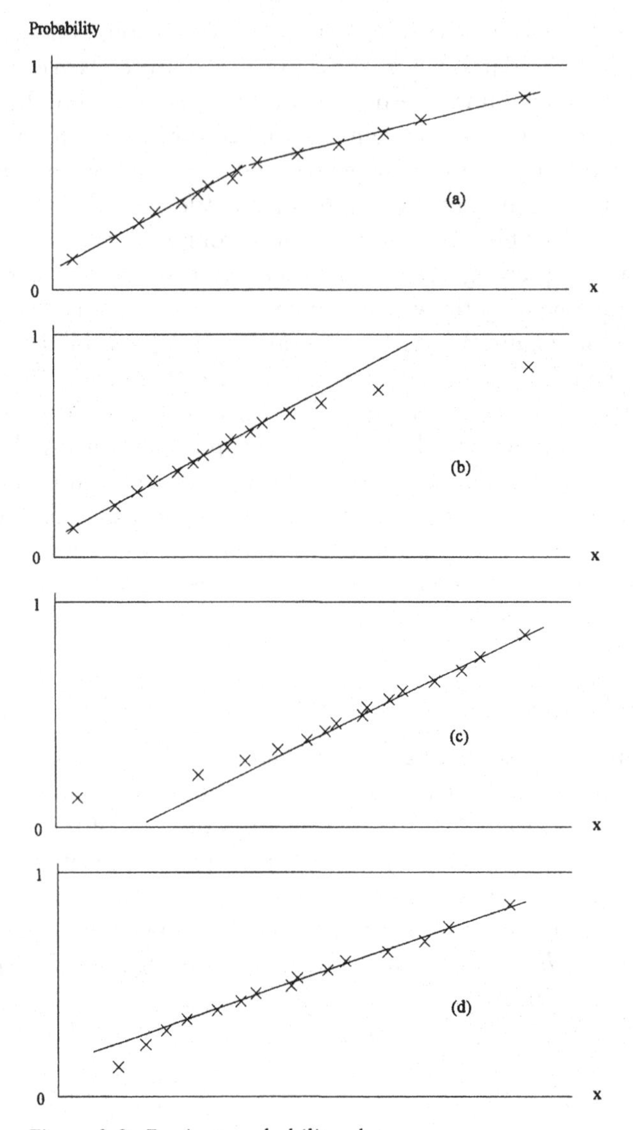

Figure 3.8. Deviant probability plots.

Formula (3.31) can now be adjusted to give the distribution-free *median* estimate of the cdf $F(x_{(k)})$. This adjusted estimate is the recommended plotting position for data from multiply censored samples:

$$p_k = 1 - \frac{n+0.7}{n+0.4} \prod_{q=1}^{k} \frac{n-q+0.7}{n-q+1.7}. \tag{3.33}$$

As before, q is the order of an *actual* observation in the *total* sample, which includes the censored items. When there is no censoring, (3.33) reduces to (3.29). See Example 12.1 for an illustration of plotting data from a censored sample.

EXAMPLE 3.1

A small hydroelectric turbine installation is operated by a resource company in a remote area. A design modification was made to the turbine runner to improve the turbine efficiency at part load. The following measurements were taken over a period of time at a certain wicket-gate setting (partially closed):

Effective hydraulic head *H* (ft)	Flow rate *Q* (cfs)	Power *P* produced (hp)
43.2	5390	19400
42.7	5280	19500
39.0	5160	17200
46.3	5620	22300
45.4	5660	21400
38.7	5100	16600
42.1	5340	19400
40.7	5190	18000
38.6	4960	16400
39.4	5170	17000

Estimate the turbine efficiency and the uncertainty (standard error) of this estimate.

Data matrix:

$$D := \begin{bmatrix} 43.2 & 5390 & 19400 \\ 42.7 & 5280 & 19500 \\ 39 & 5160 & 17200 \\ 46.3 & 5620 & 22300 \\ 45.4 & 5660 & 21400 \\ 38.7 & 5100 & 16600 \\ 42.1 & 5340 & 19400 \\ 40.7 & 5190 & 18000 \\ 38.6 & 4960 & 16400 \\ 39.4 & 5170 & 17000 \end{bmatrix}$$

$n := 10 \qquad m := 3$

$i := 1..n \qquad k := 1..m$

average values:

$$A_{j,k} := \frac{1}{n} \cdot \left(\sum_i D_{i,k} \right) \qquad \begin{aligned} H &:= A_{1,1} & H &= 41.6 \\ Q &:= A_{1,2} & Q &= 5287 \\ P &:= A_{1,3} & P &= 18720 \end{aligned}$$

mean-adjusted data matrix: $\quad M := D - A$

Convariance matrix

$$V := \frac{1}{n-1} \cdot M^T \cdot M \qquad V = \begin{pmatrix} 7.819 & 596.033 & 5643.111 \\ 596.033 & 49401.111 & 4.346 \cdot 10^5 \\ 5643.111 & 4.346 \cdot 10^5 & 4.133 \cdot 10^6 \end{pmatrix}$$

Correlation matrix

Let $T_k := \frac{1}{\sqrt{V_{k,k}}}$ and $S_{i,k} := M_{i,k} \cdot T_k$; then $C := \frac{1}{n-1} \cdot S^T \cdot S$

$$C = \begin{pmatrix} 1 & 0.959 & 0.993 \\ 0.959 & 1 & 0.962 \\ 0.993 & 0.962 & 1 \end{pmatrix}$$

Approximate average efficiency

Let $a := \frac{550}{62.4} \qquad E(H, Q, P) := a \cdot \frac{P}{H \cdot Q} \qquad E(H, Q, P) = 0.750$

Uncertainty in *E*

derivatives of E:

$$d_1 := \frac{d}{dH} E(H, Q, P) \qquad d_2 := \frac{d}{dQ} E(H, Q, P) \qquad d_3 := \frac{d}{dP} E(H, Q, P)$$

variance of E:

$$J := 1\,..\,3 \qquad \mathrm{var}E = \sum_j \sum_k d_j \cdot d_k \cdot V_{j,k} \qquad \mathrm{var}E = 0.000126$$

standard error of E:

$$\mathrm{se}E := \sqrt{\mathrm{var}E} \qquad \mathrm{se}E = 0.011$$

Note: The average efficiency estimate E can be improved by adding a correction term involving second-order partial derivatives of the decision function E:

$$DD_{1,1} := \frac{d}{dH}\frac{d}{dH} E(H, Q, P) \qquad DD_{2,2} := \frac{d}{dQ}\frac{d}{dQ} E(H, Q, P)$$

$$DD_{3,3} := \frac{d}{dP}\frac{d}{dP} E(H, Q, P) \qquad DD_{1,2} := \frac{d}{dH}\frac{d}{dQ} E(H, Q, P)$$

$$DD_{2,1} := DD_{1,2} \qquad DD_{1,3} := \frac{d}{dH}\frac{d}{dP} E(H, Q, P) \qquad DD_{3,1} := DD_{1,3}$$

$$DD_{2,3} := \frac{d}{dQ}\frac{d}{dP} E(H, Q, P) \qquad DD_{3,2} := DD_{2,3}$$

The improved estimate of E is

$$EE := E(H, Q, P) + \frac{1}{2} \sum_j \sum_k DD_{j,k} \cdot V_{j,k} \qquad EE = 0.748$$

PART TWO

DISCRETE DISTRIBUTIONS

CHAPTER FOUR

Introduction to Discrete Distributions

4.1 Discrete Probability Functions

There are many situations where the engineer measures a *discrete* quantity. That is, the measured variable X only admits discrete values x, usually integers as in the case of a *counting* process.

Examples of counting processes are: the number of components failing during a system's mission, the number of defectives in a manufactured lot, the number of production orders received at a factory, the number of earthquakes experienced in a region, the number of vehicles crossing a bridge, and the number of industrial accidents in a factory.

The uncertainty in the number of counts is measured by probability and is modeled by a mathematical function $p(x)$. This function gives the probability of occurrence of discrete values x measured in a sample space S:

$$Pr(X = x) = p(x); \qquad x \in S$$

(see also Section 1.6). The function p is termed the *probability mass function* (pmf), since probabilities are lumped at discrete values x. A related probability model is the *cumulative distribution function* (cdf):

$$Pr(X \leq x) = F(x); \qquad x \in S.$$

The two functions are connected via

$$F(x) = \sum_{t \leq x} p(t)$$

(see Fig. 4.1). The engineering interpretation of these functions is that they are *relative frequencies* of occurrence of x in a sequence of measurements on X. It follows that

$$0 \leq F(x) \leq 1$$

and

$$p(x) \geq 0.$$

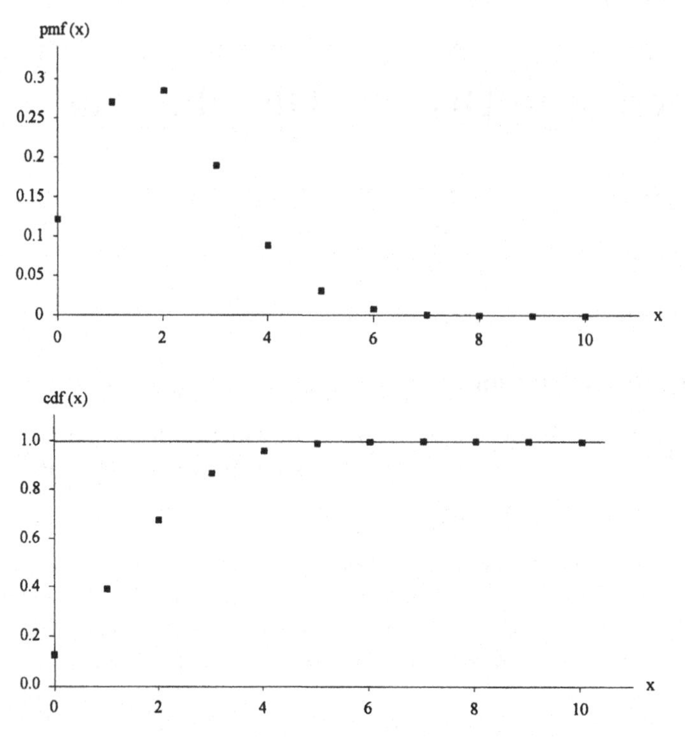

Figure 4.1. The Binomial pmf and cdf for $n = 20$ and $p = 0.1$.

The discrete probability distributions covered in this text are *indexed* by a parameter θ that takes values in a parameter space Ω:

$$p(x; \theta); \qquad x \in S, \qquad \theta \in \Omega.$$

The parameter θ can have several components. The more components there are, the more flexible is the model $p(x; \theta)$ in fitting discrete data. Although for continuous distribution models some of the parameters θ relate to the variable X in a consistent way (see Chapter 9), this is not the case for the discrete models discussed in this text. Thus, there are no location parameters, and the scale of the model is not associated with a specific parameter component. All parameter components influence, in combination, both the scale and the shape of the model.

4.2 Bernoulli Experiment

Many engineering problems can be modeled by the random occurrence or non-occurrence of a particular *event*.

For example, on inspection a randomly chosen product specimen may or may not prove to be *defective*. In a performance test a specimen device may or may not meet specifications. During its service life a structure may or may not be exposed to an earthquake of a certain magnitude. A new product may or may not meet its expected sales quota. A development project may or may not exceed its budget.

Typically there is a *sequence* of occasions, or *trials*, at which the event in question may or may not occur. If the following conditions hold, the sequence is termed a *Bernoulli sequence*, after the seventeenth-century mathematician Jakob Bernoulli. The conditions are:

1. The trials are *statistically identical*, each resulting in one of only two possible outcomes: the occurrence or nonoccurrence of a specified event.
2. For each trial the probability p of the event's occurrence is *constant*.
3. The trials are *statistically independent*.

It is customary to label the occurrence of the event in question as a "success" s and its nonoccurrence as a "failure" f. Furthermore, the event labeled s is typically so chosen that its probability p is less than f's probability $q = 1 - p$. A Bernoulli sequence, then, consists of a string of s and f events of length n; for example, $ffsfffssffff$ for $n = 12$. This sequence of events is a basic experimental scheme that underlies the discrete distribution models discussed in this text.

4.3 Randomness Test

The question of validity of a chosen discrete probability model needs to be addressed. This translates initially to checking the above conditions on which the Bernoulli process is based. That the first condition holds is usually obvious from the problem context: There are only two possible outcomes for each trial, and trials are conducted in identical circumstances.

The second and third conditions can sometimes be affirmed directly; the classical example is the repeated toss of a coin. In many practical instances, however, these conditions are *known* to not hold precisely, in which case the Bernoulli scheme may provide an *approximate* model. In particular, the probability p may not be precisely constant from trial to trial because the experiment is not stable over time. As well, the trials may not be entirely independent because experimental conditions may *drift* over time in a particular direction.

For example, there are many reasons why the characteristics of a machining process change consistently over time: cutting edges dull, gear teeth wear down, bearings loosen, sliding surfaces fret, and so forth. The result is that the probability p of producing a defective item cannot be constant over time, and that the change is likely in a consistent direction, implying the statistical dependence of successive trials. Nevertheless, the Bernoulli sequence provides a useful standard against which actual process performance can be measured to discover process changes that would call for timely corrective action.

When it is not clear whether conditions 2 and 3 hold for the particular discrete sample on hand, these conditions need to be *tested*. What is to be tested is essentially the condition of *randomness* (see Section 2.2). The following simple, but not too powerful, randomness test checks the number of *runs* observed in the data sequence. This test is "distribution-free," in that it does not depend on the choice of a particular distribution for modeling the random variable.

A *run* is defined as a subsequence of consecutive outcomes of one kind, immediately preceded and followed by outcomes of the other kind. Thus, in the preceding section the sample sequence shows five runs: ff, s, fff, ss, and $ffff$. Clearly, to obtain only one run (outcomes of one kind only) would be viewed as an unusual *clustering* and would intuitively be taken as evidence against the sample being random. Similarly, if n runs are found (outcomes alternate one by one), improbable *mixing* would be suspected, and the sample's randomness would also be questioned.

Consider a sample sequence with a successes and b failures. The sample size is $n = a + b$, and $a, b \geq 1$. If the randomness condition holds, all arrangements of elements in the sequence are equally likely. The probability that an arrangement contains exactly r runs of either kind can be shown to be as follows:
For r even:

$$p_{\mathrm{e}}(r; a, b) = \frac{2\binom{a-1}{\frac{r}{2}-1}\binom{b-1}{\frac{r}{2}-1}}{\binom{n}{a}}, \tag{4.1}$$

where $r = 2, 4, 6, \ldots, 2a$, and $a \leq b$.
For r odd:

$$p_{\mathrm{o}}(r; a, b) = \frac{\binom{a-1}{\frac{r+1}{2}-2}\binom{b-1}{\frac{r+1}{2}-1} + \binom{a-1}{\frac{r+1}{2}-1}\binom{b-1}{\frac{r+1}{2}-2}}{\binom{n}{a}}, \tag{4.2}$$

where $r = 3, 5, 7, \ldots, 2a - 1$, and $a < b$. The number of runs ranges between 2 and $r_{\max} = 2a + 1$. When $a = b$, $r_{\max} = n$. The symbol $\binom{a}{b}$ denotes the "Binomial coefficient" $\frac{a!}{b!(a-b)!}$.

The expected value of the number of runs is

$$E\{R\} = 1 + \frac{2ab}{n}, \tag{4.3}$$

and the variance of the number of runs is

$$\mathrm{Var}(R) = \frac{2ab(2ab - n)}{n^2(n-1)}. \tag{4.4}$$

A one-sided significance test on the observed number r of runs, in a sequence of size $n = a + b$, can be constructed as follows. If *clustering* is suspected, r would be substantially smaller than the expected value (4.3). A test at the significance level α requires the quantile r_q such that q equals α, or falls just below α. This

Table 4.1.[a] *The largest number r_c of runs, with a events of one kind and b events of another, such that $Pr\,(r \leq r_c) \leq 0.05$.*

	a:2	3	4	5	6	7	8	9	10	11	12	13	14	15	16	17	18	19	20
b:2																			
3																			
4			2																
5		2	2	3															
6		2	3	3	3														
7		2	3	3	4	4													
8	2	2	3	3	4	4	5												
9	2	2	3	4	4	5	5	6											
10	2	3	3	4	5	5	6	6	6										
11	2	3	3	4	5	5	6	6	7	7									
12	2	3	4	4	5	6	6	7	7	8	8								
13	2	3	4	4	5	6	6	7	8	8	9	9							
14	2	3	4	5	5	6	7	7	8	8	9	9	10						
15	2	3	4	5	6	6	7	8	8	9	9	10	10	11					
16	2	3	4	5	6	6	7	8	8	9	10	10	11	11	11				
17	2	3	4	5	6	7	7	8	9	9	10	10	11	11	12	12			
18	2	3	4	5	6	7	8	8	9	10	10	11	11	12	12	13	13		
19	2	3	4	5	6	7	8	8	9	10	10	11	12	12	13	13	14	14	
20	2	3	4	5	6	7	8	9	9	10	11	11	12	12	13	13	14	14	15

[a]Extracted by permission of the publisher from "Tables for Testing Randomness of Grouping in a Sequence of Alternatives," F. S. Swed, C. Eisenhart, *Annals of Mathematical Statistics*, Vol. 14, pp. 66–87, 1943.

critical value r_c can be found in Table 4.1 for $a, b \leq 20$ and $\alpha = 0.05$. Thus, there is a 5%, or lower, chance that the observed number of runs r is less than the tabulated value r_c even though the sequence is random. Hence, if $r \leq r_c$, the randomness hypothesis is rejected at the significance level α.

For example, the sequence $ffsfffssffff$ features $a = 3$, $b = 9$, $n = 12$, and $r = 5$. The expected value (4.3) is 5.5 and the standard deviation is 1.2 from (4.4). Thus there appears no reason to reject the randomnes hypothesis. As a check, the lower critical value of runs, at $\alpha = 5\%$, is found from Table 4.1 as $r_C = 2$, confirming the above conjecture.

If *mixing* is suspected, r would be substantially larger than the expected value (4.3). An α-level significance test can be constructed as above, with the upper 5% critical value obtained from Table 4.2. Note that Tables 4.1 and 4.2 also give the critical values for a two-sided run test at the significance level $\alpha = 0.10$.

If it is desired to construct a test at a significance level different from that of the table, the critical value r_c can be obtained by computing, and cumulating,

Table 4.2.[a] *The smallest number r_c of runs, with a events of one kind and b events of another, such that $Pr(r \geq r_c) \leq 0.05$.*

	a:2	3	4	5	6	7	8	9	10	11	12	13	14	15	16	17	18	19	20
b:2	4																		
3	5	6																	
4	5	6	7																
5	5	7	8	8															
6	5	7	8	9	10														
7	5	7	8	9	10	11													
8	5	7	9	10	11	12	12												
9	5	7	9	10	11	12	13	13											
10	5	7	9	10	11	12	13	14	15										
11	5	7	9	11	12	13	14	14	15	16									
12	5	7	9	11	12	13	14	15	16	16	17								
13	5	7	9	11	12	13	14	15	16	17	17	18							
14	5	7	9	11	12	13	15	16	16	17	18	19	19						
15	5	7	9	11	13	14	15	16	17	18	18	19	20	20					
16	5	7	9	11	13	14	15	16	17	18	19	20	20	21	22				
17	5	7	9	11	13	14	15	16	17	18	19	20	21	21	22	23			
18	5	7	9	11	13	14	15	17	18	19	20	20	21	22	23	23	24		
19	5	7	9	11	13	14	15	17	18	19	20	21	22	22	23	24	24	25	
20	5	7	9	11	13	14	16	17	18	19	20	21	22	23	24	24	25	26	26

[a] Extracted by permission of the publisher from "Tables for Testing Randomness of Grouping in a Sequence of Alternatives," F. S. Swed, C. Eisenhart, *Annals of Mathematical Statistics*, Vol. 14, pp. 66–87, 1943.

successive probabilities $p(2), p(3), p(4), \ldots$ from (4.1) and (4.2). See Example 4.1 for an illustration of this calculation.

When $a > 20$ and $b > 20$, the sampling distribution of R can be approximated by a Normal model (see Chapter 10) with mean μ equal to (4.3) and variance σ^2 equal to (4.4). One-sided critical values, for a significance level α, are

$$r_l = \mu + z_\alpha \cdot \sigma \tag{4.5}$$

for suspected clustering and

$$r_u = \mu - z_\alpha \cdot \sigma \tag{4.6}$$

for suspected mixing, where z_α is the *standard* Normal quantile of order α. See Example 4.2 for an illustration of the calculation.

4.4 Model Choice

The Hypergeometric distribution (Chapter 5) models the *number of successes* in a random sample from a *finite* dichotomous population. The Binomial distribution (Chapter 6) extends this model to Bernoulli samples from populations of *unlimited* size.

For example, the number of defective items found in a sample from a *fixed* production lot would be modeled by a Hypergeometric distribution. However, the number of defectives in a sample from a *continuous* production process would be modeled by a Binomial distribution.

The Negative Binomial distribution (Chapter 7) models the "inverse sampling" aspect of Bernoulli trials: the *number of trials* to the rth success. A special case of the Hypergeometric is the Geometric distribution, which models the number of trials to the *first* success.

For example, a flood-control structure is designed to contain a daily flow volume D. Each year there will be a maximum daily flow volume ("annual flood") that may or may not exceed D. The Geometric distribution models the number of years to the *first* exceedance of D.

The Poisson distribution (Chapter 8) models situations where an event may or may not occur at any point in time or space. By dividing the time or space continuum into small intervals such that only one event may occur in any one interval, a Bernoulli sequence of trials results, provided that the probability of occurrence of the event is constant for each trial. Thus, the Poisson distribution is the limiting case of the Binomial distribution, as the number of trials becomes large while the average occurrence rate remains constant.

For example, cracks may occur anywhere along a line of weld. Dividing the length of weld into very small segments (trials) results in a large number of segments to cover the weld length. If the rate of crack occurrence per unit length is constant, the Poisson distribution models the number of welds occurring along the weld line.

In *discrete*-distribution modeling, the problem situation usually leads to the appropriate model. Thus, distribution choice is not an issue. Similarly, a test-of-fit (see Section 3.12) is typically not a requirement in engineering applications of discrete distributions. When such a test is required, however, the classical Pearson *Chi-squared* statistic is recommended. This statistic, and its use, is well described in many introductory statistics texts.[1] Note that a large sample is needed for this statistic to be of value. Unfortunately, in many engineering problems the sample size is small.

To obtain visual feedback on how well the estimated model represents the data, simple *probability plotting* is recommended, when several measured data are on hand. That is, one plots the estimated cdf $F(x_{(i)}; \widehat{\theta})$ and a suitable sample

[1] See, for example, Devore, J. L., *Probability and Statistics for Engineering and the Sciences*, Wadsworth Inc, Belmont, Ca, 1995.

plotting position p_i versus the ordered data $x_{(i)}$. The median plotting position (see Section 3.13), is recommended:

$$p_i = \frac{i - 0.3}{n + 0.4}.$$

Although these plots cannot be linearized, they afford a visual check on the model's fit to the data. See Example 7.2 of Chapter 7 for an illustration.

EXAMPLE 4.1

For the sequence $f\,f\,s\,f\,f\,f\,s\,s\,f\,f\,f\,f$ determine the 5% critical number of runs if clustering is suspected.

$a := 3 \quad b := 9 \quad n := a + b \quad r\max := 2 \cdot a + 1 \quad r\max = 7 \quad r := 5$

expected value: $\mathrm{Er} := 1 + \dfrac{2 \cdot a \cdot b}{n} \qquad \mathrm{Er} = 5.5$

std. deviation: $\mathrm{SDr} := \sqrt{\dfrac{2 \cdot a \cdot b \cdot (2 \cdot a \cdot b - n)}{n^2 \cdot (n-1)}} \qquad \mathrm{SDr} = 1.2$

(4.1) simplified: $\mathrm{pe}(r) := \dfrac{2 \cdot (a-1)! \cdot (b-1)! \cdot a! \cdot b!}{\left(\left(\frac{r}{2} - 1\right)!\right)^2 \cdot \left(a - \frac{r}{2}\right)! \cdot \left(b - \frac{r}{2}\right)! \cdot n!}$

(4.2) simplified:

$$\mathrm{po}(r) := \frac{a! \cdot (a-1)! \cdot b! \cdot (b-1)!}{n! \cdot \left(\frac{r+1}{2} - 1\right)! \left(\frac{r+1}{2} - 2\right)!} \cdot \left(\frac{1}{\left(a - \frac{r+1}{2} + 1\right)! \cdot \left(b - \frac{r+1}{2}\right)!} + \frac{1}{\left(a - \frac{r+1}{2}\right)! \cdot \left(b - \frac{r+1}{2} + 1\right)!}\right)$$

Probability	**Cumulative probability**
pe(2) = 0.009	0.009
po(3) = 0.045	0.055
pe(4) = 0.145	0.200
po(5) = 0.291	0.491
pe(6) = 0.255	0.745

Hence, the critical value is $r_c = 2$. Since $r > r_c$, there is no evidence of clustering.

EXAMPLE 4.2

In a sample of 150 inspected highway sections, 32 sections were found to be substandard. There were 41 runs in the inspection sequence, and clustering due to assignable causes was suspected. Test for clustering at the 5% significance level.

$a := 32 \qquad n := 150 \qquad b := n - a \qquad r := 41$

expected value: $\mu := 1 + \dfrac{2 \cdot a \cdot b}{n} \qquad \mu = 51.3$

std. deviation: $\sigma := \sqrt{\dfrac{2 \cdot a \cdot b \cdot (2 \cdot a \cdot b - n)}{n^2 \cdot (n-1)}} \qquad \sigma = 4.1$

The standard Normal 5-percentile is

$z05 := -1.645$

From (4.5) the critical number of runs for clustering is

$r_c := \mu + z05 \cdot \sigma$ so $r_c = 45$

Since $r < r_c$, the randomness hypothesis is rejected: There is evidence of clustering.

CHAPTER FIVE

Hypergeometric Distributions

INTRODUCTION

5.1 Definition

In engineering, an experimental event (here called a *trial*) is often constrained to admit only two possible outcomes, usually labeled *success* s and *failure* f.

For example, a randomly chosen product specimen is classified, upon inspection, as defective or nondefective. In a destructive performance test, a prototype survives or it fails. (See also Section 4.2.)

Suppose the engineer contemplates a sequence of n such trials. If the population, from which the sample sequence is randomly chosen, is of finite size N, then it will contain some number M of items that would each produce a trial success s. The number x of successes s that could turn up in the sample sequence may then be of interest to the engineer.

Suppose, for example, a product shipment consists of $N = 100$ pieces. There will be $0 \leq M \leq N$ defectives in that shipment. If a random sample of size $n = 15$ specimens is inspected, the number x of defectives found in that sample gives the engineer information on the quality of the shipment.

The above experimental situation is characterized by the following conditions:

1. The sampled population is of finite size N.
2. A sample of $n < N$ trials is randomly selected.
3. Each trial admits only two outcomes: s or f.
4. There are $0 \leq M \leq N$ successes s in the population.

The number X of successes in such a sample is a discrete random variable with a pmf defined by the preceding combinatorial conditions as

$$h(x; n, M, N) = \frac{\binom{M}{x}\binom{N-M}{n-x}}{\binom{N}{n}}. \tag{5.1}$$

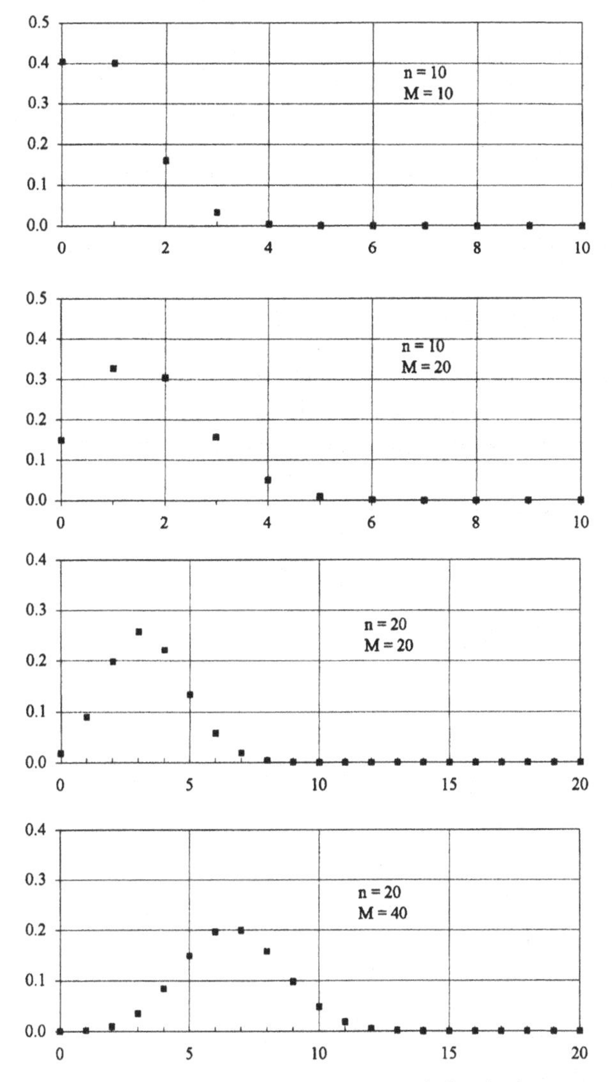

Figure 5.1. Several Hypergeometric pmfs for the lot size $N = 120$.

It is understood that x can only range between $\max(0, n - N + M)$ and $\min(n, M)$. The quantities N, M, and n are parameters, and $\binom{a}{b} = \frac{a!}{b!(a-b)!}$ is a Binomial coefficient. Variables with pmf (5.1) are termed *Hypergeometric.* Figure 5.1 shows several Hypergeometric pmfs.

The cdf (see Section 1.6) is defined for the Hypergeometric variable X as

$$H(x; n, M, N\} = \sum_{i=0}^{x} \frac{\binom{M}{i}\binom{N-M}{n-i}}{\binom{N}{n}}. \tag{5.2}$$

Both (5.1) and (5.2) are easily evaluated with a modern computational aid such as Mathcad.

5.2 Background

Historically, the Hypergeometric distribution arose as the probability model of certain compound events in games of chance.

For example, the probability of finding 3 aces in a fairly dealt hand of 13 cards is

$$h(x=3; n=13, M=4, N=52) = \frac{\binom{4}{3}\binom{52-4}{13-3}}{\binom{52}{13}} = \frac{4!\,48!\,13!\,39!}{3!\,1!\,10!\,38!\,52!} = 0.041,$$

since there are 4 aces in a deck of 52 cards.

Analogous situations arise in engineering, chiefly in the assessment of product quality by attributes, when the sampled population is of finite size. When the population size is large, the Hypergeometric distribution is approximated by the Binomial distribution (see Chapter 6 and the following section).

PROPERTIES

5.3 Hypergeometric Distribution

The expected value of a Hypergeometric variable X is

$$\mu_1'(X) = \frac{nM}{N} \tag{5.3}$$

and the variance is

$$\mu_2(X) = \frac{nM}{N^2}\frac{(N-M)(N-n)}{N-1}. \tag{5.4}$$

The third central moment is

$$\mu_3(X) = \frac{nM(N-2M)(N-M)(N-2n)(N-n)}{N^3(N-1)(N-2)}. \tag{5.5}$$

The fourth central moment is best calculated from

$$\mu_4(X) = \mu_4'(X) - 4\mu_3'(X)\mu_1'(X) + 6\mu_2'(X)[\mu_1'(X)]^2 - 3[\mu_1'(X)]^4 \tag{5.6}$$

and the following moments about the origin:

$$\mu_2'(X) = \frac{nM}{N}\left(1 + \frac{(n-1)(M-1)}{N-1}\right), \tag{5.7}$$

$$\mu_3'(X) = \frac{nM}{N}\left(1 + \frac{(n-1)(M-1)}{N-1}\left[3 + \frac{(n-2)(M-2)}{N-2}\right]\right), \tag{5.8}$$

$$\mu_4'(X) = \frac{nM}{N}\left(1 + \frac{(n-1)(M-1)}{N-1} \cdot \left\{7 + \frac{(n-2)(M-2)}{N-2}\left[6 + \frac{(n-3)(M-3)}{N-3}\right]\right\}\right). \tag{5.9}$$

The first shape factor (see Section 1.10) is

$$\gamma_1 = \frac{(N-2M)(N-2n)\sqrt{N-1}}{(N-2)\sqrt{nM(N-n)(N-M)}}. \tag{5.10}$$

The second shape factor is best calculated from its definition:

$$\gamma_2 = \frac{\mu_4(X)}{[\mu_2(X)]^2}. \tag{5.11}$$

As the population size N becomes large, while $M/N = p$ remains constant, the Hypergeometric moments and shape factors approach their Binomial equivalents. Hence, the Binomial distribution can be regarded as a limiting case of the Hypergeometric distribution.

The mode value x_m is the largest integer k satisfying the relation

$$k \le \frac{(n+1)(M+1)}{N+2}. \tag{5.12}$$

If the right-hand side of (5.12) is an integer, then x_m has two values: k and $(k-1)$.

5.4 Simulation

Simulation of a Hypergeometric variable X with known parameters M, N, and sample size n, follows directly from the conditions of the experiment (see Section 5.1). Let $C_S = M$ and $C_N = N$ be counters for the number of successes s in the population and the remaining population size, respectively, for successive trials. Generate n Uniform random numbers u_i on (0,1). If $u_1 \le C_S/C_N$, one success was simulated: Decrement C_S to $M-1$ and C_N to $N-1$. If $u_1 > C_S/C_N$, a failure was simulated: Decrement C_N to $N-1$. Repeating the process to the nth trial gives the cumulative count $x = M - C_S$ as one simulated Hypergeometric observation.

INFERENCES

5.5 Estimation

In engineering applications of the Hypergeometric distribution the lot size N and the sample size n are usually known, while the number M of *nonconforming* items in the lot needs to be estimated. The likelihood function of a Hypergeometric experiment is the pmf itself:

$$L(M) = \binom{M}{x}\binom{N-M}{n-x}\binom{N}{n}^{-1}. \tag{5.13}$$

To find the value of M that maximizes the above likelihood (see Section 3.6) consider $L(M+1)$ expressed in terms of $L(M)$:

$$L(M+1) = \frac{(M+1)(N-M-n+x)}{(N-M)(M+1-x)} L(M).$$

For L to achieve its maximum at M, the above fraction must be just less than 1. That is,

$$\frac{(M+1)(N-M-n+x)}{(N-M)(M+1-x)} \leq 1$$

or

$$M \geq \frac{x(N+1)}{n} - 1. \tag{5.14}$$

Thus, the maximum likelihood estimator of M is the smallest integer $\widehat{M}$ satisfying (5.14). When (5.14) is an equality, $\widehat{M}+1$ is also maximum likelihood.

The sampling variance of $\widehat{M}$ is obtained, using (5.4), as

$$\begin{aligned} \text{Var}(\widehat{M}) &= \left(\frac{N+1}{n}\right)^2 \text{Var}(X) \\ &= \frac{M}{nN^2} \frac{(N+1)^2(N-M)(N-n)}{N-1}. \end{aligned} \tag{5.15}$$

A confidence interval on M at the level of at least $(1-\alpha)$ (see Section 3.8), is given by (M_L, M_U), where M_L is the largest integer such that

$$H(x; n, M_L, N) \geq 1 - \frac{\alpha}{2} \tag{5.16}$$

and M_U is the smallest integer such that

$$H(x; n, M_U, N) \leq \frac{\alpha}{2}. \tag{5.17}$$

The value x is the observed number of successes in the experiment. The exact confidence level is the difference between the above cdfs. See Example 5.1 for an illustration of the computations.

5.6 Test

Suppose it is desired to discriminate between two conflicting claims on the number M of successes in a lot of size N. Denote the null hypothesis by $H_0: M = M_0$ and the alternative hypothesis by $H_a: M = M_a$, where $M_a < M_0$. Given a sample of size n and the significance level α of the test, the critical number of successes is the integer x_c such that

$$Pr(X \leq x_c \mid n, M_0, N) = H(x_c; n, M_0, N) \leq \alpha. \tag{5.18}$$

The *rejection* region is then $x \leq x_c$, resulting in the acceptance of the alternative H_a.

The *power* of the test is calculated from

$$Pr(X \leq x_c \mid n, M_a, N) = H(x_c; n, M_a, N) = 1 - \beta. \qquad (5.19)$$

If the calculated power is considered insufficient, the sample size n needs to be increased. See Example 5.2 for an illustration.

APPLICATIONS

5.7 Quality Control

To *improve* the quality of his products, the engineer uses *statistical experiments* to determine the influence on product quality of various design and production factors under his control. To *monitor* the quality of his production, he uses *control charts* that tell him when he needs to act to maintain acceptable quality levels. A third use of statistics in quality control is to *decide* the acceptance of incoming materials that feed the production process. These materials (e.g., manufactured components) are usually purchased in finite lots whose quality level is not under the direct control of the receiving engineer. The statistical aid for this decision is *acceptance sampling*. When quality is measured by an *attribute* (i.e., an item conforms to specifications or not), the Hypergeometric distribution models the situation exactly, whereas the Binomial distribution (see Chapter 6) approximates the situation when the lot size N is large.

The central problem here is to design a sampling plan with desirable characteristics. These characteristics are usually expressed as type I and II error probabilities (see Sections 3.10 and 3.11), at stipulated quality levels. In this context, the type I error probability α is called the *producer's risk*: the probability of rejecting a lot of acceptable quality level (AQL). The type II error probability β is termed the *consumer's risk*: the probability of accepting a lot with the worst acceptable (i.e., limiting), quality level (LQL).

Given the lot size N, the quality level $M_1 = AQL$ with risk α, and the limiting quality level $M_2 = LQL$ with risk β, one needs to find the sample size n and the critical number of defectives x_c in the sample. Since n and x_c are integers, an exact solution is rarely obtained. Rather, one chooses a combination (n, x_c) such that α and β do not exceed their stipulated values. The defining relations are

$$1 - H(x_c; n, M_1, N) \leq \alpha \qquad (5.20)$$

and

$$H(x_c; M_2, N) \leq \beta. \qquad (5.21)$$

The appropriate values for (n, x_c) can be obtained by trial and error with a computational aid such as Mathcad. See Example 5.3 for an illustration.

The characteristics of a chosen sampling plan are displayed by its *operating characteristic (OC) curve*, which plots the probability of accepting the lot as a function of the lot quality level M or the fraction defective M/N. It may then be instructive to also plot the OC curve for the limiting Binomial sampling plan, to see how closely the Binomial solution approximates the exact Hypergeometric solution. See Example 5.3.

5.8 Population Size

An application of the Hypergeometric distribution, outside engineering, concerns estimating the size N of an animal species on the basis of recapture data. For example, suppose that M members of an endangered species are caught, tagged, and released. In a later sample of size n, x tagged specimens are found. Clearly the probability of x is given by the Hypergeometric pmf.

The likelihood of this experiment, in terms of N, is therefore given by (5.13). To find the value of N that maximizes that likelihood, express $L(N+1)$ in terms of $L(N)$ as

$$L(N+1) = \frac{(N+1-M)(N+1-n)}{(N+1-M-n+x)(N+1)} L(N).$$

For L to be maximum at N requires that the above fraction be just less than or equal to 1. This condition reduces to

$$N \geq \frac{nM}{x} - 1. \tag{5.22}$$

Thus, the ML estimator of N is the smallest integer $\hat{N}$ satisfying (5.22). When (5.22) is an equality, $\hat{N}+1$ is also maximum likelihood. However, the moments of (5.22) are usually infinite. A more suitable estimator is given by

$$\tilde{N} \doteq \frac{(n+1)(M+1)}{x+1} - 1 \tag{5.23}$$

with approximate variance

$$\text{Var}(\tilde{N}) \doteq kN^2(1+2k+6k^2), \tag{5.24}$$

where $k = N/nM$.

EXAMPLE 5.1

Consider a shipment of 150 components. A sample of 20 components was inspected and 4 defectives were found. Estimate the number M of defectives in the shipment and calculate a 90% or higher confidence interval on M. What is the actual confidence level on that interval?

$N := 150 \qquad n := 20 \qquad x := 4 \qquad \alpha := 0.10$

1. Estimate

From (5.14): $\dfrac{x \cdot (N+1)}{n} - 1 = 29.2$ Hence the estimate is $M := 30$

standard error of the estimate:

$$SE_M := \frac{N+1}{N} \cdot \sqrt{\frac{(N-M) \cdot (N-n) \cdot M}{(N-1) \cdot n}} \qquad SE_M = 13$$

2. Confidence interval

Expressing the Hypergeometric cdf in terms of factorial functions:

$$H(M) := \sum_{i=0}^{x} \frac{M!}{i! \cdot (M-i)!} \cdot \frac{(N-M)!}{(n-i)! \cdot (N-M-n+i)!} \cdot \frac{n! \cdot (N-n)!}{N!}$$

Evaluating $H(M)$ for several values of M gives:

$H(15) = 0.968$
$H(16) = 0.958$ Lower confidence limit is $M_L = 16$
$H(17) = 0.946$

Similarly for the upper confidence limit:

$H(56) = 0.067$
$H(57) = 0.059$
$H(58) = 0.052$
$H(59) = 0.045$ Upper confidence limit is $M_U = 59$

The exact confidence level for this interval is $(0.958 - 0.045) = 0.913$.

EXAMPLE 5.2

A purchased component is received in lots of 160. The defect rate has been running at 10%. The supplier claims that he has reduced that rate to 5%. Every other component from an improved lot is to be checked for defects. Specify a hypothesis test at the significance level of 10% or less, and determine the power of this test.

$N := 160 \qquad n := 80 \qquad Ho: \quad Mo := 16 \qquad Ha: \quad Ma := 8$

The Hypergeometric cdf is expressed in terms of factorial functions as

$$H(x, M) := \sum_{i=0}^{x} \frac{M!}{i! \cdot (M-i)!} \cdot \frac{(N-M)!}{(n-i)! \cdot (N-M-n+i)!} \cdot \frac{n! \cdot (N-n)!}{N!}$$

To find the critical number of defects x, evaluate H for several values of x, according to (5.18):

$H(6, Mo) = 0.22$
$H(5, Mo) = 0.09$
$H(4, Mo) = 0.03$

The critical number is $x_c = 5$. Thus, if 5 or fewer defects are found in the sample, the supplier's claim is accepted.
Power of the test: From (5.19) the power is $H(x_c, Ma) = 0.86$.

EXAMPLE 5.3

Shipments of a manufactured component are received for assembly into a product. The shipment lot size is 160. Determine an attribute acceptance sampling plan with the producer's risk limited to 5% at 10 defectives/lot and the consumer's risk limited to 10% at 20 defectives/lot.

$N := 160 \quad \alpha := 0.05 \quad AQL: \; M1 := 10$
$\beta := 0.10 \quad LQL: \; M2 := 20$

Define the Hypergeometric cdf in terms of factorial functions as

$$H(x, n, M) := \sum_{i=0}^{x} \frac{M!}{i! \cdot (M-i)!} \cdot \frac{(N-M)!}{(n-i)! \cdot (N-M-n+i)!} \cdot \frac{n! \cdot (N-n)!}{N!}$$

Let $n := 75..90$

$x1 := 6 \quad x2 := 7 \quad x3 := 8$

n	$H(x1,n,M1)$	$H(x1,n,M2)$	$H(x2,n,M1)$	$H(x2,n,M2)$	$H(x3,n,M1)$	$H(x3,n,M2)$
75	0.88	0.08	0.97	0.18	0.99	0.34
76	0.87	0.07	0.97	0.17	0.99	0.32
77	0.86	0.07	0.96	0.15	0.99	0.3
78	0.86	0.06	0.96	0.14	0.99	0.28
79	0.85	0.05	0.95	0.13	0.99	0.26
80	0.84	0.05	0.95	0.12	0.99	0.24
81	0.83	0.04	X 0.95	X 0.1	0.99	0.22
82	0.81	0.04	0.94	0.09	0.99	0.2
83	0.8	0.03	0.94	0.08	0.99	0.18
84	0.79	0.03	0.93	0.08	0.99	0.17
85	0.78	0.02	0.93	0.07	0.99	0.15
86	0.77	0.02	0.92	0.06	0.98	0.14
87	0.75	0.02	0.91	0.05	0.98	0.13
88	0.74	0.02	0.91	0.05	0.98	0.12
89	0.73	0.01	0.9	0.04	0.98	0.1
90	0.71	0.01	0.89	0.04	0.98	0.09

A suitable sampling plan is therefore $n = 81$ and $x = 7$.

Plot the operating characteristic of this plan and compare with the Binomial approximation:

let $\quad M := 7..30 \quad$ the Binomial OC is:

$$B(x, n, M) := \sum_{i=0}^{x} \frac{n!}{i! \cdot (n-i)!} \cdot \left(\frac{M}{N}\right)^{i} \cdot \left(1 - \frac{M}{N}\right)^{n-i}$$

$B(x2, 81, M1) = 0.87$

$B(x2, 81, M2) = 0.19$

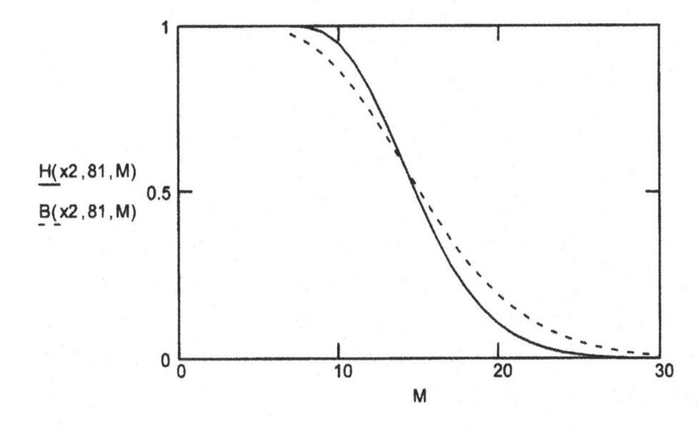

It is seen that the Binomial approximation is quite inaccurate in this example.

CHAPTER SIX

Binomial Distributions

INTRODUCTION

6.1 Definition

Consider a continuous manufacturing process that mass produces a certain component. In contrast to a finite manufactured lot (see Chapter 5), the population of possible components produced is large. Even for a well-designed and operated production process, there will be a small proportion of components that do not meet all manufacturing specifications. The production engineer would want to monitor the *rate* p at which these defectives are produced. To obtain information on p, he would randomly sample n components from the process, have them inspected, and have the number x of defectives counted. If the process is stable during sampling, the probability p of a component being defective ("success" s) is constant, and the sampling experiment generates a Bernoulli sequence of "trials" (see Section 4.2). The quantity of interest here is the number X of successes s in n Bernoulli trials.

There are $\binom{n}{x} = \frac{n!}{x!(n-x)!}$ distinct arrangements in which x successes can appear in a sequence of n trials. Since the probability of a success is p at any one trial, each arrangement occurs with probability $p^x(1-p)^{n-x}$. The probability of obtaining x successes in any order in n Bernoulli trials is therefore given by the pmf

$$b(x; p, n) = \binom{n}{x} p^x(1-p)^{n-x}, \qquad x = 0, 1, 2, 3 \ldots, n, \tag{6.1}$$

where p and n are parameters, and $\binom{n}{x} = \frac{n!}{x!(n-x)!}$ is a Binomial coefficient. The value of n is usually known from the circumstances of the experiment, whereas p is unknown and needs to be estimated from data. Discrete random variables X that are distributed according to (6.1) are called *Binomial* variables. Figure 6.1 shows several Binomial pmfs, indicating the range of shapes this pmf can take.

The cdf (see Section 1.6) is defined for the Binomial variable by

$$B(x; p, n) = \sum_{i=0}^{x} \binom{n}{i} p^i(1-p)^{n-i}. \tag{6.2}$$

Although this cdf is widely tabulated for various combinations of n and p, it is

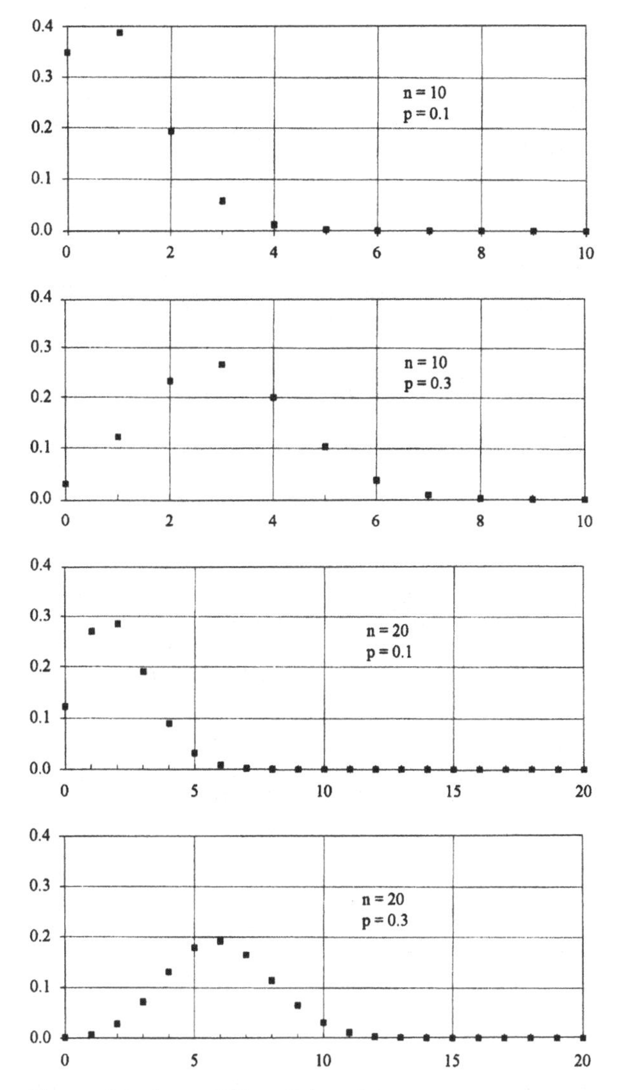

Figure 6.1. Several Binomial pmfs.

perhaps more convenient to evaluate that sum directly with a modern computational tool. See Example 6.1 for a Mathcad illustration. Note that Mathcad 6+ has a built-in function that gives the Binomial cdf.

6.2 Background

The Binomial distribution was derived by Jakob Bernoulli in the early seventeenth century. It is one of the oldest distributions studied systematically. Interest in it arose originally in connection with games of chance. The term "Binomial" derives from the fact that the pmf (6.1) is also the $(x+1)$th term in the expansion of the binomial $(p+q)^n$, with $p+q=1$ and n a positive integer.

PROPERTIES

6.3 Binomial Distribution

The expected value of the Binomial variable X is given by

$$\mu_1'(X) = np, \tag{6.3}$$

and the variance of X is

$$\mu_2(X) = np(1 - p), \tag{6.4}$$

which is less than the expected value. With $q = 1 - p$, higher central moments are

$$\mu_3(X) = npq(q - p) \tag{6.5}$$

and

$$\mu_4(X) = 3(npq)^2 + npq(q - p). \tag{6.6}$$

All of the above quantities can, of course, take noninteger values. The most likely (mode) value of X is the integer x_m satisfying the relation

$$(n + 1)p - 1 < x_m \leq (n + 1)p. \tag{6.7}$$

The first and second shape factors (see Section 1.10) are

$$\gamma_1 = \frac{q - p}{\sqrt{npq}} \tag{6.8}$$

and

$$\gamma_2 = 3 + \frac{1 - 6pq}{npq}. \tag{6.9}$$

We see that the distribution shape is symmetrical with respect to p and q and that the skew is positive if $p < q$.

As $n \to \infty$, $\gamma_1 \to 0$ and $\gamma_2 \to 3$. These values are the corresponding shape factors for the Normal model (see Section 10.3). Thus, for large n the Normal distribution with mean value $\mu = np$ and variance $\sigma^2 = npq$ provides a (crude) approximation to the Binomial distribution. Other approximations to the Binomial distribution are available. For example, as $n \to \infty$ and $p \to 0$, while $np = \lambda$ remains constant, the Binomial pmf approaches

$$\frac{\lambda^x \exp\{-\lambda\}}{x!}.$$

This is the Poisson pmf (see Section 8.1). Such limiting distributions were traditionally used to simplify the calculation of the Binomial cdf, but they are no longer required for this purpose because modern computational tools efficiently evaluate that cdf for all n.

The Binomial distribution has a *reproductive property*. That is, the *sum* of n independent Binomial variables X_i with parameters n_i and common p is also Binomial with parameters $\sum_i n_i$ and p:

$$S = \sum_{i=1}^{n} X_i \sim b\left(s;\, p, \sum_i n_i\right), \tag{6.10}$$

where the symbol $\sim$ denotes "distributed as."

The minimum-variance-bound (see Section 3.2) of unbiased estimators of the parameter p is given by

$$MVB_p = \frac{p(1-p)}{n}. \tag{6.11}$$

6.4 Simulation

There are many instances where an industrial system component can be modeled by a Binomial distribution. Often such systems are studied and optimized by *simulating* the system's operation. Thus, the Binomial variable X, given p and n, needs to be simulated (see Sections 2.7 and 2.8).

A Bernoulli trial is simulated quite simply by a single random number u. The trial is a success if $u < p$ and a failure if $u \geq p$. A Binomial realization x is then simulated by counting the number of successes in n such Bernoulli trials. This requires n random numbers for one simulated Binomial observation and a counting procedure. Alternatively, one could compute a table of the Binomial cdf and then search the table for the condition $B(x;\, p, n) \leq u < B(x+1;\, p, n)$ to give one simulated Binomial value. This requires only one random number and a search procedure for each simulated value x.

When the required number of simulations is large, as it often is, it is preferable to approximate the Binomial cdf by a Normal cdf, as in the preceding section, provided $n > 20$ and the value of the p is near 0.5. The simulated Normal observation (see Section 10.5) is then rounded to the nearest integer to provide an approximate simulated observation on the Binomial variable X. Note that Mathcad 6+ has a built-in function that generates Binomial random observations.

INFERENCES

6.5 Estimation

The likelihood function of a Binomial experiment is the pmf itself:

$$L(p) = \binom{n}{x} p^x (1-p)^{n-x}. \tag{6.12}$$

Differentiating the log of (6.12) with respect to p and equating to zero gives the maximum likelihood estimator of the parameter p as the *success ratio*

$$\hat{p} = \frac{x}{n} \tag{6.13}$$

(see also Section 3.6). The estimator is unbiased and minimum-variance-bound for all n (see Section 3.2). Its standard error is the square root of (6.11):

$$se(\hat{p}) = \sqrt{\frac{p(1-p)}{n}}. \tag{6.14}$$

A $(1-\alpha)$-level confidence interval on p (see Section 3.8) is given by (p_L, p_U), where p_L is the solution of

$$B(x; p_L, n) = \sum_{i=0}^{x} \binom{n}{i} p_L^i (1-p_L)^{n-i} = 1 - \frac{\alpha}{2} \tag{6.15}$$

and p_U is the solution of

$$B(x; p_U, n) = \sum_{i=0}^{x} \binom{n}{i} p_U^i (1-p_U)^{n-i} = \frac{\alpha}{2}. \tag{6.16}$$

Here, x is the observed number of successes in the Binomial experiment. See Example 6.2 for an illustration of the computations.

When the sample size n is large, a simpler, approximate confidence interval on p is obtained from the Normal approximation to the sampling distribution of the maximum likelihood estimator $\hat{p}$ (see Section 3.7):

$$(p_L, p_U) = \hat{p} \pm z_\alpha \cdot \sqrt{\frac{\hat{p}(1-\hat{p})}{n}},$$

where z_α is the *standard* Normal quantile of order α (see Section 10.3).

6.6 Test

In the industrial context the Binomial parameter p is often a process characteristic, such as a defect rate, that is desired to be small. Process improvements aim to reduce it, and a statistical test is required (see Section 3.11) to discriminate between the null hypothesis $H_0: p = p_0$ (existing condition) and an alternative hypothesis $H_a: p = p_a < p_0$ (process improvement). Given a sample size n and the significance level α of the test, the critical number of successes is given by the integer x_c such that

$$Pr(X \le x_c \mid p_0, n) = B(x_c; p_0, n) \le \alpha. \tag{6.17}$$

The *rejection region* is therefore $x < x_c$, resulting in the acceptance of H_a.

The *power* of the test is obtained from

$$Pr(X \leq x_c \mid p_a, n) = B(x_c; p_a, n) = 1 - \beta. \tag{6.18}$$

If the calculated power is inadequate, the sample size must be increased and x_c adjusted for α. See Example 6.3 for an illustration.

APPLICATIONS

6.7 Quality Control

In engineering work the Binomial distribution appears most often in the context of quality control for attributes. That is, when precise measurements of a product characteristic are unnecessary or infeasible, an inspected specimen is simply classified as acceptable or not in terms of a required attribute.

For example, a locating pin on a casting needs to be small enough to fit a hole in a mating part. The inspection of a specimen casting would attempt to fit a go-/no-go gauge to the pin. If the gauge fits, the casting is classified as acceptable. If it does not fit, the casting is classified as a reject.

When the population from which specimens are sampled can be considered very large (e.g., a continuous production process), the removal of the sample will not alter the sample space of the population. In that case the sample represents, at least approximately, a Bernoulli sequence, and the number X of successes in n trials is a Binomial variable. The *control* of an in-house production process is the usual application, where the process defect rate p is of central interest.

A related application is *acceptance sampling* by attributes, where the quality level of incoming material is of concern. Here one samples from lots of finite size N. It is usually impractical to return an inspected item indistinguishably to the lot, so that inspection is *without replacement*. The exact model of the number of successes in n such trials is the Hypergeometric distribution (see Chapter 5). However, if N is large compared to the sample size n ($N > 10n$), the Binomial distribution provides a reasonable approximation to the Hypergeometric distribution (see also Section 5.7).

To illustrate the approximate nature of the Binomial model for finite lots, consider a lot of 100 items, containing 5 defects. Suppose the first inspected item is defective, at the probability 5/100. Unless this item is indistinguishably returned to the lot, the next inspected item has a probability of 4/99 of being defective, and so forth. Thus p is not constant and the trials are not independent, so that inspections do not conform to the Bernoulli sequence. However, the Binomial model for the number of defects found in the inspection sample would provide a satisfactory (two-decimal) approximation for sample sizes less than ten.

In addition to applications in quality control, the Binomial distribution is used for distribution-free inferences on distribution values, distribution quantiles, and *tolerance limits* for distributions. By *distribution-free* inference is meant a procedure that does not assume a specific distributional form for the measurement variable in question. Such procedures are required when a distributional postulate is not warranted and the data base is deemed too small to fit a distribution to the data. The remainder of this chapter briefly introduces these inferences.

6.8 Distribution-Free Estimation of a cdf

Consider an ordered sample of measurements $t_{(1)}, t_{(2)}, \ldots, t_{(n)}$, on a variable T. The cdf of T is to be estimated at a specified value $t = h$. The *number* of observations less than or equal to h can be regarded as a Binomial random variable X, with probability of success equal to $p = F(h)$. The distribution-free estimate of $F(h)$ is thus given by the success ratio (6.13), and a distribution-free confidence interval on $F(h)$ is provided by (6.15) and (6.16). See Example 6.4 for an illustration.

6.9 Distribution-Free Estimation of a Quantile

Given an ordered sample of measurements $t_{(1)}, t_{(2)}, \ldots, t_{(n)}$, consider estimating the quantile t_q, where q is specified. A distribution-free estimate of t_q is clearly the order statistic $t_{(k)}$ where $\frac{k}{n}$ equals q or is just less than it.

A distribution-free confidence interval on t_q is given by a pair of order statistics $t_{(r)}, t_{(s)}$ such that $t_{(r)} < t_{(q)} < t_{(s)}$. The *number* of observations less than t_q is a Binomial random variable with parameter q. Hence, the probability of obtaining fewer than s observations t is

$$B(s; q, n) = \sum_{i=0}^{s-1} \binom{n}{i} q^i (1-q)^{n-i},$$

and for fewer than r observations that probability is

$$B(r; q, n) = \sum_{i=0}^{r-1} \binom{n}{i} q^i (1-q)^{n-i}.$$

Hence, the probability of the order statistics $t_{(r)}, t_{(s)}$ bracketing the unknown value t_q is given by the difference of the above probabilities:

$$Pr\left(t_{(r)} \le t_q < t_{(s)}\right) = \sum_{i=r}^{s-1} \binom{n}{i} q^i (1-q)^{n-i} = 1 - \alpha. \tag{6.19}$$

Thus, given the values n, r, s, and q, with $\frac{r}{n} < q < \frac{s}{n}$, the above expression determines the corresponding confidence level $(1 - \alpha)$.

Alternatively, given n, q, and a desired confidence level $(1-\alpha)$, the problem is to find the closest order statistics $t_{(r)}, t_{(s)}$ such that

$$Pr\left(t_{(r)} \leq t_q < t_{(s)}\right) = \sum_{i=r}^{s-1} \binom{n}{i} q^i (1-q)^{n-i} \geq 1-\alpha. \tag{6.20}$$

With the additional constraint that r/n and s/n be symmetrically close to q, the problem is easily solved by evaluating (6.20) for different symmetrical combinations of r and s. See Example 6.5 for an illustration.

6.10 Tolerance Limits for Distributions

In the manufacture of products, the engineer is often concerned with the range of measurements on a critical product characteristic X that he can expect for a *portion* of the distribution of X. An acceptable range of measurements may be stated as the *physical tolerance limits* (t_L, t_U) within which an inspected product specimen is acceptable. The engineer may then want to know the proportion d of acceptable products he can expect from the production process *on average*. This proportion is obtained from the cdf $F(x)$ as $d = F(t_U) - F(t_L)$.

For example, suppose the diameter of a stubshaft at the bearing seat is of critical importance. To produce acceptable assemblies, that diameter must be between two *physical tolerance limits* specified by the design engineer. Stubshafts with diameters outside these limits are classified as defective.

Since the estimated model $F(x)$ will be based on a finite sample of measurements on X, the calculated proportion d will feature statistical uncertainty. That is, repeated samples of size n produce slightly different Fs, and therefore slightly different calculated values d. Hence, d is associated with a sampling distribution.

In distinction to *estimating* the proportion d of acceptable specimens between the known physical tolerance limits t_L, t_U, the engineer may wish to know the *statistical* tolerance limits l_1, l_2 that bracket a *specified* proportion d of measurements X, at a *specified* level of statistical confidence $(1-\alpha)$.

Continuing the preceding example, we consider material hardness of the stubshaft as another important product characteristic. Its nominal value is also specified by the design engineer. Deviations from this value are undesirable but can be tolerated. Production personnel may be concerned with knowing what range $[l_1, l_2]$ of hardness values can be expected for a specified portion d, say 90%, of the production. The quantities l_1 and l_2, as well as d, are realizations of statistical variables L_1, L_2, and D, since they will vary from sample to sample.

Statistical tolerance limits are succinctly defined by the probability statement

$$Pr\left(D = \int_{L_1}^{L_2} f(x)\,dx \geq d\right) \geq 1-\alpha.$$

This problem has been solved for several distributions f. For example, Normal tolerance limits are tabulated.[1] When the distributional form f is unspecified, the well-known solution to this problem identifies two sample order statistics $x_{(r)} < x_{(s)}$, with the orders r and s defined by a Beta cdf F_B (see Chapter 14):

$$Pr(D \geq d) = 1 - F_B(d; s - r, n - s + r + 1) \geq 1 - \alpha.$$

Using the identity (14.7), this result is more conveniently expressed in terms of a Binomial sum as

$$Pr(D \geq d) = \sum_{i=0}^{s-r-1} \binom{n}{i} d^i (1 - d)^{n-i} \geq 1 - \alpha. \tag{6.21}$$

There are five parameters in the above expression: d, n, s, r, and α. To simplify matters, the orders r and s are usually assumed to be symmetric: $r = c$ and $s = n - c + 1$, which gives

$$Pr(D \geq d) = \sum_{i=0}^{n-2c} \binom{n}{i} d^i (1 - d)^{n-i} \geq 1 - \alpha. \tag{6.22}$$

Given any three of the four parameters d, n, c, or α, the fourth parameter is easily calculated numerically from (6.22). If α is to be calculated, then (6.22) is considered as an equality. If one of n or c is to be determined, then (6.22) is taken as an inequality and the solution is the smallest integer value n (or the largest integer value c) to equal or just exceed the specified confidence level $(1 - \alpha)$. See Example 6.6 for an illustration of the computation.

It is sometimes required to find the *lower* statistical tolerance limit l_1 that a portion d of measurements X can be expected to *exceed*, at a confidence level of at least $(1 - \alpha)$.

For example, in reliability engineering one often wishes to know the *time-to-failure* l_1 that a portion d of the equipment population in question is expected to *survive*, at a confidence level of at least $(1 - \alpha)$. Similarly, in the area of material strength, the design engineer is concerned with the lower tail of the strength distribution, since the weak specimens are the ones likely to fail. He therefore may consider the 5-percentile of strength as the appropriate design input. He may then wish to know the strength value l_1 that $d = 95\%$ of specimens can be expected to *exceed*, at a confidence level of at least $(1 - \alpha)$.

When the underlying distributional form f is not specified, the required lower tolerance limit is given by a sample order statistic $x_{(r)}$. The order r is determined from the following well-known expression involving a Beta distribution:

$$Pr(D \geq d) = 1 - F_B(d; n - r + 1, r) \geq 1 - \alpha.$$

[1] *CRC Basic Statistical Tables*, CRC Press, Boca Raton, FL.

This probability is more conveniently expressed as a Binomial sum, using identity (14.7), as

$$Pr(D \geq d) = \sum_{i=0}^{n-r} \binom{n}{i} d^i (1-d)^{n-i} \geq 1 - \alpha. \tag{6.23}$$

Four parameters are involved: d, n, r, and α. Given any three parameter values, the fourth is readily computed numerically; see Example 6.7 for an illustration.

EXAMPLE 6.1

A certain production process generates defective items at the rate of $p = 6\%$. What is the probability that a shipment of 200 products contains no more than 8 defectives?

$n := 200 \qquad x := 8 \qquad p := 0.06$

Answer: $Pr(X$ is less than or equal to $8) = B(8; 0.06, 200)$

To avoid numerical overflow in the factorial function for large n, the common factors in the Binomial coefficient can be cancelled and the coefficient can be defined for $j = 0$ to x as follows, provided x is small:

$$BC_0 := 1 \qquad j := 1 \,..\, x \qquad BC_j := \frac{\prod_{k=n-j+1}^{n} k}{j!}$$

The required probability is

$$B := \sum_{i=0}^{x} BC_i \cdot p^i \cdot (1-p)^{n-i} \qquad \text{with result:} \quad B = 0.147$$

Alternatively, express the factorials in the Binomial coefficient as the exponential of the sum of logs:

$$bc_0 := 1 \qquad bc_j := \exp\left(\sum_{i=1}^{n} \ln(i) - \sum_{i=1}^{j} \ln(i) - \sum_{i=1}^{n-j} \ln(i)\right)$$

Hence, the required probability is

$$B := \sum_{i=0}^{x} bc_i \cdot p^i \cdot (1-p)^{n-i} \qquad \text{resulting in} \quad B = 0.147$$

EXAMPLE 6.2

A sample of 200 items was chosen randomly from a production process. Each item was checked for conformity to specifications. Six items did not pass this inspection. Estimate the process defect rate and its standard error, and calculate the 90% confidence interval on that rate.

$n := 200 \qquad x := 6$

Estimate

success ratio $p := \frac{x}{n}$ $\qquad p = 0.030$

std. error $SEp := \sqrt{\frac{p \cdot (1-p)}{n}}$ $\qquad SEp = 0.012$

Exact confidence interval

To avoid numerical overflow in the factorial function for large n, the Binomial coefficient can be defined for $j = 0$ to x as follows, provided x is small:

$$BC_0 := 1 \qquad j := 1..x \qquad BC_j := \frac{\prod_{k=n-j+1}^{n} k}{j!}$$

lower limit: $pl := \text{root}\left[\sum_{i=0}^{x} BC_i \cdot p^i \cdot (1-p)^{n-i} - 0.95,\ p\right]$ $\qquad pl = 0.017$

upper limit: $pu := \text{root}\left[\sum_{i=0}^{x} BC_i \cdot p^i \cdot (1-p)^{n-i} - 0.05,\ p\right]$ $\qquad pu = 0.058$

Approximate confidence interval

$z05 := -1.645$

$pl := p + z05 \cdot \sqrt{\frac{p \cdot (1-p)}{n}}$ $\qquad pl = 0.010$

$pu := p - z05 \cdot \sqrt{\frac{p \cdot (1-p)}{n}}$ $\qquad pu = 0.058$

EXAMPLE 6.3

A production process has been operating at the defect rate $Po = 0.05$. A process improvement has aimed to reduce that rate to $Pa = 0.025$. Three-hundred and fifty specimens from the improved process are ready for inspection. Design a hypothesis test on Po versus Pa with a 5% significance level or lower. What is the power of this test?

$n := 350 \qquad po := 0.05 \qquad pa := 0.025$

To find the critical number of defects, evaluate (6.17) for several values of x, defining the Binomial coefficient as in preceding examples:

$$BC_0 := 1$$

$$x := 11 \qquad j := 1..x \qquad BC_j := \frac{\prod_{k=n-j+1}^{n} k}{j!}$$

$$B := \sum_{i=0}^{x} BC_i \cdot po^i \cdot (1 - po)^{n-i} \qquad B = 0.064$$

$$x := 10 \qquad j := 1..x \qquad BC_j := \frac{\prod_{k=n-j+1}^{n} k}{j!}$$

$$B := \sum_{i=0}^{x} BC_i \cdot po^i \cdot (1 - po)^{n-i} \qquad B = 0.035$$

The critical number of defects is 10, giving a significance level of 3.5%. If 10 or fewer defects are found in the sample, the null hypothesis is rejected and it is concluded that the process has improved.

Power of this test: $P := \sum_{i=0}^{x} BC_i \cdot pa^i \cdot (1 - pa)^{n-i}$ $\qquad P = 0.737$

EXAMPLE 6.4

Consider a sample of 40 observations on a variable Y. A distribution $F(y)$ has not been postulated. What is the probability that Y is less than, or equal to, the fourth ordered observation? Calculate a 90%, or higher, confidence interval on that probability.

$$n := 40 \qquad x := 4$$

Distribution estimate

success ratio $p := \frac{x}{n}$ $\qquad p = 0.100,$

std. error $SEp := \sqrt{\frac{p \cdot (1 - p)}{n}}$ $\qquad SEp := 0.047$

Exact confidence interval

lower limit: $pl := \text{root}\left[\sum_{i=0}^{x} \frac{n!}{i! \cdot (n-i)!} \cdot p^i \cdot (1-p)^{n-i} - 0.95,\ p\right]$ $\qquad pl = 0.051$

upper limit: $pu := \text{root}\left[\sum_{i=0}^{x} \frac{n!}{i! \cdot (n-i)!} \cdot p^i \cdot (1-p)^{n-i} - 0.05,\ p\right]$ $\qquad pu = 0.214$

Approximate confidence interval

$z05 := -1.645$

$pl := p + z05 \cdot \sqrt{\frac{p \cdot (1-p)}{n}}$ $\qquad pl = 0.022$

$pu := p - z05 \cdot \sqrt{\frac{p \cdot (1-p)}{n}}$ $\qquad pu = 0.214$

EXAMPLE 6.5

Consider a sample of 50 observations on a variable X for which a distribution $F(x)$ has not been postulated. Provide a distribution-free confidence interval on the 80-percentile, at the 90% or higher level of confidence.

$n := 50$

$q := 0.80 \qquad k := q \cdot n \qquad k = 40$

A point estimate would be given by the 40th observation.

$$i := 1..8 \quad r_i := 30 + i \quad s_i := 50 - i \quad P_i := \sum_{j=r_i}^{s_i - 1} \frac{n!}{j! \cdot (n-j)!} \cdot q^j \cdot (1-q)^{n-j}$$

r_i	s_i		P_i
31	49		0.999
32	48		0.996
33	47		0.988
34	46		0.967
35	45	X	0.921
36	44		0.836
37	43		0.699
38	42		0.507

Thus the 35th- and 45th-order statistics bracket the unknown 80-percentile at the 92.1% level of confidence.

EXAMPLE 6.6

Consider a sample of 50 observations. A distribution $F(x)$ has not been estimated yet. Provide symmetrical, distribution-free tolerance limits on 75% of the distribution, at a level of confidence of 95% or higher.

$n := 50 \qquad d := 0.75$

$$i := 1..4 \qquad c_i := 2 + i \qquad P_i := \sum_{j=0}^{n-2 \cdot c_i} \frac{n!}{j! \cdot (n-j)!} \cdot d^j \cdot (1-d)^{n-j}$$

c_i		P_i
3		0.993
4	X	0.955
5		0.836
6		0.618

The solution is $c = 4$, so that the required tolerance limits are given by the sample order statistics of order $r = 4$ and $s = n - c + 1 = 47$. The statistical confidence level for this interval is 95.5%.

EXAMPLE 6.7

A sample of 80 observations on the operating life to failure of a component has become available from a life test. Suppose that, before the estimation of a life model is attempted, it is required to produce a lower tolerance limit on 75% of the component population, at a confidence level of 90% or higher.

$n := 50 \qquad d := 0.75$

$$i := 1..6 \qquad r_i := 6 + i \qquad P_i := \sum_{j=0}^{n-r_i} \frac{n!}{j! \cdot (n-j)!} \cdot d^j \cdot (1-d)^{n-j}$$

r_i		P_i
7		0.981
8		0.955
9	X	0.908
10		0.836
11		0.738
12		0.618

The solution is $r = 9$; thus the required lower tolerance limit is given by the ninth sample order statistic. The statistical confidence level for this tolerance limit is 90.8%.

CHAPTER SEVEN

Negative Binomial Distributions

INTRODUCTION

7.1 Definition

Consider a sequence of *Bernoulli trials* (see Section 4.2). That is, the events of that sequence are independent, admit only one of two outcomes (success or failure), and the probability p of a failure in any one trial is constant. The quantity of interest is the number of successes X *before* the rth failure.

For example, suppose an engineered system comprises several standby-redundant components. During each of repeated system missions (trials) the on-line component may fail with probability p. Each component failure degrades the system reliability, and after r component failures the system is considered to have failed. Assuming that only the on-line component may fail during a system mission, a question of engineering interest is the number $(X+r-1)$ of missions that precede the rth component failure.

In contrast to a Binomial experiment (see Chapter 6) the number of failures is fixed at r, whereas the number X of successes before the rth failure event is a discrete random variable. Thus, in the first $(X+r-1)$ trials, $(r-1)$ failures occur with Binomial probability $b(r-1; p, X+r-1)$, and the rth trial is a failure with probability p. The pmf of X is therefore

$$nb(x; p, r) = \binom{x+r-1}{r-1} p^{r-1}(1-p)^{(x+r-1)-(r-1)} p$$

$$= \binom{x+r-1}{r-1} p^{r}(1-p)^{x}, \quad x = 0, 1, 2, 3, \ldots, \tag{7.1}$$

where $\binom{a}{b} = \frac{a!}{b!(a-b)!}$ is a Binomial coefficient. A discrete random variable with the above pmf is called *Negative Binomial*, since the pmf can also be generated as the $(x+1)$th term in the expansion of the negative binomial $p^r(1-q)^{-r}$, where $p+q=1$. The cdf (see Section 1.6) of a Negative Binomial variable X is

$$NB(x; p, r) = \sum_{i=0}^{x} \binom{i+r-1}{r-1} p^{r}(1-p)^{i}. \tag{7.2}$$

Both the pmf and the cdf are easily computed directly. Note that Mathcad 6+ gives these functions directly.

The Negative Binomial distribution is one of the oldest in the probability literature, dating back to the work of Pascal. When the parameter r is an integer, as it usually is for engineering applications, the Negative Binomial is often termed the *Pascal* distribution.

When the number of failures is $r = 1$, the variable X becomes the number of trials before the *first* failure. The Negative Binomial distribution then specializes to the *Geometric* distribution, with pmf

$$g(x;\, p) = p(1-p)^x, \quad x = 0, 1, 2, 3, \dots \tag{7.3}$$

and cdf

$$G(x;\, p) = 1 - (1-p)^{x+1}. \tag{7.4}$$

PROPERTIES

7.2 Negative Binomial Distribution

The expected value of a Negative Binomial variable X is

$$\mu_1'(X) = \frac{r(1-p)}{p}. \tag{7.5}$$

The variance is

$$\mu_2(X) = \frac{r(1-p)}{p^2}, \tag{7.6}$$

which is greater than the expected value. The coefficient of variation is

$$cv(X) = \frac{1}{\sqrt{r(1-p)}}. \tag{7.7}$$

The third and fourth central moments are

$$\mu_3(X) = \frac{r(1-p)(2-p)}{p^3} \tag{7.8}$$

and

$$\mu_4(X) = \frac{r(1-p)}{p^4}[3(1-p)(2+r) + p^2]. \tag{7.9}$$

The first shape factor (see Section 1.10) is

$$\gamma_1 = \frac{2-p}{\sqrt{r(1-p)}}, \tag{7.10}$$

indicating that the pmf is always positively skewed. The second shape factor is

$$\gamma_2 = 3 + \frac{1}{r}\left(6 + \frac{p^2}{1-p}\right). \tag{7.11}$$

As r becomes large, $\gamma_1 \to 0$ and $\gamma_2 \to 3$. These values are the corresponding shape factors for the Normal model (see Section 10.3). Thus, the Negative Binomial pmf tends toward the Normal "bell curve" as r increases. Furthermore, as $r \to \infty$ and $p \to 0$ while the expected value $r(1-p)/p = \theta$ remains constant, the Negative Binomial pmf approaches the Poisson pmf $\theta^x \exp\{-\theta\}/x!$ (see Chapter 8). Figure 7.1 shows several Negative Binomial pmfs.

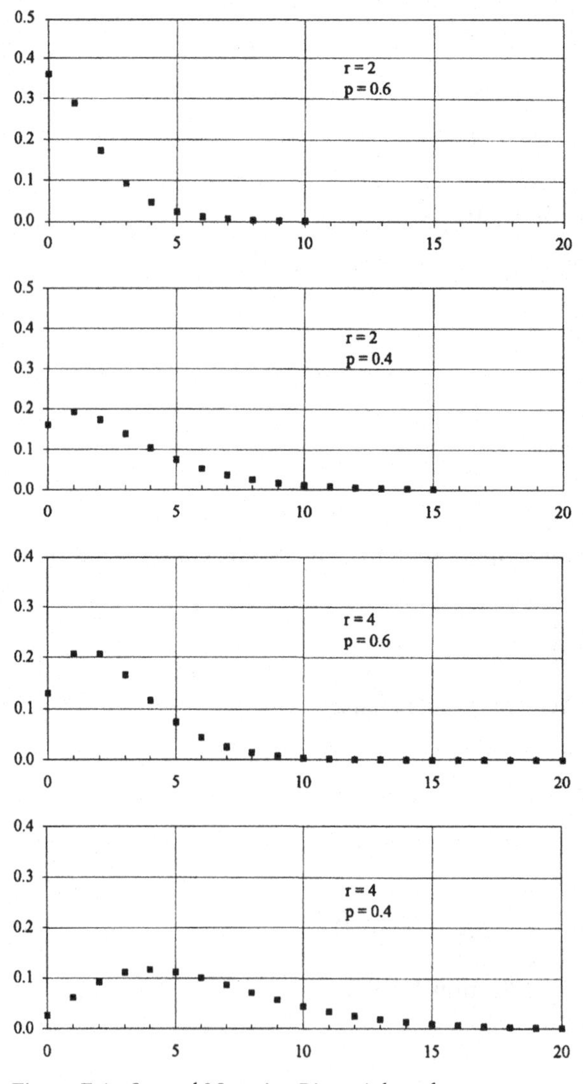

Figure 7.1. Several Negative Binomial pmfs.

The mode value x_m is the smallest integer k satisfying the relation

$$k \geq \frac{r(1-p)-1}{p}. \tag{7.12}$$

If the right side of (7.12) is an integer, then x_m has two values: k and $(k+1)$.

The Negative Binomial distribution has a *reproductive property*. That is, the *sum* of n independent variables X_i with parameters (p, r_i) is also Negative Binomial, with parameters $(p, \sum_{i=1}^{n} r_i)$:

$$S = \sum_{i=1}^{n} X_i \sim nb\left(s;\, p, \sum_{i=1}^{n} r_i\right), \tag{7.13}$$

where the symbol $\sim$ stands for "distributed as."

The *local* information matrix (see Section 3.7) for the estimation of the parameters p and r is given by

$$\begin{bmatrix} I_{pp} = \frac{nr}{p^2(1-p)} & I_{pr} = -\frac{n}{p} \\ I_{pr} = -\frac{n}{p} & I_{rr} = n\psi'(r) - \sum_{i=1}^{n} \psi'(x_i + r) \end{bmatrix}, \tag{7.14}$$

where ψ' is the *trigamma* function (see Section 13.2).

7.3 Geometric Distribution

For the Geometric distribution, we can simplify the above properties, noting that the Geometric variable X is the number of trials prior to the first failure. The expected value is

$$\mu_1' = \frac{1-p}{p}, \tag{7.15}$$

and the variance is

$$\mu_2 = \frac{1-p}{p^2}, \tag{7.16}$$

giving a coefficient of variation that is always greater than 1:

$$cv(X) = \frac{1}{\sqrt{1-p}}. \tag{7.17}$$

The third central moment is

$$\mu_3(X) = \frac{(1-p)(2-p)}{p^3}. \tag{7.18}$$

The fourth central moment is

$$\mu_4(X) = \frac{1-p}{p^4}[9(1-p) + p^2]. \tag{7.19}$$

The first and second shape factors are

$$\gamma_1 = \frac{2-p}{\sqrt{1-p}} \tag{7.20}$$

and

$$\gamma_2 = 9 + \frac{p^2}{1-p}. \tag{7.21}$$

Figure 7.2 shows several Geometric pmfs. Note how widely dispersed the pdf becomes as p decreases.

The Geometric distribution has a *memoryless* property, similar to the Exponential distribution (see Section 12.4). That is, the conditional probability of an

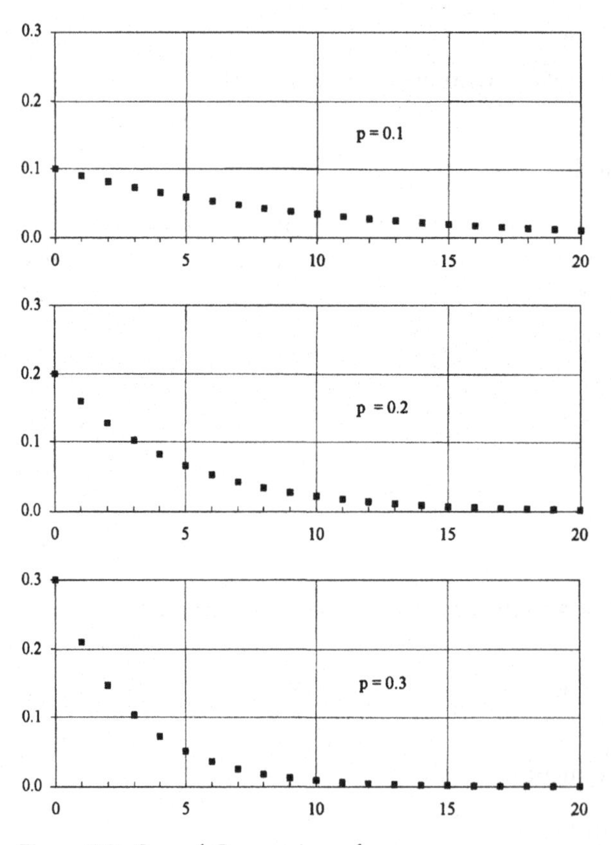

Figure 7.2. Several Geometric pmfs.

event, given some related information on that event, equals its unconditional probability:

$$Pr(X = a + b \mid X \geq b) = Pr(X = a).$$

This means, for example, that past events do not influence the probability of future events.

7.4 Simulation

To simulate observations from a Geometric distribution, the cdf (7.4) can be inverted directly (see Section 2.8):

$$x_i = \text{int}\left[\frac{\ln(1 - u_i)}{\ln(1 - p)}\right], \qquad (7.22)$$

where u_i is a Uniform random number on (0, 1). A Negative Binomial variable from $NB(x; p, r)$ can be simulated as the sum of r simulated values from the Geometric distribution $G(x; p)$. This makes use of the reproductive property (7.13). Note that Mathcad 6+ has a built-in function that generates random observations from a Negative Binomial distribution.

INFERENCES

7.5 Negative Binomial Distribution: *r* Known

In engineering applications of the Negative Binomial distribution, the integer r is usually chosen as a decision quantity, so that only the parameter p may have to be estimated from the observation x. The likelihood function (see Section 3.1) is the pmf itself:

$$L(p) = \binom{x + r - 1}{r - 1} p^r (1 - p)^x.$$

Equating the derivative of $L(p)$ with respect to p to zero (see Section 3.6) gives the maximum likelihood (ML) estimator:

$$\hat{p} = \frac{r}{x + r}.$$

This estimator is highly biased, particularly for small p, which is the region of interest in engineering. The alternate estimator

$$\hat{p}_u = \frac{r - 1}{x + r - 1}, \qquad r \geq 2 \qquad (7.23)$$

is unbiased, since

$$\begin{aligned}
E\{\widehat{p}_u\} &= \sum_{i-0}^{\infty} \frac{r-1}{i+r-1} \binom{i+r-1}{r-1} p^r (1-p)^i \\
&= p^r \sum_{i=0}^{\infty} \binom{i+r-1}{r-2} (1-p)^i \\
&= p^r p^{-r+1} \\
&= p.
\end{aligned}$$

An unbiased estimate of the sampling variance of $\widehat{p}_u$ is

$$\text{Var}(\widehat{p}_u) = \frac{p^2(1-p)}{r-1-p}, \quad r \geq 2. \tag{7.24}$$

If n observation on X are available, (7.23) becomes

$$\widehat{p}_u = \frac{r-1}{\bar{x}+r-1}, \quad r \geq 2, \tag{7.25}$$

where $\bar{x}$ is the average of the observations, and (7.24) becomes

$$\text{Var}(\widehat{p}_u) = \frac{1}{n} \cdot \frac{p^2(1-p)}{r-1-p}, \quad r \geq 2. \tag{7.26}$$

A confidence interval on p, with confidence level approximately $(1-\alpha)$, can be obtained from the reproductive property (see Section 7.2). Since $S = \sum_{i=1}^{n} x_i \sim nb(s; p, nr)$, consider the probability

$$Pr(s_L \leq S \leq s_U) \geq 1 - \alpha.$$

Here s_L is the quantile of order $\alpha/2$ or just less, and s_L is the quantile of order $(1-\alpha/2)$ or just greater. Replacing $S = n\bar{X}$ by $n(r-1)(1/\widehat{p}_u - 1)$ from (7.25), the event in the above probability can be reexpressed (see Section 3.8) to give

$$Pr\left(\frac{r-1}{\frac{s_U}{n}+r-1} \leq \widehat{p}_u \leq \frac{r-1}{\frac{s_L}{n}+r-1}\right) \geq 1 - \alpha.$$

Hence, the required confidence interval is

$$\left(\frac{r-1}{\frac{s_U}{n}+r-1}, \frac{r-1}{\frac{s_L}{n}+r-1}\right), \quad r \geq 2. \tag{7.27}$$

See Example 7.1 for an illustration.

7.6 Negative Binomial Distribution: r and p Unknown

When *fitting* the Negative Binomial distribution to discrete data, it becomes necessary to estimate both p and r. It is then convenient to generalize the parameter r to include *noninteger* values. This is easily done by expressing the Binomial coefficient in the pmf (7.1) in terms of Gamma functions (see Section 13.1), which admit noninteger arguments:

$$\binom{x+r-1}{r-1} = \frac{(x+r-1)!}{(r-1)!x!} = \frac{\Gamma(x+r)}{\Gamma(r)\Gamma(x+1)}. \tag{7.28}$$

The log-likelihood function can then be written as

$$\begin{aligned} LL(p,r) = {} & \sum_{i=1}^{n} \ln[\Gamma(x_i+r)] - n\ln[\Gamma(r)] - \sum_{i=1}^{n} \ln[\Gamma(x_i+1)] \\ & + nr\ln(p) + n\bar{x}\ln(1-p). \end{aligned} \tag{7.29}$$

Differentiating this function with respect to p and r, and equating to zero, gives two maximum likelihood estimating equations:

$$\hat{p} = \frac{\hat{r}}{\bar{x}+\hat{r}} \tag{7.30}$$

and

$$\ln(\hat{p}) = \psi(\hat{r}) - \frac{1}{n}\sum_{i=1}^{n} \psi(x_i+\hat{r}), \tag{7.31}$$

where ψ is the *digamma* function (see Section 13.2). Combining these expressions gives a single equation in $\hat{r}$:

$$\ln(\hat{r}) - \ln(\bar{x}+\hat{r}) - \psi(\hat{r}) + \frac{1}{n}\sum_{i=1}^{n} \psi(x_i+\hat{r}) = 0. \tag{7.32}$$

It is straightforward to find the root $\hat{r}$ numerically, using a computational aid such as Mathcad. The unbiased estimate (7.25) is recommended for p instead of (7.30), since the resulting fit tends to be superior.

The asymptotic sampling variances of these estimators can be calculated from the inverse of the information matrix (7.14). See Example 7.2 for an illustration. Note that the solution equation (7.32) may turn out to be ill-conditioned, in which case the Negative Binomial distribution would not be an appropriate model for the data.

7.7 Geometric Distribution

In engineering applications of the Geometric distribution ($r = 1$), the parameter p is usually known from the context of the problem, so that nothing needs estimating. However, should p need to be estimated from n observation on X, the ML estimator is

$$\hat{p} = \frac{1}{\bar{x} + 1}, \tag{7.33}$$

where $\bar{x}$ is the sample mean. The asymptotic sampling variance of this estimator is

$$\text{Var}(\hat{p}) = \frac{p^2(1 - p)}{n}. \tag{7.34}$$

The approximate $(1 - \alpha)$-level confidence interval on p is

$$\left(\frac{1}{\frac{s_U}{n} + 1}, \frac{1}{\frac{s_L}{n} + 1} \right), \tag{7.35}$$

where s_L and s_U are the quantiles of $nb(s; p, n)$ close to the orders $\alpha/2$ and $(1 - \alpha/2)$, respectively.

APPLICATIONS

7.8 Quality Control

In the context of quality control, the Negative Binomial distribution models "inverse binomial sampling plans" for the acceptance sampling by attributes (see also Section 6.7). That is, items are sampled one by one, and checked for conformity to specifications. Sampling continues until r nonconforming items are found and the lot is rejected, or until a conforming items are counted and the lot is accepted. The sampling plan is designated by the values of r and a, which are termed the rejection- and acceptance-number, respectively. The model assumes that sampled items are returned to the lot indistinguishably (not practical) or that the lot is sufficiently large so that sampling does not materially change the quality level p of the unsampled lot.

The characteristics of a specified sampling plan are given by the type I and II error probabilities (see Sections 3.10 and 3.11) at specific product quality levels. Here the type I probability α is the *producer's risk* of rejecting a lot with an acceptable quality level (AQL). The type II probability β is the *consumer's risk* of accepting a lot with a limiting (worst acceptable) quality level (LQL). The challenge is to design a sampling plan with desirable characteristics α, β at specified quality levels AQL and LQL. In terms of the Negative Binomial model, the defining equations are

$$NB(a + r; p = AQL, r) \leq \alpha \tag{7.36}$$

and

$$NB(a + r;\ p = LQL, r) \geq 1 - \beta. \tag{7.37}$$

Suitable values for a and r are obtained numerically by trial and error. The characteristics of the designated sampling plan are displayed by its *operating characteristic* (OC) curve, which plots the probability β as a function of the quality level p. See Example 7.3 for an illustration.

7.9 Accident Process

In the study of accidents, and similar processes, the number of occurrences X is often modeled by a Poisson pmf (8.1):

$$p(x; \theta) = \frac{\theta^x \exp\{-\theta\}}{x!}.$$

However, it is plausible that the mean occurrence rate θ is a variable depending on the "accident proneness" of the individual involved. When θ is modeled by a Gamma density (13.1):

$$f(\theta; \sigma, \lambda) = \frac{1}{\sigma^\lambda \Gamma(\lambda)} \theta^{\lambda - 1} \exp\left\{-\frac{\theta}{\sigma}\right\},$$

the marginal probability of X is

$$Pr(X = x) = \int_{\theta=0}^{\infty} p(x; \theta)\, f(\theta; \sigma, \lambda)\, d\theta.$$

This reduces to

$$Pr(X = x) = \frac{(x - \lambda - 1)!}{(\lambda - 1)!x!} \left(\frac{1}{1+\sigma}\right)^\lambda \left(1 - \frac{1}{1+\sigma}\right)^x. \tag{7.38}$$

Thus, the number of occurrences X has an exact Negative Binomial distribution with parameters $r = \lambda$ and $p = 1/(1 + \sigma)$.

The above property could also be used to simulate a Negative Binomial variable: First one simulates a Gamma observation θ_i (see Section 13.4) with scale parameter $\sigma = (1 - p)/p$ and shape parameter $\lambda = r$; then one simulates the Negative Binomial observation x_i from a Poisson distribution with parameter θ_i (see Section 8.5).

7.10 Discrete Data Model

The Poisson distribution (see Chapter 8) is a most important model of counting processes of rare events that occur randomly at a fixed rate. When the randomness condition is questioned, one searches for a somewhat more flexible alternative model. The Negative Binomial distribution is then a first choice, and the problem becomes one of estimating the parameters (p, r) to fit this distribution to the given counting data. See Section 7.6 and Example 7.2.

7.11 Waiting Time

Suppose one begins with the Negative Binomial model of a situation where each trial occupies an interval Δt, with the probability of a failure proportional to $\Delta t: p = \Delta t/\sigma$. The total time $t = (x + r)\Delta t$ required to the rth occurrence of a failure then approaches a Gamma variable with pdf

$$f(t; \sigma, r) = \frac{t^{r-1}}{\sigma^r \Gamma(r)} \exp\left\{-\frac{t}{\sigma}\right\}, \tag{7.39}$$

as $\Delta t \to 0$. The Gamma variable T is interpreted as the *waiting time* to the rth failure event; see also Section 13.16. Thus, the Negative Binomial distribution can be viewed as the discrete analogue of the Gamma distribution.

Similarly, when $r = 1$, T is the waiting time to the first failure, with a Geometric distribution that approaches the Exponential:

$$f(t; \sigma) = \frac{1}{\sigma} \exp\left\{-\frac{t}{\sigma}\right\}. \tag{7.40}$$

Thus, the Geometric distribution is the discrete analogue of the Exponential distribution. Note that the Geometric and Exponential models share the memoryless property (see Sections 7.3 and 12.4).

Applications of the Geometric distribution usually involve waiting-time situations where Bernoulli trials cover *fixed* intervals. The number of trials to the first failure event is then equivalent to the waiting time to that event.

An example from civil engineering is the *return period* of extreme load events such as the "annual floods" of a watershed. The concept arises when one considers the annual probability p of a load event that exceeds the design capacity of a structure. That probability value is, in fact, a *decision* quantity that is incorporated in the design of the structure. It relates to the *risk* associated with an economical design. The smaller the design value p_d, the larger (and more costly) the designed structure. Instead of specifying an acceptable design value p_d directly, it is common practice to quote, instead, the number of years that, *on average*, precede the first occurrence of the exceedance event. That measure is termed the *return period* t of the exceedance and is given by the average of the Geometric distribution with failure probability p_d:

$$t = \frac{1 - p_d}{p_d},$$

according to (7.15). Because for well-designed engineering structures p_d is small, the usual definition of the return period is stated as

$$t = \frac{1}{p_d}, \tag{7.41}$$

which is interpreted as the number of years to, and including, the first exceedance event. Recall from (7.17) that the coefficient of variation of the Geometric distribution is close to 1 for small p, indicating the wide dispersion of this model. Thus, although the return period t may be comfortably large, the yearly probability of an exceedance event remains at p.

For example, consider a flood-control structure designed to a 1,000-year flood. Here $t = 1{,}000$ and so $p = 10^{-3}$. The probability that the first exceedance of the design value occurs in less than t years is therefore

$$G\left(t;\ p = \frac{1}{t}\right) = 1 - \left(1 - \frac{1}{t}\right)^{t+1} = 0.633,$$

which is a sobering thought.

EXAMPLE 7.1

Three systems were observed for the number of failure-free missions X prior to the mission with the third component failure. The average of observations was 14. Estimate the failure probability of the components in the system environment, and obtain an 80% confidence interval on that probability.

$n := 3 \qquad r := 3 \qquad xb := 14 \qquad \alpha := 0.20$

1. Estimate

$$p := \frac{r - 1}{xb + r - 1} \qquad\qquad p = 0.125$$

standard error: $\quad \mathrm{VARp} := \dfrac{1}{n} \cdot \dfrac{p^2 \cdot (1 - p)}{r - 1 - p} \qquad \mathrm{SEp} := \sqrt{\mathrm{VARp}} \qquad \mathrm{SEp} = 0.049$

2. Confidence Interval

The distribution of the sum s is

$$P(s) := p^{n \cdot r}\left[1 + \sum_{i=1}^{s} \frac{(i + n \cdot r - 1)!}{(n \cdot r - 1)! \cdot i!} \cdot (1 - p)^i\right]$$

Trying a few values of s gives

$P(34) = 0.081 \qquad P(92) = 0.898$

$P(35) = 0.091 \qquad P(93) = 0.904$

$P(36) = 0.102$

Hence $\quad sL := 35 \quad$ and $\quad sU := 93$

The confidence limits are

$$pL := \frac{r-1}{\frac{sU}{n}+r-1} \qquad pL = 0.061$$

$$pU := \frac{r-1}{\frac{sL}{n}+r-1} \qquad pU = 0.146$$

The actual confidence level is $0.904 - 0.091 = 0.813$.

EXAMPLE 7.2

Five observations on a discrete random variable are on hand: 19, 22, 13, 17, 29. Fit a Negative Binomial distribution to these data.

$n := 5 \qquad i := 1..n$

$x_i :=$ $\begin{bmatrix} 19 \\ 22 \\ 13 \\ 17 \\ 29 \end{bmatrix}$ $\qquad xb := \frac{1}{n} \cdot \sum_i x_i \qquad xb = 20 \qquad \psi(a) := \frac{d}{da} \ln(\Gamma(a))$

1. Solution equation for *r*

$$g(a) := \ln(a) - \ln(xb+a) - \psi(a) + \frac{1}{n} \cdot \sum_i \psi(x_i + a)$$

guess value: $a := 2 \qquad r := \text{root}(g(a), a) \qquad r = 46.98$

2. Solution equation for *p*

$$p := \frac{r}{xb+r} \qquad p = 0.701$$

unbiased: $p_u := \frac{r-1}{xb+r-1} \qquad p_u = 0.697$

3. Check on solution

Plot the log-likelihood function, expressed in terms of r:

$$LL(r) := \sum_{i=1}^{n} \ln(\Gamma(x_i+r)) - n \cdot \ln(\Gamma(r)) + n \cdot r \cdot \ln\left(\frac{r}{xb+r}\right) + n \cdot xb \cdot \ln\left(1 - \frac{r}{xb+r}\right)$$

$rt := 40..60$

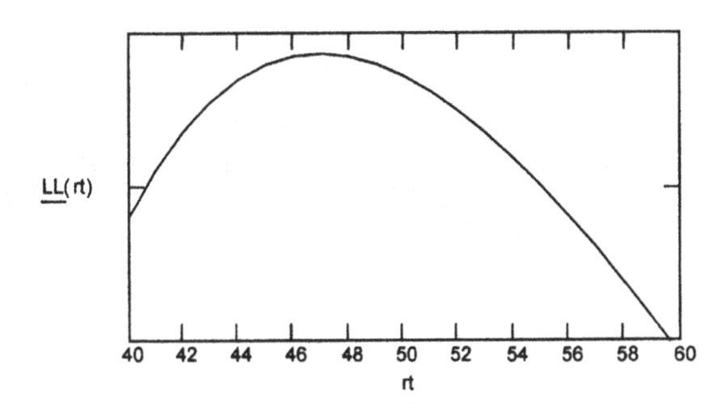

4. Standard errors of the estimates

local information matrix:

$$\psi d(a) := \frac{d}{da}\psi(a) \qquad I := \begin{bmatrix} \frac{n \cdot r}{p^2 \cdot (1-p)} & -\frac{n}{p} \\ -\frac{n}{p} & n \cdot \psi d(r) - \sum_{i=1}^{n} \psi d(x_i + r) \end{bmatrix}$$

covariance matrix: $V := I^{-1}$
standard errors:

$$\mathrm{SEp} := \sqrt{V_{0,0}} \qquad \mathrm{SEp} = 0.44$$

$$\mathrm{SEr} := \sqrt{V_{1,1}} \qquad \mathrm{SEr} = 98.61$$

Note: It is evident from the large standard errors that there is not enough information in the sample to estimate the parameters reliably.

graphical check of estimated model:
Plot sample cdf as

$$pl_i := \frac{i - 0.3}{n + 0.4}$$

ordered observations: $z := \mathrm{sort}(x)$
Plot model cdf:

$$NB_i := \sum_{k=0}^{z_i} \frac{\Gamma(k+r)}{\Gamma(r) \cdot \Gamma(k+1)} \cdot p_u^r \cdot (1 - p_u)^k$$

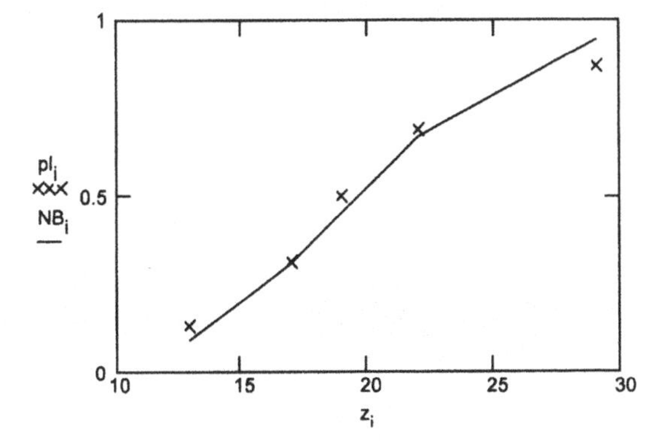

EXAMPLE 7.3

Design an inverse Binomial sampling plan with producer's risk no greater than 5% at an *AQL* of 3% and consumer's risk no greater than 12% at a *LQL* of 10%.

AQL: $p1 := 0.03 \qquad \alpha := 0.05$
LQL: $p2 := 0.10 \qquad \beta := 0.12$

The Negative Binomial cdf is

$$NB(x, p, r) := p^r \cdot \left[1 + \sum_{i=1}^{x} \frac{(i + r - 1)!}{(r-1)! \cdot i!} \cdot (1-p)^i \right]$$

Try: $x := 70..85$ and $r = 5$ and 6:

x	$NB(x, p1, 5)$		$NB(x, p1, 6)$	$NB(x, p2, 5)$		$NB(x, p2, 6)$
70	0.075		0.027	0.881		0.784
71	0.078		0.028	0.888		0.794
72	0.082		0.03	0.894		0.804
73	0.085		0.032	0.901		0.814
74	0.089		0.033	0.906		0.823
75	0.093		0.035	0.912		0.832
76	0.097		0.037	0.917		0.841
77	0.1		0.039	0.922		0.849
78	0.104		0.041	0.927		0.857
79	0.108		0.043	0.931		0.864
80	0.113		0.045	0.936		0.871
81	0.117		0.047	0.94		0.878
82	0.121	X	0.049	0.943	X	0.885
83	0.125		0.052	0.947		0.891
84	0.13		0.054	0.95		0.897
85	0.134		0.056	0.953		0.902

Thus the rejection number is $r = 6$, and the acceptance number is $82 - 6 = 76$.

OC curve:

$j := 1..20 \qquad p_j := 0.005 + \dfrac{0.05 \cdot j}{10}$

$1 - NB(82, p_j, 6)$

p_j

CHAPTER EIGHT

Poisson Distributions

INTRODUCTION

8.1 Definition

The engineer often encounters problems where important information derives from the random occurrence of critical events during an extended period or over a lengthy segment of space.

For example, the possible occurrence of an earthquake exceeding a specified magnitude influences the risk inherent in a designed structure. The failure of a redundant component in an engineered system degrades its reliability. The occurrence of a flaw in a length of optical material compromises its performance.

Suppose that the occurrence of the event in question, during a *short* time (or space) interval, conforms to a Bernoulli scheme (see Section 4.2). That is, only one event may occur during that interval, the event's occurrence probability p is constant for all such intervals, and the events are statistically independent. The probability of x events occurring in n intervals is given by the Binomial pmf (5.1):

$$b(x; p, n) = \binom{n}{x} p^x (1-p)^{n-x},$$

where $\binom{n}{x} = \frac{n!}{x!(n-x)!}$ is a Binomial coefficient. In many instances the engineer deals with *rare* events, so that p is small and n is large, while the average rate of occurrence $np = \lambda$ is constant. This constant rate characterizes the event phenomenon. In the limit, as the length of the measurement interval decreases, $p \to 0$ and $n \to \infty$ while np remains at λ. To obtain the limiting form of the Binomial pmf, express it in n and λ as

$$\begin{aligned} b(x; n, \lambda) &= \frac{n!}{x!(n-x)!} \left(\frac{\lambda}{n}\right)^x \left(1 - \frac{\lambda}{n}\right)^{n-x} \\ &= \frac{\lambda^x}{x!} \left(1 - \frac{\lambda}{n}\right)^n \left\{ \frac{n!}{(n-x)!} \left(1 - \frac{\lambda}{n}\right)^{-x} \right\}. \end{aligned}$$

In the limit, the term in braces becomes 1 and the preceding term becomes $\exp\{-\lambda\}$. Thus, the Binomial pmf approaches the limit

$$p(x;\lambda) = \frac{\lambda^x \exp\{-\lambda\}}{x!}; \quad x = 0, 1, 2, 3\ldots; \quad \lambda > 0, \tag{8.1}$$

where λ is the occurrence rate of the events and x is the number of observed events, in a space or time *window* of observation. Discrete random variables X that are distributed according to (8.1) are termed *Poisson* variables. Figure 8.1 shows several Poisson pmfs, indicating the range of shapes this pmf can take.

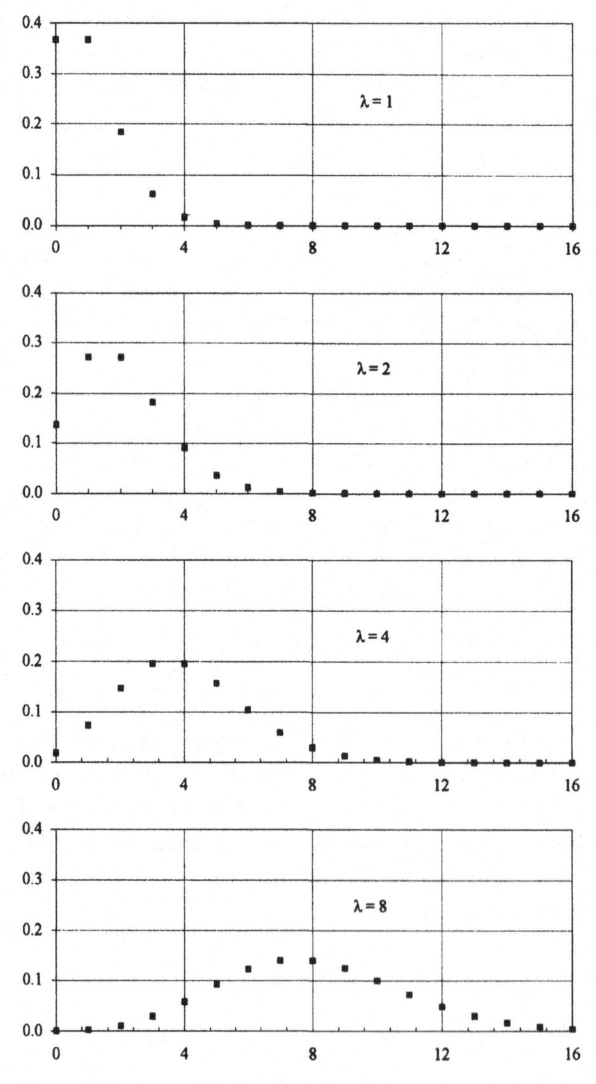

Figure 8.1. Several Poisson pmfs.

For example, suppose that from records the yearly average number of significant industrial accidents in a large factory is 2.6. Assuming that such accidents occur randomly, the probability of exactly 2 accidents occurring in a year is then

$$p(2; 2.6) = 2.6^2 \exp\{-2.6\}/2! = 0.251.$$

Note that, although the above derivation implies rare events, this need not be so for the Poisson distribution to model a counting phenomenon. The randomness of event occurrences, and their constant rate, are the essential features.

The cdf (see Section 1.6) is defined for the Poisson variable as

$$P(x; \lambda) = \exp\{-\lambda\} \sum_{i=0}^{x} \frac{\lambda^i}{i!}. \tag{8.2}$$

This cdf is widely tabulated for different values of the parameter λ. However, with a modern computational tool, such as Mathcad, (8.2) is easily evaluated directly.

This distribution was derived by the French scientist Poisson about 150 years ago. At the turn of the nineteenth century, Bortkiewicz introduced the Poisson distribution as a statistical model of rare events, calling it the "law of small numbers." In the meantime this distribution has become the most prominent model for a wide range of discrete random variables.

8.2 Poisson Process

An important generalization of the Poisson distribution is obtained by considering the window of observation as a function of time t, so that the Poisson parameter λ becomes νt, where ν is a unit rate. The Poisson pmf (8.1) then reads

$$p(x; \nu, t) = \frac{(\nu t)^x \exp\{-\nu t\}}{x!}. \tag{8.3}$$

In this formulation the pmf models the probability of a *sample function* $X(t)$ for any given time t, if that sample function is generated by a Poisson *counting process*. For a counting process to be Poisson, it must meet the following conditions:

I. The probability of an event occuring in a short interval Δt is approximately $\nu \Delta t$.
II. The probability of more than one event occurring in Δt is approximately zero.
III. The number of events that occur in a time span t_1 is independent of the number of events that occur in any other (nonoverlapping) time span t_2.

Because these assumptions are not particularly restrictive, the Poisson distribution is a widely applicable model for general counting processes. Note also the analogy to the conditions underlying the Bernoulli scheme (see Section 4.2).

For example, suppose that knots occur in a certain species of lumber at $\nu = 0.042$ knots per centimeter board length, and suppose that the above assumptions are approximately satisfied for this defect process. The number of knots that occur in 4-meter boards would then be modeled by a Poisson distribution with parameter $\lambda = \nu t = 0.042(400) = 16.8$.

PROPERTIES

8.3 Poisson Distribution

The moment generating function (see Section 1.8) is

$$M(t) = \exp\{\lambda[\exp(t) - 1]\}, \tag{8.4}$$

so that the expected value of X is obtained as

$$\mu_1'(X) = \lambda \tag{8.5}$$

and the next few moments about the origin are

$$\begin{aligned} \mu_2'(X) &= \lambda + \lambda^2, \\ \mu_3'(X) &= \lambda + 3\lambda^2 + \lambda^3, \\ \mu_4'(X) &= \lambda + 7\lambda^2 + 6\lambda^3 + \lambda^4. \end{aligned} \tag{8.6}$$

Using expression (1.4), the variance of X is obtained as

$$\mu_2(X) = \lambda, \tag{8.7}$$

and the next two moments about the mean are

$$\begin{aligned} \mu_3(X) &= \lambda, \\ \mu_4(X) &= \lambda + 3\lambda^2. \end{aligned} \tag{8.8}$$

The first and second shape factors (see Section 1.10) are therefore

$$\gamma_1 = \frac{1}{\sqrt{\lambda}} \tag{8.9}$$

and

$$\gamma_2 = 3 + \frac{1}{\sqrt{\lambda}}. \tag{8.10}$$

Hence, the Poisson pmf is always skewed to the right. As λ increases, the shape factors approach the Normal values 0 and 3 (see Section 10.3). Thus, for large λ the Poisson distribution approaches the shape of the Normal distribution.

The mode value is the integer portion of the parameter λ:

$$x_m = \text{int}[\lambda]. \tag{8.11}$$

When λ is itself an integer, there are two mode values: λ and $(\lambda - 1)$.

The Poisson distribution has a *reproductive property*; namely, the *sum* of n independent Poisson variables X_i with parameters λ_i is also Poisson, with parameter $\lambda = \sum_{i=1}^{n} \lambda_i$:

$$S = \sum_{i=1}^{n} X_i \sim p\left(s; \sum_{i=1}^{n} \lambda_i\right), \tag{8.12}$$

where the symbol $\sim$ stands for "distributed as." Thus, the average $\bar{X} = S/n$ of n observations from a Poisson process with parameter λ is itself Poisson with parameter $n\lambda$:

$$\bar{X} \sim p(\bar{x}; n\lambda), \qquad \bar{x} = 0, \frac{1}{n}, \frac{2}{n}, \frac{3}{n}, \ldots. \tag{8.13}$$

The minimum-variance-bound (see Section 3.2) of unbiased estimators for λ, given a single observation x, is

$$MVB_\lambda = \lambda. \tag{8.14}$$

Given n observations x_i, that bound is

$$MVB_\lambda = \frac{\lambda}{n}. \tag{8.15}$$

8.4 Relation to Exponential Distribution

A characterization of a Poisson process is that the distribution of *distances* D between consecutive Poisson events is Exponential:

$$f(d; \nu) = \nu \exp\{-\nu d\}.$$

From expression (12.3) it is recognized that $\sigma = 1/\nu$ can here be regarded as a *scale parameter*. Thus, the Exponential distribution of distances,

$$f_{\text{Exp}}(d; \nu) = \frac{1}{\sigma} \exp\left\{-\frac{d}{\sigma}\right\}, \tag{8.16}$$

is an equivalent description of the Poisson process $p(x; t/\sigma)$. Thus the *randomness* of Poisson events takes expression in the *memoryless* property of the Exponential variable D (see Section 12.4).

In the context of *waiting line theory*, the distance D is termed the *inter-arrival time* (IAT) between Poisson events (see also Section 13.16). The parameter σ is then interpreted as the mean arrival time for the events, and the sum of n Exponential IATs is the *waiting time* to the $(n-1)$th Poisson event. In Section 12.4 it is indicated that the exact distribution of the sum of n Exponential variables is Gamma distributed with parameters σ and n. Since the shape parameter n is here integer valued, the Gamma distribution specializes to the *Erlang* distribution (see Chapter 13). Thus, the Erlang distribution of the waiting time to the $(n-1)$th Poisson event is the third equivalent description of a Poisson process.

8.5 Simulation

To simulate Poisson events, the above relation to an Exponential process (8.16) can be used. Thus, if u_i is a Uniform random number on the interval (0, 1), an observation d_i from (8.16) is simulated as

$$d_i = -\sigma \ln(u_i)$$

(see Section 12.5). A simulated Poisson observation x is then obtained by cumulating Exponential observations d_i until the following condition is satisfied:

$$\sum_{i=1}^{x} d_i \leq \lambda < \sum_{i=1}^{x+1} d_i. \tag{8.17}$$

When λ is large, a faster scheme is to use the Normal approximation with mean and variance λ. The integer portion of the Normal random deviate (see Section 10.5) serves as an approximate Poisson simulation. Note that Mathcad 6+ has a built-in function that generates random observations from a Poisson distribution.

INFERENCES

8.6 Point Estimates

The likelihood function of a sample of n observations x_i on a Poisson process is

$$L(\lambda) = \exp\{-n\lambda\} \prod_{i=1}^{n} \frac{\lambda^{x_i}}{x_i!}, \tag{8.18}$$

and the log-likelihood function is

$$LL(\lambda) = \ln(\lambda) \sum_{i=1}^{n} x_i - \sum_{i=1}^{n} \ln(x_i!) - n\lambda. \tag{8.19}$$

Differentiating (8.19) with respect to λ and equating to zero gives the maximum-likelihood estimator for the parameter λ:

$$\hat{\lambda} = \frac{1}{n} \sum_{i=1}^{n} x_i = \bar{x}. \tag{8.20}$$

This estimator is unbiased and has minimum variance for all n. The standard error of the estimate $\widehat{\lambda}$ is the square root of (8.15):

$$se(\widehat{\lambda}) = \sqrt{\frac{\lambda}{n}}. \tag{8.21}$$

It is sometimes required to estimate individual Poisson probabilities $p(y; \lambda)$ directly from data. A minimum-variance unbiased estimate of $p(y; \lambda)$ is given by:[1]

$$\widetilde{P}_y = \binom{n\widehat{\lambda}}{y}\left(1 - \frac{1}{n}\right)^{n\widehat{\lambda}}(n-1)^{-y}. \tag{8.22}$$

An unbiased estimate of the sampling variance of (8.22) is

$$\begin{aligned}\mathrm{Var}(\widetilde{P}) = &\left\{\frac{(n\widehat{\lambda})!}{y!(n\widehat{\lambda} - y)!}\right\}^2 n^{-2y}\left(1 - \frac{1}{n}\right)^{2(n\widehat{\lambda} - y)} \\ &- \frac{(n\widehat{\lambda})!}{(n\widehat{\lambda} - 2y)!(y!)^2} n^{-2y}\left(1 - \frac{2}{n}\right)^{n\widehat{\lambda} - 2y}.\end{aligned} \tag{8.23}$$

See Example 8.1 for an illustration.

8.7 Interval Estimates

Given an observed number x of Poisson events, a $(1-\alpha)$-level confidence interval on the parameter λ (see Section 3.8) is given by $[\lambda_L, \lambda_U]$, where λ_L is the solution of

$$P(x; \lambda_L) = \exp\{-\lambda_L\}\sum_{i=0}^{x}\frac{\lambda_L^i}{i!} = 1 - \frac{\alpha}{2}, \tag{8.24}$$

and λ_U is the solution of

$$P(x; \lambda_U) = \exp\{-\lambda_U\}\sum_{i=0}^{x}\frac{\lambda_U^i}{i!} = \frac{\alpha}{2}. \tag{8.25}$$

See Example 8.2 for an illustration of the calculations. When there are n observations x_i on a Poisson process, the preceding expressions generalize to

$$P(s; n\lambda_L) = \exp\{-n\lambda_L\}\sum_{i=0}^{s}\frac{(n\lambda_L)^i}{i!} = 1 - \frac{\alpha}{2} \tag{8.26}$$

[1] Glasser, G. J., "Minimum Variance Unbiased Estimators for Poisson Probabilities," *Technometrics*, Vol. 4, 409–418, 1962.

and

$$P(s; n\lambda_U) = \exp\{-n\lambda_U\} \sum_{i=0}^{s} \frac{(n\lambda_U)^i}{i!} = \frac{\alpha}{2}, \tag{8.27}$$

where

$$s = \sum_{i=1}^{n} x_i.$$

8.8 Probability Plot

Probability plotting (see Section 3.13) of Poisson data is acomplished on the basis of the characterization (8.16): The distances d_i between Poisson events are Exponentially distributed. Such an Exponential probability plot (see Section 12.6) follows a linear trend if the data came from a Poisson process. Similarly, a test-of-fit of a Poisson model to counting data can be based on an Exponential test of the distances d_i; see Section 12.7.

APPLICATIONS

8.9 Counting Phenomena

As the most prominent model of counting phenomena, the Poisson distribution spans a wide range of applications, including particle counts in physics, spatial distributions of plants in agriculture, distributions of animals in ecology, cell or bacteria counts in biology, counts of disease victims in medicine, and demand on inventoried goods in business management.

Of particular interest to engineers are the following applications: In reliability engineering of electronic systems one encounters component failures, which tend to occur randomly. Thus, the Poisson distribution models the number of component failures during a period, thereby supplying information to forecast spare parts requirements. In quality control, the number of defectives in production, the number of flaws in a manufactured part, and the number of machines requiring service in a factory, are examples of Poisson applications. However, most engineering applications of the Poisson distribution relate to *waiting line* phenomena, where it models the number of *calling units* entering a congested service system. Examples include the number of cars arriving at a toll booth, the number of production orders arriving at a factory, the number of calls at a service facility, and the number of lines in use at a telephone exchange.

All these applications assume that the events in question occur randomly, at a constant rate. In several situations, the rate itself may be modeled as a random variable. The resulting distribution of the event count X is then a *mixed Poisson distribution.* For example, in the occurrence of industrial accidents, the occurrence rate often varies as a function of the "accident proneness" of individual workers. If that rate is modeled by a Gamma distribution, the mixed Poisson distribution of the number of accidents becomes Negative Binomial (see Section 7.9).

When the randomness assumption of the Poisson model is not justified, it is prudent to consider a more flexible discrete distribution to obtain an improved data fit. In this situation the Negative Binomial distribution is a good alternative. Its two parameters provide the flexibility that often produces a surprisingly good data fit. See Example 7.2 of Chapter 7.

8.10 Return Period

In Section 7.11 the *return period* of extreme events was introduced for Bernoulli trials, each covering a fixed time period. Annual flood values for a watershed are an example. When the event in question occurs randomly at any time (for example an earthquake), its return period is obtained from the Poisson model of the number of event occurrences during a given period. That is, the recurrence time between Poisson events is Exponentially distributed (see Section 8.4), so that the return period is the mean value of recurrence time:

$$t = \frac{1}{\nu} \text{ time units,} \tag{8.28}$$

where ν is the unit rate of the Poisson process. See Example 8.2.

EXAMPLE 8.1

Over a 14-day period, a total of 63 distress calls were received by a Coast Guard station. Assuming that calls follow a Poisson distribution, estimate the probability that exactly 6 distress calls arrive during any one day.

$$n := 14 \qquad x := 63 \qquad xb := \frac{x}{n} \qquad y := 6$$

1. ML estimate of parameter

$$\lambda := xb \qquad \lambda := 4.5$$

standard error of estimate:

$$SE\lambda := \sqrt{\frac{\lambda}{n}} \qquad SE\lambda = 0.567$$

2. ML estimate of probability

$$p(\lambda) := \frac{\lambda^{y} \cdot \exp(-\lambda)}{y!} \qquad p(\lambda) = 0.128$$

standard error of ML estimate:

$$SEp := \frac{d}{d\lambda} p(\lambda) \cdot SE\lambda \qquad SEp = 0.024$$

3. MVU estimate of probability

$$pu := \frac{x!}{y! \cdot (x-y)!} \cdot \left(1 - \frac{1}{n}\right)^{x} \cdot (n-1)^{-y}$$

standard error of MVU estimate:

$$SEpu := \left(\left(\frac{x!}{y! \cdot (x-y)!} \right)^{2} \cdot n^{-2 \cdot y} \cdot \left(1 - \frac{1}{n}\right)^{2 \cdot (x-y)} - \frac{x!}{(x - 2 \cdot y)! \cdot (y!)^{2}} \cdot n^{-2 \cdot y} \cdot \left(1 - \frac{2}{n}\right)^{x - 2 \cdot y} \right)^{\frac{1}{2}}$$

$$SEpu = 0.026$$

EXAMPLE 8.2

During a 20-year period, 3 earthquakes of magnitude greater than Richter-scale 5 were recorded in a certain geographical area. Assuming that earthquake occurrences of this magnitude follow a Poisson process, calculate the 90% confidence interval on the mean occurrence rate for a 20-year period. What is the return period of this event?

Confidence interval

$$x := 3 \qquad \alpha := 0.10 \qquad P(\lambda) := \exp(-\lambda) \cdot \sum_{i=0}^{x} \frac{\lambda^{i}}{i!}$$

$$t := 20$$

lower limit: $a := 2$ $\quad \lambda l := \text{root}(P(a) - 0.95, a)$ $\quad \lambda l = 1.4$
upper limit: $b := 4$ $\quad \lambda u := \text{root}(P(b) - 0.05, b)$ $\quad \lambda u = 7.8$

Return period

Yearly occurrence rate is $\nu := \frac{x}{t}$ $\quad \nu = 0.15$ earthquakes per year.

Return period is $T := \frac{1}{\nu}$ $\quad T = 6.7$ years.

PART THREE

CONTINUOUS DISTRIBUTIONS

CHAPTER NINE

Introduction to Continuous Distributions

9.1 Continuous Probability Functions

Most measurable quantities in engineering and the sciences admit a continuous range of measured values. Such quantities are therefore modeled by a *continuous random variable* X, whose realizations x can occur anywhere on some continuum of points called the *sample space* S.

For example, the time-to-failure of a piece of equipment could be any positive real number. The cost of an engineering project of a given scope and complexity could be any positive dollar-value between a feasible minimum and a plausible maximum. The pressure in a pipeline could be any positive or negative number within the strength limits of the material.

The uncertainty of occurrence of a particular value x is measured by probability. It is modeled by a mathematical function $f(x)$ that describes the *density* of probability for possible values x over its sample space S:

$$Pr\left(x - \frac{dx}{2} \leq X \leq x + \frac{dx}{2}\right) = f(x)\,dx, \quad x \in S.$$

(See also Section 1.6.) The function f is termed a *probability density function* (pdf). A particular cumulation of the above probabilities is the *cumulative distribution function* (cdf):

$$Pr(X \leq x) = F(x); \quad x \in S.$$

The two functions are of course related as

$$f(x) = \frac{dF(x)}{dx}$$

and are pictured in Fig. 9.1. They can be interpreted as *relative frequencies* of occurrence of x in a sequence of measurements on X. To accord with this interpretation, we have

$$0 \leq F(x) \leq 1$$

and

$$f(x) \geq 0.$$

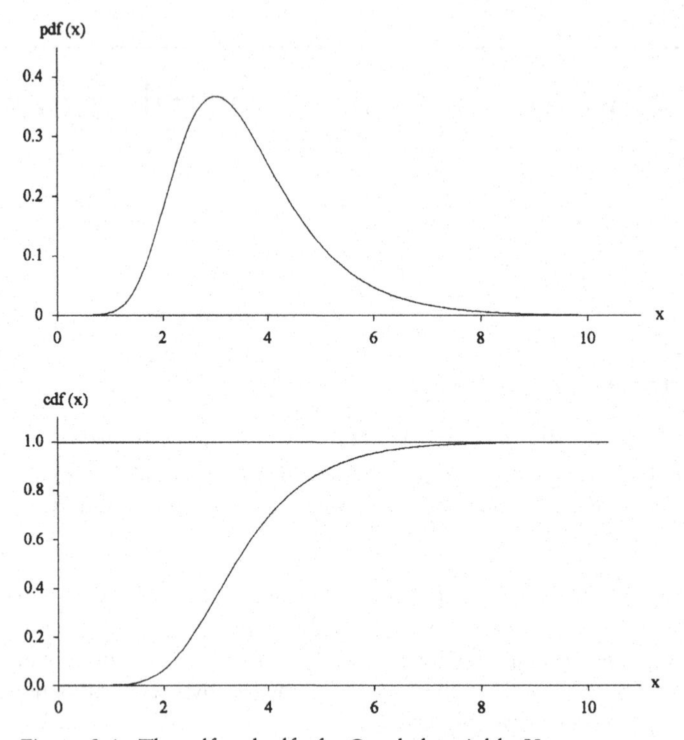

Figure 9.1. The pdf and cdf of a Gumbel variable X.

Chapters 10 to 17 introduce the continuous probability models f and F, which are of practical importance in the applied sciences.

9.2 Distribution Structure

To provide flexibility in modeling an assortment of measurement variables X, the probability model $f(x)$ is indexed by a *parameter* θ that takes values in a parameter space Ω:

$$f(x; \theta); \qquad x \in S, \qquad \theta \in \Omega.$$

In the presence of a parameter θ, the function $f(x; \theta)$ can be termed a *family of distributions*. A particular *value* of θ identifies a specific *member* of that family.

Often, θ is a vector of two or more parameters. The more parameters there are, the more flexible is the model to fit data. There are three types of parameters, permitting three kinds of flexibility for the model f:

1. Location parameters: These *locate* the model f on its measurement axis. This operation can be thought of as choosing the origin of measurements on X. Location parameters are denoted in this text by μ or δ. They are

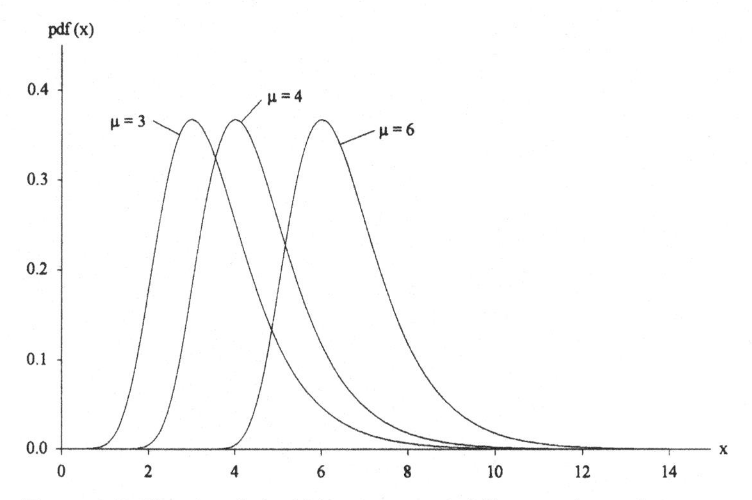

Figure 9.2. The Gumbel pdf for $\sigma = 1$ and different values of μ.

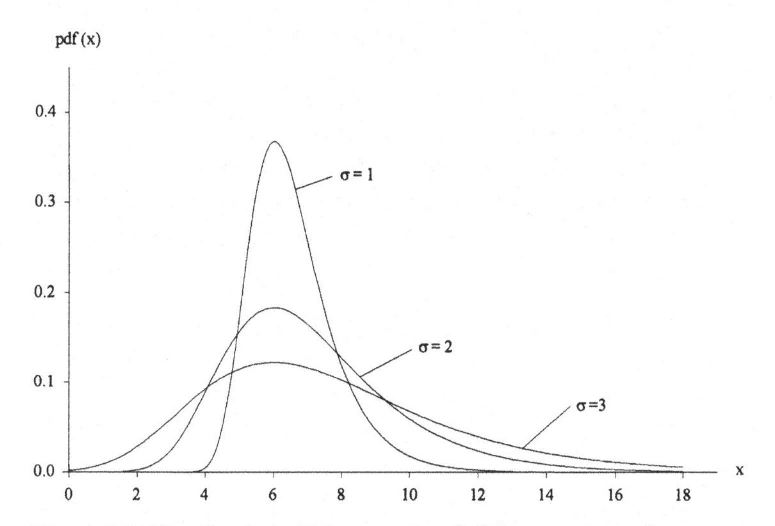

Figure 9.3. The Gumbel pdf for $\mu = 6$ and different values of σ.

recognized by their relation to x in the function f:

$$(x - \mu) \quad \text{or} \quad (x - \delta).$$

Different values of μ (or δ) shift the otherwise unchanged model f along the measurement axis x (see Fig. 9.2).

2. Scale parameters: These *scale* the model f on the measurement axis. This operation can be likened to choosing the units of measurement for X (e.g., inches or centimeters). The result is a horizontal stretching or contracting of the model f (see Fig. 9.3). Scale parameters are denoted by the symbol σ

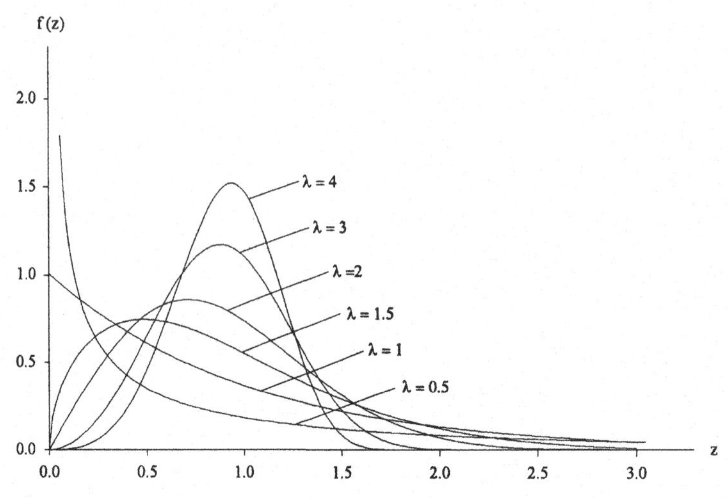

Figure 9.4. Weibull pdfs for $\sigma = 1$ and several values of λ.

in this text. They always appear as the configurations $\frac{x-\mu}{\sigma}$ and $\frac{dx}{\sigma}$ in the expression $f(x; \theta)\, dx$.

3. Shape parameters: These determine the *basic shape* of the function f (see Fig. 9.4). They are denoted by the symbol λ in this book. The flexibility of f to model a wide variety of variables X is primarily controlled by the shape parameter λ, which is therefore a most important value to determine. This parameter does not relate to x in a set configuration common to all models f.

Because μ and σ always appear in the configurations $\frac{x-\mu}{\sigma}$ and $\frac{dx}{\sigma}$, consider transforming the random variable x to a new random variable z by $x = \mu + \sigma z$, $dx = \sigma\, dz$. The transformed variable

$$z = \frac{x - \mu}{\sigma}$$

is termed *reduced*. When the model f features only a location parameter μ and a scale parameter σ, it has *location-scale* structure. Its reduced pdf $f(z)$ is then *parameter free*. The Normal distribution (see Chapter 10) is an example. For such models, $f(z)$ can be computed and tabulated once for all. When the model f features a shape parameter λ, its reduced version $f(z; \lambda)$ will show λ as the only parameter. This simplifies the problem of distinguishing the essential shapes admitted by the model f. The Weibull distribution is an example (see Chapter 17 and Fig. 9.4).

9.3 Overview of Models

The continuous distributions discussed in this text cover a multitude of shapes to model most random variables encountered in engineering and the physical sciences. The following provides a brief overview of these distributions.

Chapter 10 treats the *Normal distribution*, which is undoubtedly the "standard" to which other distributions are compared, principally because this model and its methods are generally well known and widely used. The Normal pdf has the familiar, symmetrical bell shape, and its sample space extends from minus to plus infinity. Since most engineering variables admit only positive values, and since engineering data typically show skewed frequency patterns, the Normal distribution must be approached with caution, particularly if inferences will focus on the tails of the distribution, as is often the case in engineering.

However, on the basis of a central limit theorem (see Section 2.4) the Normal distribution is the model of choice if it can be argued that the random variable under consideration is the aggregate *sum* of many underlying causes. Thus, a variety of error processes and cost functions are well modeled by this distribution. In addition, if interest focuses on the *average* value of a variable, the Normal distribution often provides a reasonable approximation to the *central* region of the true distribution. An advantage of considerable importance is that many inferential methods of practical interest have been worked out for the Normal distribution, in terms of exact results for all sample sizes.

Chapter 11 treats the *Log-Normal distribution*, which is logarithmically related to the Normal model. The Log-Normal pdf shows considerable flexibility of shape, which is always skewed to the right (longer right tail). Its sample space admits all positive values. This distribution is therefore admirably suited to model many engineering variables. Because of its close relation to the Normal model, many of the exact Normal inferential procedures transfer directly to the Log-Normal distribution. Applications cover chiefly natural growth and decay processes. In engineering, the Log-Normal distribution has been extensively applied to model material strength phenomena.

When it can be argued that the random variable under observation is the aggregate *multiplicative* result of many underlying causes, the Log-Normal model is the appropriate distributional postulate. This follows from a central limit theorem. An example is fatigue strength of steels, based on a proportional damage theory.

Chapter 12 treats the *Exponential distribution*, which models random phenomena in space or time, occurring at a *constant* rate. The shape of the pdf is fixed, rising abruptly from zero to its maximum value at the lower end of the sample space. As a general measurement model, the Exponential distribution is therefore not recommended. However, there are two important application areas where the Exponential distribution is the appropriate model.

One application is the time or space between adjoining Poisson events, since in this formulation the Exponential distribution models a Poisson counting process exactly. The other application is the time-to-failure of equipment with a *constant* failure hazard. Constant failure hazard implies that the equipment's operating history does not affect the likelihood of future failure. Hence, this application is largely restricted to electronic components. When in doubt of the constant hazard condition, it is best to assume a more general model (Gamma or Weibull), for which the Exponential is a special case. If the estimated model turns out to be Exponential, or close to it, it is advantageous to use it because of the simplicity of its exact inferential procedures.

Chapter 13 treats the *Gamma distribution*. The Gamma pdf shows considerable shape flexibility, with positive skew. Hence, the Gamma distribution is an excellent candidate to model engineering variables. Special cases include the Chi-squared, Exponential, and Erlang distributions.

There are two application areas where the Gamma distribution is prominent. One is the waiting time to the nth Poisson event. In this formulation the Gamma distribution becomes the Erlang distribution, which is an exact model of a Poisson counting process. Note that for $n = 1$, the Erlang reduces to the Exponential, and the waiting time variable becomes inter-arrival time between adjoining Poisson events. The second application is the time-to-failure of equipment, since the Gamma distribution admits failure hazards that decrease, increase, or are constant over time. Thus, the Gamma model generalizes the constant-hazard Exponential distribution.

When it can be argued that the observed random variable is the *sum* of a few underlying causes that are plausibly Gamma distributed, the Gamma model becomes a strong distribution candidate because the sum of Gamma variables is also Gamma distributed.

Chapter 14 treats the *Beta distribution*. This pdf has exceptional shape flexibility. Its sample space is limited to a finite range of values. Since many engineering variables are restricted to take values between a minimum and a maximum, the Beta distribution is an important measurement model. When the limited range is restricted to the unit interval, applications include efficiency measures and proportions. More generally, the Beta distribution is applied to cost variables, load variables with restricted maxima, task completion times, and the like. The Uniform distribution is a special case from which "random numbers" are produced; it is therefore at the core of Monte Carlo simulation work.

Chapter 15 treats the *Gumbel distribution*. The Gumbel pdf is of fixed shape, slightly skewed to the right. This distribution is therefore not recommended as a general measurement model. However, the Gumbel is the first of three "extreme value" distributions, which are of particular importance in engineering.

There are many engineering variables that are the result of an underlying extreme value process, that is, the variable under consideration is determined by the maximum or the minimum extreme of other variables. An example of a minimum extreme is the shortest-lived turbine blade, among those assembled on a turbine disk, that determines the life time of the disk. An example of maximum extreme is the largest load, placed on a structure during its service life, that determines the structure's reliability. Under fairly general conditions (see Section 15.2), a maximum extreme is asymptotically modeled by the Gumbel distribution. Thus, prominent applications include design loads and the magnitude of events that result from natural maximum extreme processes such as floods, storm winds, extreme ocean waves, earthquakes, and extreme atmospheric events.

Chapter 16 treats the *Frechet distribution*, which is the second of the extreme value models. Like the Gumbel distribution, it models maximum extremes. Although not all of its moments necessarily exist, the Frechet pdf is highly flexible and always skewed to the right. Its sample space comprises all positive values. This distribution is well suited as a general measurement model in engineering. In particular, it is employed to model maximum load variables that produce unusually large values in the upper tail. Thus, in a sense the Frechet distribution extends the Gumbel model's upper tail.

Finally, Chapter 17 treats the *Weibull distribution*. This is the third of the extreme value distributions, asymptotically modeling the minima of extreme value processes. It is perhaps the most prominent distribution in engineering statistics. The Weibull pdf is extremely flexible, and its sample space comprises all nonnegative values. The Weibull distribution is therefore recommended as a suitable general model for many engineering variables. However, because so many engineering quantities are in fact minimum extremes, the Weibull distribution is often the *a priori* distributional postulate. That is, whenever the "chain link" model applies (the weakest link determines the strength of the chain), the Weibull model is generally appropriate. The most prominent areas of engineering application are therefore material strength and the life time of equipment. In the latter situation the Weibull, like the Gamma, generalizes the Exponential life distribution.

9.4 Model Choice

It should be clearly understood that, although the inferential methods presented for each distribution produce excellent results, these results do depend on the distribution chosen to represent the data. Thus, if the true distribution differs markedly in shape, scale, or location from the true model of the random variable, inferences may be substantially in error, particularly if the inferences focus on a distribution tail (e.g., the 95-percentile). Model choice is therefore an important consideration, not to be taken lightly.

If at all possible, the physics of the process generating the random variable under consideration should be examined to see if an *a priori* model choice is indicated (see the preceding section). If this cannot be done, one should analyze the given data by assuming several plausible distributions in turn and study the effect of model choice on the inferences of concern. With the computational tools currently available, this strategy has now become feasible. If inferences are insensitive to the choice of model, there is no problem. However, sensitive inferences would justify a call for more data! If decisions need to be made on the basis of given data, engineering judgment must be applied to determine the most plausible inferences produced by the various models. Additional (visual) information may be garnered from the probability plots for the assumed models.

Sometimes statisticians will use the probability value of the test statistic from a test-of-fit to rank different models. The highest p value selects the model of choice. This is a sensible approach in general. However, when distribution tails are of importance, it is recommended to supplement this strategy with a close examination of the corresponding probability plot to ensure that tail discrepancies have not been overlooked.

9.5 Data Randomness

The statistical methods presented in this text rest on the assumption that the data constitute a simple random sample (see Section 2.2). This means that 1. the data are statistically independent of one another and 2. they all came from the same population. When these conditions are not met, the inferences drawn from the data are in error, sometimes seriously so. In practice it is therefore important to check these conditions, first at the level of the data generating process and then at the level of the data values obtained.

Lack of independence may be caused by serial correlation among data. For example, as redundant components fail in a system, the system load may redistribute over the remaining components, which face an increasingly severe load environment. Component lives are then not independent. Such a sample is no longer a simple random one, and standard inferential methods do not apply directly.

The question of whether data came from a common, homogeneous population can be a vexing one, requiring a careful investigation of the process that generated the data.

For example, time-to-failure data may pertain to an identifiable type of equipment, but the failed specimens may have experienced distinctly different operating environments during their service lives. As another example, specimens in a sample of failed components may have experienced different failure modes. As a further example, specimens for a material strength test may have been machined from portions of the steel ingot that have different strength properties. More generally, the variable under investigation may be influenced by a "factor" not accounted for. In each case the sample data did not come from a single population but represent a mixture of populations.

When the engineer has only the data before him, and he cannot examine the conditions under which they were generated, he will want to test the data directly for randomness. One very simple, distribution-free scheme for doing so is to apply a *run test* to the sample (see Section 4.3). To set up the sample for a run test of randomness, consider the *signs* of differences in magnitude of consecutive observations. Thus, the sequence of observations $x_1, x_2, x_3, \ldots$ converts to the sequence $\text{sign}(x_2 - x_1), \text{sign}(x_3 - x_2), \ldots$, where the sign function is $+1$ if the argument is positive and -1 if the argument is negative. When $x_{i+1} = x_i$, the sign is zero; such ties are ignored.

For example, consider the following sample of 17 tensile strength results of a new thermoplastic material (1,000 psi), in the order in which the specimens were prepared and tested:

5.25, 8.44, 10.09, 8.95, 8.23, 6.32, 7.19, 6.54, 7.88,
10.74, 8.65, 9.46, 10.02, 8.16, 8.02, 7.05, 10.74

The engineer wishes to test the randomness of these data. Converting to a sequence of signs gives:

$$+ + - - - + - + + - + + - - - +$$

Thus, there are $a = 8$ plus signs and $b = 8$ minus signs, and $r = 9$ runs. From Table 4.1, the 5-percentile of the run statistic R is 5; Table 4.2 gives the 95-percentile of R as 12. Therefore, at the 10% significance level, the randomness hypothesis is not rejected.

CHAPTER TEN

Normal Distributions

INTRODUCTION

10.1 Definition

A continuous random variable X has a *Normal distribution* if its pdf has the form

$$f(x; \mu, \sigma) = \frac{1}{\sigma\sqrt{2\pi}}\exp\left\{-\frac{1}{2}\left(\frac{x-\mu}{\sigma}\right)^2\right\}; \quad \sigma > 0, \quad -\infty < x,\ \mu < \infty. \tag{10.1}$$

This model occupies a central position in statistics, since much of statistical theory was developed around this distribution, and since problems with its applications have been worked out. The work of Laplace and Gauss introduced the Normal distribution to the sciences, principally astronomy, where it still dominates as a model for continuous random variables.

The pdf (10.1) shows that the Normal model is symmetrical about the value $x = \mu$ and that its tails diminish exponentially with the square of x. This distribution often serves as a good approximation to the *central* region of data frequency functions. In engineering, however, decisions frequently focus on the *tails* of distributions, so that an acceptable model must provide good tail-fits. Because engineering random variables often follow highly *skewed* frequency patterns, the symmetrical Normal model can easily lead to erroneous decisions and must be applied with caution. There are, however, engineering applications (see Sections 10.21 and 10.22) where the Normal model is appropriate. In addition, this model serves as a *standard* with which other models can be compared.

PROPERTIES

10.2 Standard Normal Distribution

The Normal distribution is of location-scale structure (see Section 9.2), so that the pdf of the *reduced* variable $z = (x - \mu)/\sigma$ is parameter-free:

$$f(z) = \frac{1}{\sqrt{2\pi}}\exp\left\{-\frac{z^2}{2}\right\}. \tag{10.2}$$

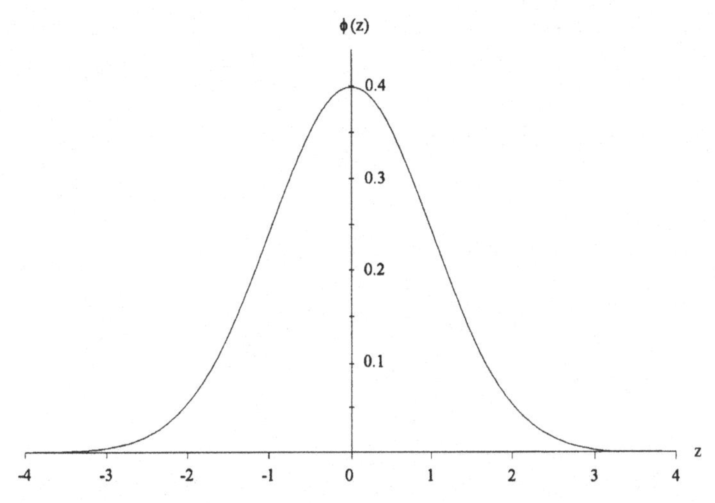

Figure 10.1. The standard Normal pdf.

This pdf is often denoted as $\phi(z)$. Figure 10.1 shows the familiar symmetrical bell shape of this pdf. The corresponding cdf,

$$F(z) = \frac{1}{\sqrt{2\pi}} \int_{-\infty}^{z} \exp\left\{-\frac{t^2}{2}\right\} dt, \tag{10.3}$$

cannot be expressed in closed form, but it is widely tabulated as the *standard Normal* cdf and is denoted as $\Phi(z)$. In many computer routines $\Phi(z)$ is available as a numerical subroutine. For example, in Mathcad this function is invoked as *cnorm*(z). The general Normal cdf,

$$F(x; \mu, \sigma) = \frac{1}{\sigma\sqrt{2\pi}} \int_{-\infty}^{x} \exp\left\{-\frac{1}{2}\left(\frac{t-\mu}{\sigma}\right)^2\right\} dt, \tag{10.4}$$

is *equal* to the standard Normal cdf:

$$F(x; \mu, \sigma) = \Phi\left(\frac{x-\mu}{\sigma}\right), \tag{10.5}$$

whereas the pdfs are related as

$$f(x; \mu, \sigma) = \frac{1}{\sigma}\phi\left(\frac{x-\mu}{\sigma}\right). \tag{10.6}$$

The expected value of Z is zero and the variance of Z is 1.

10.3 General Normal Distribution

The moment generating function of X (see Section 1.8) is

$$M(t) = \exp\left\{t\mu + \frac{\sigma^2 t^2}{2}\right\}. \tag{10.7}$$

Thus, the expected value of X is

$$\mu_1'(X) = \mu, \tag{10.8}$$

which is also the mode and the median. The variance of X is

$$\mu_2(X) = \sigma^2. \tag{10.9}$$

The coefficient of variation of X is therefore

$$cv(X) = \frac{\sigma}{\mu}. \tag{10.10}$$

The quantile of order q is obtained from

$$x_q = \mu + z_q\sigma, \tag{10.11}$$

where z_q is the corresponding *standard* Normal quantile, obtained from standard Normal tables, or numerically as the inverse of the standard Normal cdf: $\Phi(z_q) = q$. Note that Mathcad 6+ has a built-in function that computes the quantile (10.11).

The first shape factor (see Section 1.10) is

$$\gamma_1 = 0, \tag{10.12}$$

signifying that the Normal model is symmetrical, and the second shape factor is

$$\gamma_2 = 3. \tag{10.13}$$

The matrix of minimum-variance-bounds (see Section 3.2) for estimators of μ and σ is

$$\begin{bmatrix} V_{\mu\mu} & V_{\mu\sigma} \\ V_{\mu\sigma} & V_{\sigma\sigma} \end{bmatrix} = \begin{bmatrix} \frac{\sigma^2}{n} & 0 \\ 0 & \frac{\sigma^2}{2n} \end{bmatrix}, \tag{10.14}$$

giving the lower bound on the variances of estimators for μ and σ.

10.4 Reproductive Property

The Normal distribution features an important reproductive property: If k independent Normal variables X_i with location parameters μ_i and scale parameters σ_i are summed as

$$T = \sum_{i=1}^{k} a_i X_i,$$

where the a_i are constants, then T is again Normal with parameters

$$\mu = \sum_{i=1}^{k} a_i \mu_i \quad \text{and} \quad \sigma = \sqrt{\sum_{i=1}^{k} a_i^2 \sigma_i^2}. \tag{10.15}$$

Consider the case $\mu_i = \mu, \sigma_i = \sigma$, and $a_i = 1/n$. This implies that $T = \bar{X}$ is the average of n observations from the *same* Normal process. Thus, $\bar{X}$ is exactly Normally distributed with parameters μ and $\sigma/\sqrt{n}$.

10.5 Simulation

Random observation z_i from a *standard* Normal process are efficiently simulated by the Box/Muller[1] technique as

$$z_i = \sqrt{-2\ln(u_i)}\sin(2\pi v_i)$$

and

$$z_{i+1} = \sqrt{-2\ln(u_i)}\cos(2\pi v_i). \tag{10.16}$$

Here u_i and v_i are Uniform random numbers on the interval (0, 1) (see Section 14.6). Random Normal observations are thus simulated in pairs. Random observations x_i from a *general* Normal process with known parameters μ and σ are then obtained as

$$x_i = \mu + z_i\sigma. \tag{10.17}$$

Note that Mathcad 6+ has a built-in function that generates random Normal variates x_i directly.

One should check the adequacy of a simulated sample by comparing at least its first two sample moments with those of the given model. For system simulation work it is recommended to mix up the generated random observations by a shuffling procedure (see the footnote of Section 2.8), because it is known that the *linear congruential* generators used on many computers produce dependent pairs (z_i, z_{i+1}).

SAMPLING DISTRIBUTIONS

There are three important distributions used to model the random behavior of various *statistics*: the *Chi-squared*, the *t*-, and the *F*-distributions. These *sampling* distributions are of particular importance for the Normal measurement process, as they model the sampling behavior of certain Normal statistics exactly.

10.6 Chi-Squared Distribution

In Section 13.6 the *Chi-squared* (χ^2) distribution is introduced as a special case of the Gamma distribution. The χ^2 pdf is defined by (13.25), and Fig. 13.2 shows

[1] Box, G. E. P., Muller, M. E., "A Note on the Generation of Normal Deviates," *Annals of Mathematical Statistics*, Vol. 28, 610–611, 1958.

this pdf for several degrees of freedom ν. As ν increases, the χ^2 pdf approaches a Normal pdf. The mean value and variance are given by (13.26). From the reproductive property of Gamma variables (see Section 13.3), the sum of $k\ \chi^2$ variables with ν_i degrees of freedom is also a χ^2 variable, with $\nu = \sum \nu_i$ degrees of freedom.

The importance of the χ^2 distribution as a sampling model for Normal statistics arises from the following result:

Let $X_i\ i=1,\ldots,n$, be random observations from a Normal process with parameters (μ, σ). Then the statistic

$$V = \frac{1}{\sigma^2}\sum_{i=1}^{n}(X_i - \overline{X})^2 \tag{10.18}$$

is exactly χ^2 distributed with $\nu = n - 1$ degrees of freedom, where $\overline{X}$ is the sample mean.

The statistic (10.18) can be expressed in terms of the sample variance

$$S^2 = \frac{1}{n-1}\sum_{i=1}^{n}(X_i - \overline{X})^2 \tag{10.19}$$

as

$$V = \frac{1}{\sigma^2}(n-1)S^2. \tag{10.20}$$

Since V involves only the parameter σ^2, its χ^2 distribution is used to construct hypothesis tests and confidence intervals on σ^2.

Tables of selected χ^2-quantiles are widely available. However, with modern computational aids it is straightforward to compute the χ^2 cdf or its inverse directly from the definition of the pdf. See Example 10.1. Note that Mathcad 6+ has a built-in function that produces χ^2-quantiles.

10.7 Central *t*-Distribution

A variable T has a *central t-distribution* if its pdf is

$$f(t;\nu) = \frac{\Gamma\left(\frac{\nu+1}{2}\right)}{\sqrt{\nu\pi}\,\Gamma\left(\frac{\nu}{2}\right)}\left(1+\frac{t^2}{\nu}\right)^{-\frac{\nu+1}{2}}; \qquad \nu > 0, \quad -\infty < t < \infty, \tag{10.21}$$

where the parameter ν is again the degrees of freedom. Figure 10.2 shows this pdf for several values of ν. We see that the central t pdf is symmetrical, centered at the origin, and bell shaped. As ν increases, $f(t;\nu)$ approaches the standard

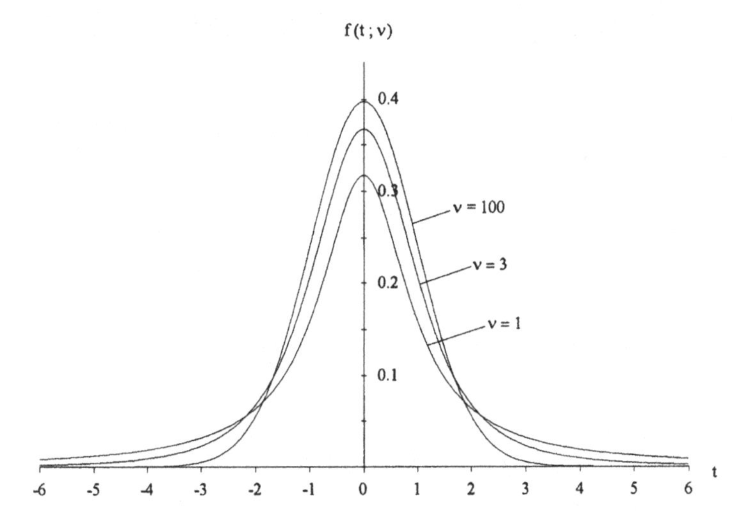

Figure 10.2. The t-pdf for several degrees of freedom ν.

Normal pdf $\phi(z)$; as ν decreases, $f(t;\,\nu)$ becomes more widely dispersed than $\phi(z)$. The mean value of t is zero and its variance is

$$\mu_1'(T) = \frac{\nu}{\nu - 1}. \tag{10.22}$$

Consider a Normal random variable X with parameters (μ, σ). The reduced variable

$$Z = \frac{X - \mu}{\sigma}$$

is, of course, standard Normal, and the reduced sample mean $\bar{Z}$ is distributed with mean zero and standard deviation $1/\sqrt{n}$. Next, consider the χ^2 variable V of the preceding section, expressed as (10.20). It can be shown that:

1. **The statistics $\bar{Z}$ and S^2 are statistically independent.**
2. **The statistic**

$$W = \frac{\bar{Z}}{\sqrt{V/(n-1)}} = \frac{\bar{X} - \mu}{S/\sqrt{n}} \tag{10.23}$$

is exactly distributed as a central t variable with $\nu = n - 1$ degrees of freedom.

Because W involves only the parameter μ, its t-distribution is used to construct hypothesis tests and confidence intervals on μ, when σ is not known (the usual case).

Tables of selected t-quantiles are widely available. These quantiles are, however, easily computed as the inverse of the t cdf, directly from the definition

(10.21) of the t pdf. See Example 10.1. Note that Mathcad 6+ has a built-in function that gives t-quantiles.

10.8 Noncentral t-Distribution

A variable T' has a *noncentral t-distribution* if its cdf is

$$F(t'; \delta, \nu) = \frac{2^{1-\frac{\nu}{2}}}{\Gamma\left(\frac{\nu}{2}\right)\sqrt{2\pi}} \int_0^{\infty} x^{\nu-1} \exp\left\{-\frac{x^2}{2}\right\} \int_0^{\frac{xt'}{\sqrt{\nu}}} \exp\left\{-\frac{(\mu-\delta)^2}{2}\right\} du\, dx, \tag{10.24}$$

where δ is a *noncentrality* parameter and the parameter ν is the degrees of freedom. This distribution arises as an exact sampling model of the statistic (10.23) when the mean value μ of the Normal variable X is displaced by δ. Thus, the *power function* of tests on μ (see Section 3.11) can be determined from (10.24) as a function of δ. Also, tests and confidence intervals on linear combinations of μ and σ are based on this distribution.

Although tables of noncentral t-quantiles t'_q are available, they are somewhat inaccessible. Direct computation of these quantiles is straightforward, provided care is taken to define the numerical integration limits on the variable x in (10.24). A convenient approximation to the quantile t'_q is given by

$$t'_q \doteq \frac{\delta + z_q\sqrt{1 + \frac{\delta^2 - z_q^2}{2\nu}}}{1 - \frac{z_q^2}{2\nu}}, \tag{10.25}$$

where z_q is the standard Normal quantile and the square root must be real. This approximation may be used as a starting value for the exact computation. See Example 10.1 for an illustration.

It is useful to note that the noncentral t pdf is symmetrical about the origin for $\pm\delta$. This means that

$$t'_q(\delta) = -t'_{1-q}(-\delta), \tag{10.26}$$

which simplifies the computation of the quantile when δ is negative.

10.9 F-Distribution

A variable Y has an F-distribution if its pdf is

$$f(y; \nu_1, \nu_2) = \frac{\Gamma\left(\frac{\nu_1+\nu_2}{2}\right)}{\Gamma\left(\frac{\nu_1}{2}\right)\Gamma\left(\frac{\nu_2}{2}\right)} \nu_1^{\frac{\nu_1}{2}} \nu_2^{\frac{\nu_2}{2}} y^{\frac{\nu_1}{2}-1} (\nu_2 + \nu_1 y)^{-\frac{\nu_1+\nu_2}{2}}; \quad y \geq 0; \quad \nu_1, \nu_2 > 0, \tag{10.27}$$

where the parameters ν_1 and ν_2 are both degrees of freedom. Figure 10.3 shows this pdf for several combinations of ν_1 and ν_2.

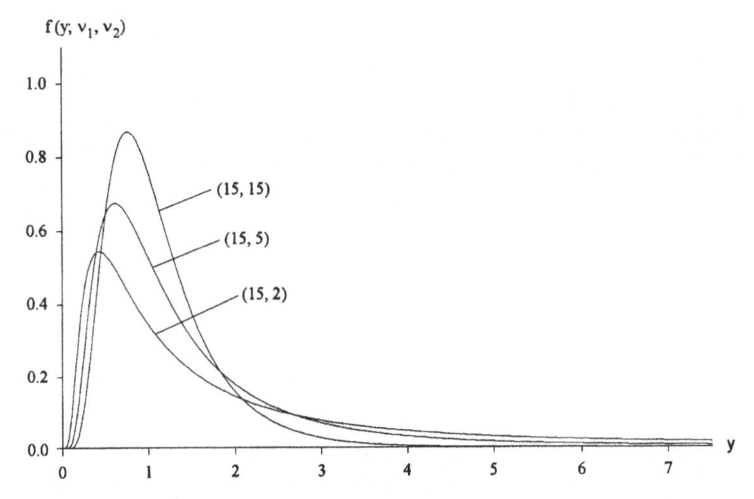

Figure 10.3. The F-pdf for several pairs of degrees of freedom.

The expected value and variance of Y are

$$\mu_1'(Y) = \frac{\nu_2}{\nu_2 - 2}, \qquad \nu_2 > 2, \tag{10.28}$$

$$\mu_2(Y) = \frac{2\nu_2^2(\nu_1 + \nu_2 - 2)}{\nu_1(\nu_2 - 2)^2(\nu_2 - 4)}, \qquad \nu_2 > 4. \tag{10.29}$$

Consider two independent χ^2 variables X_1 and X_2 with ν_1 and ν_2 degrees of freedom, respectively. It can be shown that:

The ratio $Y = \frac{X_1/\nu_1}{X_2/\nu_2}$ is F distributed with ν_1, ν_2 degrees of freedom.

Since the Normal sample variance S^2 is χ^2 distributed (see expression (10.20)), the F-distribution is used to construct hypothesis tests and confidence intervals on the ratio σ_1^2/σ_2^2 of variances from two Normal processes.

Tables of selected upper quantiles $F_{\nu_1,\nu_2,1-\alpha}$ of the F-distribution are widely available. Lower quantiles are obtained from such tables, using the relation

$$F_{\nu_1,\nu_2,1-\alpha} = (F_{\nu_2,\nu_1,\alpha})^{-1}. \tag{10.30}$$

Note the reversal of the degrees of freedom. Quantiles are also easily computed directly from the definition of the pdf; see Example 10.1. Note that Mathcad 6+ has a built-in function to compute F-quantiles.

PROBABILITY PLOT

10.10 Complete Samples

A probability plot of ordered observations $x_{(i)}$ (see Sections 3.13–3.15) should always be used to check the distributional assumption of the Normal model.

Since the standard Normal cdf (10.3) is parameter free, it can be linearized between any two points on it. If these points are the 1- and 99-percentiles, up to 69 data points can be plotted when the median plotting position is used. Normal *probability paper* is constructed with the probability scale thus linearized.

Because the data are linearly related to the standard variate by $x_{(i)} = \mu + \sigma z_{(i)}$ the data $x_{(i)}$ can be plotted directly, preserving their linear relation. Thus, plotting the approximate median position

$$p_i = \frac{i - 0.3}{n + 0.4} \tag{10.31}$$

on the probability scale against the data $x_{(i)}$ will produce a linear trend if the data came from a Normal process. A substantial departure from linearity would put the Normal assumption in question (see Section 3.14).

Clearly, the parameters (μ, σ) could be estimated from the slope and intercept of a fitted straight line. However, since the statistical properties of such estimates are not known, and since optimal estimates are easily computed, this procedure is not recommended.

Although Normal probability paper is widely available, a computer plot is usually more convenient. With z_i denoting a standard Normal quantile, the linear probability scale is given by

$$y_i = 0.500 + 0.210 z_i \tag{10.32}$$

for the endpoints $z_{0.01}$ and $z_{0.99}$. For complete samples, or those that are censored in the right tail only, the plotting positions p_i can be calculated. The standard Normal quantiles z_i are then calculated from $\Phi(z_i) = p_i$, and y_i is plotted versus the ordered data $x_{(i)}$. See Example 10.2 for an illustration.

10.11 Censored Samples

A Normal distribution is sometimes used as a life model for operating equipment. In this application area, the engineer frequently deals with data sets that are *multiply censored* (see Sections 12.2 and 17.11). Consider, then, a multiply right-censored sample of size n, with r failure times x_i and $m = n - r$ survival times c_j. Label the combined and *ordered* set of failure and survival data as $\{t_{(k)}\}_n$. Thus, some of the $t_{(k)}$ are failure data x_i and the rest are survival data c_j.

For multiply right-censored samples, the plotting position (10.31) needs to be modified to include the information implicit in the survival times c_j. A suitable modification is based on the distribution-free *product limit* estimate of the reliability function (see Section 3.15). The modified median plotting position is then

$$p_k = 1 - \frac{n + 0.7}{n + 0.4} \prod_{q=1}^{k} \frac{n - q + 0.7}{n - q + 1.7}, \tag{10.33}$$

where $q \leq k$ is the subscript of a *failure* time in the combined, ordered set $\{t_{(k)}\}_n$. When no censored data are present, (10.33) reduces to the complete-sample median plotting position (10.31). As in the preceding section, y_i is plotted versus the ordered failure data $x_{(i)}$. See Example 10.3 for an illustration of a Normal probability plot for a censored sample.

POINT ESTIMATES

10.12 Complete Samples

The likelihood function of a complete sample of n independent observations from a Normal process is

$$L(\mu, \sigma) = \sigma^{-n}(2\pi)^{-\frac{n}{2}} \exp\left\{ -\frac{1}{2} \sum_{i=1}^{n} \left(\frac{x_i - \mu}{\sigma} \right)^2 \right\}. \tag{10.34}$$

The maximum likelihood (ML) estimates (see Section 3.6) are

$$\widehat{\mu} = \frac{1}{n} \sum_{i=1}^{n} x_i = \bar{x} \tag{10.35}$$

and

$$\widehat{\sigma} = \sqrt{\frac{1}{n} \sum_{i=1}^{n} (x_i - \bar{x})^2}. \tag{10.36}$$

The estimator $\widehat{\mu}$ is unbiased, with minimum variance. From (10.14) its standard error is

$$se(\widehat{\mu}) = \frac{\sigma}{\sqrt{n}} \tag{10.37}$$

for all n. The estimator $\widehat{\sigma}$ is biased. Its unbiased version is

$$\widehat{\sigma}_u = \sqrt{\frac{1}{n-1} \sum_{i=1}^{n} (x_i - \bar{x})^2} = S_x \tag{10.38}$$

with standard error

$$se(\widehat{\sigma}_u) = \frac{\sigma}{\sqrt{2(n-1)}}. \tag{10.39}$$

The standard errors of parameter functions $g(\mu, \sigma)$ are estimated from the error propagation formula (see Section 3.3), which is evaluated at the ML estimates $\widehat{\mu}$ and $\widehat{\sigma}_u$. Using the above standard errors for complete samples we get:

$$\text{Var}(g) \doteq \left(\frac{\partial g}{\partial \mu} \right)^2 \frac{\sigma^2}{n} + \left(\frac{\partial g}{\partial \sigma} \right)^2 \frac{\sigma^2}{2(n-1)}. \tag{10.40}$$

For a statistical test-of-fit the *Anderson–Darling* statistics is recommended:

$$A = -n - \frac{1}{n}\sum_{i=1}^{n}(2i-1)[\ln(w_i) + \ln(1 - w_{n-i+1})], \tag{10.41}$$

where w_i is the standard Normal cdf $\Phi(\frac{x_{(i)}-\widehat{\mu}}{\widehat{\sigma}})$; note that the *biased* estimate $\widehat{\sigma}$ is used here. The value A is then modified to account for sample size:

$$A_{\mathrm{m}} = A\left(1 + \frac{0.75}{n} + \frac{2.25}{n^2}\right). \tag{10.42}$$

Critical values[2] for A_{m} are:

α	0.100	0.050	0.025	0.010
A_{c}	0.631	0.752	0.873	1.035

Example 10.4 demonstrates the estimation process.

10.13 Censored Samples

The likelihood function of a multiply censored sample, defined in Section 10.11, from a Normal process is

$$L(\mu,\sigma) = \sigma^{-r}(2\pi)^{-\frac{r}{2}}\exp\left\{-\frac{1}{2}\sum_{i=1}^{r}\left(\frac{x_i-\mu}{\sigma}\right)^2\right\}\prod_{j=1}^{m}[1 - F(c_j;\mu,\sigma)], \tag{10.43}$$

where F is the Normal cdf (10.4). The maximum likelihood equations (see Section 3.6) can be written compactly as

$$\sum_{i=1}^{r} z_i + \sum_{j=1}^{m} h(y_j) = 0 \tag{10.44}$$

and

$$\sum_{i=1}^{r} z_i^2 + \sum_{j=1}^{m} y_j h(y_j) = r, \tag{10.45}$$

where $z_i = (x_i - \widehat{\mu})/\widehat{\sigma}$ and $y_j = (c_j - \widehat{\mu})/\widehat{\sigma}$. The function $h(z)$ is the standard Normal *hazard* function:

$$h(t) = \frac{\phi(t)}{1 - \Phi(t)} \tag{10.46}$$

[2] Extracted from R. B. D'Agostino, M. A. Stephens, *Goodness-of-Fit Techniques*, Marcel Dekker Inc., New York, 1986, by courtesy of the publisher.

(see Section 10.22). Here, $\phi(t)$ is the standard Normal pdf (10.2), and $\Phi(t)$ is the standard Normal cdf (10.3). The simultaneous solution of these two equations is straightforward with an equation solver, although good starting values should be used for the solution process to converge. The mean and standard deviation of the failure data often provide adequate starting values.

Standard errors of these estimates can be obtained from the *local* information matrix, defined by the terms:

$$I_{\mu\mu} = \frac{1}{\sigma^2}\left\{r + \sum_{j=1}^{m} h(y_j)[h(y_j) - y_j]\right\},$$

$$I_{\mu\sigma} = I_{\sigma\mu} = \frac{1}{\sigma^2}\left\{2\sum_{i=1}^{r} z_i + \sum_{j=1}^{m} h(y_j)(1 + y_j[h(y_j) - y_j])\right\},$$

$$I_{\sigma\sigma} = \frac{1}{\sigma^2}\left\{3\sum_{i=1}^{r} z_i^2 + 2\sum_{j=1}^{m} y_j h(y_j) + \sum_{j=1}^{m} y_j^2 h(y_j)[h(y_j) - y_j] - r\right\}, \qquad (10.47)$$

where $h(t)$ is the standard Normal hazard function (10.46). The variance–covariance matrix V is then the inverse of the matrix I, evaluated at the ML estimates $(\widehat{\mu}, \widehat{\sigma})$. The square roots of the diagonal terms in V give the standard errors of the parameter estimates. See Example 10.5 for an illustration of the calculations.

Statistical tests-of-fit for multiply censored Normal samples are not available. Probability plotting is recommended as a reliable check on the fit of the estimated model to the data.

INTERVAL ESTIMATES AND TESTS: COMPLETE SAMPLES

10.14 Mean Value

A $(1 - \alpha)$-level confidence interval on the Normal mean μ (σ unknown) is obtained from the t-distribution of the statistic (10.23) as

$$\left(\bar{x} \pm t_{n-1,\frac{\alpha}{2}} \cdot \frac{s}{\sqrt{n}}\right). \qquad (10.48)$$

Here $t_{n-1,\alpha/2}$ is the $\alpha/2$-quantile of the central t-distribution with $\nu = n - 1$ degrees of freedom, $\bar{x}$ is the sample mean, and s is the sample standard deviation.

Consider the null hypothesis $H_0 : \mu = \mu_0$ and the alternative $H_a : \mu > \mu_0$ (see Chapter 3 on *Tests*). The test statistic is $T = \bar{X}$. The null hypothesis H_0 is rejected when $\bar{x} \geq t_c$, where the critical test value t_c is given by

$$t_c = \mu_0 + t_{n-1,1-\alpha} \cdot \frac{s}{\sqrt{n}} \qquad (10.49)$$

for a significance level α of the test. The power of the test for a specified μ_a is obtained from the noncentral t-distribution of the quantity

$$T' = \frac{\bar{X} - \mu_0}{s/\sqrt{n}} \tag{10.50}$$

with noncentrality factor

$$\delta = \frac{\mu_a - \mu_0}{\sigma/\sqrt{n}} \tag{10.51}$$

and degrees of freedom $\nu = n - 1$. The power is the complement of the noncentral t cdf (10.24) at t' with $\bar{x} = t_c$. Example 10.6 illustrates the computations.

10.15 Standard Deviation

A $(1 - \alpha)$-level confidence interval on the Normal standard deviation σ (μ unknown) is obtained from the χ^2 distribution of the statistic (10.18) as

$$\left(s \cdot \sqrt{\frac{n-1}{\chi^2_{n-1,1-\frac{\alpha}{2}}}}, s \cdot \sqrt{\frac{n-1}{\chi^2_{n-1,\frac{\alpha}{2}}}} \right). \tag{10.52}$$

Here $\chi^2_{n-1,q}$ is the q-quantile of the χ^2 distribution with $\nu = n - 1$ degrees of freedom, and s is the sample standard deviation.

Consider the null hypothesis $H_0 : \sigma^2 = \sigma_0^2$ and the alternative $H_a : \sigma_a^2 > \sigma_0^2$. The test statistic is the sample variance $T = S^2$. The null hypothesis is rejected when $s^2 \geq t_c$, where the critical test value is

$$t_c = \frac{\sigma_0^2}{n-1} \cdot \chi^2_{n-1,1-\alpha} \tag{10.53}$$

for a significance level α of the test. The power of the test for a specified σ_a^2 is obtained from the complement of the χ^2 distribution of V (see (10.18)), given that H_a is true. It is evaluated as the integral of the χ^2 pdf (13.25) with upper limit

$$\left(\frac{\sigma_0}{\sigma_a}\right)^2 \chi^2_{n-1,1-\alpha} \tag{10.54}$$

and degrees of freedom $\nu = n - 1$. Example 10.6 illustrates the calculation.

10.16 Differences in Mean Values

It is frequently of practical interest to compare the properties of two Normal processes X and Y. For example, if a process is deliberately changed to increase its mean value, the effect of this change needs to be verified on the basis of a sample drawn from the changed process. Under the restrictive assumption that the (unknown) process variance has *not* changed, the statistic

$$T = \frac{\bar{X} - \bar{Y} - (\mu_x - \mu_y)}{S_p\sqrt{\frac{1}{n_x} + \frac{1}{n_y}}} \tag{10.55}$$

has a t-distribution with $(n_x + n_y - 2)$ degrees of freedom, where S_p^2 is the *pooled* variance estimate

$$S_p^2 = \frac{(n_x - 1)S_x^2 + (n_y - 1)S_y^2}{n_x + n_y - 2}. \tag{10.56}$$

The $(1-\alpha)$-level confidence interval on the difference in means $(\mu_x - \mu_y)$ is thus

$$\left(\bar{x} - \bar{y} \pm t_{n_x+n_y-2,\frac{\alpha}{2}} \cdot s_p\sqrt{\frac{1}{n_x} + \frac{1}{n_y}}\right). \tag{10.57}$$

Consider the null hypothesis $H_0 : \mu_x - \mu_y = 0$ and the alternative $H_a : \mu_x - \mu_y > 0$. The test statistic is $T = \bar{X} - \bar{Y}$. The null hypothesis is rejected when $\bar{x} - \bar{y} > t_c$, where the critical test value t_c is

$$t_c = t_{n_x+n_y-2,1-\alpha} \cdot s_p\sqrt{\frac{1}{n_x} + \frac{1}{n_y}} \tag{10.58}$$

for an α-level test significance. The power of the test for a specified alternative $\mu_x - \mu_y = \Delta M$ is obtained from the noncentral t-distribution of the quantity

$$T' = \frac{\bar{X} - \bar{Y}}{\sigma\sqrt{\frac{1}{n_x} + \frac{1}{n_y}}} \tag{10.59}$$

with noncentrality factor

$$\delta = \frac{\Delta M}{\sigma\sqrt{\frac{1}{n_x} + \frac{1}{n_y}}} \tag{10.60}$$

and $\nu = n_x + n_y - 2$ degrees of freedom. The power is computed as the complement of the corresponding noncentral t cdf (10.24) at t' with $\bar{x} - \bar{y} = t_c$ and $\sigma = S_p$. See Example 10.7 for an illustration of the computations.

10.17 Ratio of Variances

It is also often of practical interest to compare the variances of two Normal processes X and Y. For example, a process may be changed deliberately in an attempt to reduce the process variability. This change then needs to be verified by a sample from the changed process. Since the statistics

$$V_x = \frac{(n_x - 1)S_x^2}{\sigma_x^2}$$

and

$$V_y = \frac{(n_y - 1)S_y^2}{\sigma_y^2}$$

are χ^2 distributed with degrees of freedom $\nu_x = n_x - 1$ and $\nu_y = n_y - 1$ according to (10.18), the ratio

$$\frac{V_x/\nu_x}{V_y/\nu_y} = \frac{S_x^2/\sigma_x^2}{S_y^2/\sigma_y^2} \tag{10.61}$$

is F distributed with degrees of freedom ν_x and ν_y (see Section 10.9).

The $(1 - \alpha)$-level confidence interval on the ratio σ_x/σ_y is obtained from the probability

$$Pr\left(F_{n_x-1,n_y-1,\frac{\alpha}{2}} \leq \frac{S_x^2/\sigma_x^2}{S_y^2/\sigma_y^2} \leq F_{n_x-1,n_y-1,1-\frac{\alpha}{2}}\right) = 1 - \alpha$$

as

$$\left(\frac{S_x/S_y}{\sqrt{F_{n_x-1,n_y-1,1-\frac{\alpha}{2}}}}, \frac{S_x/S_y}{\sqrt{F_{n_x-1,n_y-1,\frac{\alpha}{2}}}}\right). \tag{10.62}$$

Here $F_{n_x-1,n_y-1,q}$ is the q-quantile of the F-distribution with $(n_x - 1, n_y - 1)$ degrees of freedom, and S_x, S_y are the sample standard deviations.

Consider the null hypothesis $H_0 : \sigma_x^2/\sigma_y^2 = 1$ and the alternative $H_a : \sigma_x^2/\sigma_y^2 > 1$. The test statistic is $T = S_x^2/S_y^2$. The null hypothesis is rejected when $s_x^2/s_y^2 > t_c$, where the critical test value t_c is

$$t_c = F_{n_x-1,n_y-1,1-\alpha} \tag{10.63}$$

for an α-level test significance. The power of the test for a specified alternative $\sigma_x^2/\sigma_y^2 = R$ is obtained from the F-distribution of (10.61), given that H_a is true. It is evaluated as the integral of the F pdf (10.27) at $(n_x - 1, n_y - 1)$ degrees of freedom, with upper integration limit t_c/R. See Example 10.7 for an illustration.

10.18 Quantiles

In engineering practice a q-quantile of the measurement variable X is often the decision quantity of interest. Exact confidence intervals on the Normal quantile X_q are based on the noncentral t-distribution of the statistic

$$T = \frac{X_q - \bar{X}}{S/\sqrt{n}} \tag{10.64}$$

with $\nu = n - 1$ degrees of freedom and noncentrality parameter

$$\delta = z_q\sqrt{n}, \tag{10.65}$$

where $\bar{X}$ and S are the sample mean and standard deviation and z_q is the standard Normal q-quantile. A $(1 - \alpha)$-level confidence interval on X_q is thus given by

$$\left(\bar{x} + \frac{s}{\sqrt{n}} \cdot t'_{n-1,\frac{\alpha}{2}}, \bar{x} + \frac{s}{\sqrt{n}} \cdot t'_{n-1,1-\frac{\alpha}{2}}\right). \tag{10.66}$$

See Example 10.8 for an illustration.

A related decision quantity of engineering interest is the *lower* $(1 - \alpha)$-level confidence limit on the reliability $R(x) = 1 - F(x)$, where x is a given value. That is, one wants to determine the value q such that

$$Pr((R \mid x) \geq q) = 1 - \alpha.$$

Equivalently, since x is the $(1 - q)$-quantile,

$$Pr(X_{1-q} \leq x) = 1 - \alpha.$$

Substituting from (10.66) for X_{1-q} one gets

$$Pr\left(\bar{x} + \frac{s}{\sqrt{n}} \cdot t'_{n-1,q}(\delta) \leq x\right) = 1 - \alpha,$$

or

$$Pr\left(t'_{n-1,q}(\delta) \leq \frac{x - \bar{x}}{s}\sqrt{n}\right) = 1 - \alpha, \tag{10.67}$$

where δ is the noncentrality parameter (10.65). The value of δ must be found such that (10.67) is satisfied. The required lower $(1 - \alpha)$-level confidence limit on $R \mid x$ is then the order q of the standard Normal quantile z_q, which is determined by δ according to (10.65). See Example 10.8 for an illustration of the computations.

INTERVAL ESTIMATES: CENSORED SAMPLES

10.19 Large Samples

For large samples, the asymptotic sampling distributions of the MLEs can be used for constructing approximate confidence intervals on parameters. Let θ stand for μ or σ, and recall from Section 3.7 that the sampling pdf of the MLE $\widehat{\theta}$ is asymptotically Normal, with mean value θ and variance MVB_θ:

$$f_N(\widehat{\theta}; \theta, \sqrt{MVB_\theta}). \tag{10.68}$$

Thus, the approximate $(1-\alpha)$-level confidence interval on θ is obtained as

$$(l_1, l_2) \doteq \theta \pm z_{\frac{\alpha}{2}} \cdot \sqrt{MVB_\theta}, \tag{10.69}$$

such that

$$Pr(l_1 \le \theta \le l_2) \doteq 1 - \alpha. \tag{10.70}$$

The MVBs are obtained from the inverse of the information matrix (10.47).

For parameter functions $g(\theta)$, the invariance property of MLEs is used (see Section 3.7). That is, the sampling pdf of $\widehat{g} = g(\widehat{\theta})$ is asymptotically Normal, with mean value g and variance approximated by the error propagation formula (3.11).

10.20 Small Samples

For small samples, the preceding Normal approximation produces confidence limits that are somewhat in error, particularly for highly censored samples. The likelihood-ratio method (see Section 3.9), gives superior results that are more accurate for small samples.

Recall that the statistic

$$LR(\theta) = 2\ln[L(\widehat{\theta})] - 2\ln[L(\theta)] \tag{10.71}$$

is approximately χ^2 distributed with $\nu = 1$ degree of freedom. Thus a $(1-\alpha)$-level confidence interval on θ comprises those values θ for which $LR(\theta) \le \chi^2_{1,1-\alpha}$. For a two-parameter model, one parameter in (10.71) needs to be expressed in terms of the other parameter by its ML equation. Thus, a confidence interval on μ is obtained from

$$LR(\mu) = 2\ln[L(\widehat{\mu}, \widehat{\sigma})] - 2\ln[L(\mu, \sigma\{\mu\})], \tag{10.72}$$

where $\sigma\{\mu\}$ is defined by the ML equation (10.45).

Similarly, a confidence interval on σ is obtained from

$$LR(\sigma) = 2\ln[L(\widehat{\mu}, \widehat{\sigma})] - 2\ln[L(\mu\{\sigma\}, \sigma)], \tag{10.73}$$

where $\mu\{\sigma\}$ is defined by the ML equation (10.44). See Example 10.9 for an illustration.

A confidence interval on a parameter function $g(\theta)$ is more difficult to obtain since the relation among parameters, implied by $g(\theta)$, constrains the solution process. As an illustration, consider the quantile $x_q = \mu + z_q\sigma$, so that

$$\mu\{x_q, \sigma\} = x_q - z_q\sigma. \tag{10.74}$$

Substituting (10.74) into the likelihood function (10.43) and differentiating with respect to σ gives the constraint relation:

$$\sum_{i=1}^{r}\left(\frac{x_i - x_q}{\sigma}\right)\left(\frac{x_i - x_q}{\sigma} + z_q\right) + \sum_{j=1}^{m}\frac{\left(\frac{c_j - x_q}{\sigma}\right)\phi\left(\frac{c_j - x_q}{\sigma} + z_q\right)}{1 - \Phi\left(\frac{c_j - x_q}{\sigma} + z_q\right)} = r. \qquad (10.75)$$

A confidence interval on x_q is then obtained from

$$LR(x_q) = 2\ln[L(\widehat{\mu}, \widehat{\sigma})] - 2\ln[L(\mu\{x_q, \sigma\{x_q\}\}, \sigma\{x_q\})], \qquad (10.76)$$

where $\mu\{x_q, \sigma\}$ is given by (10.74) and $\sigma\{x_q\}$ is the solution of (10.75). See Example 10.9 for an illustration of the computations.

Because censored samples are often encountered in reliability engineering, the estimation of a *lower* confidence limit on the reliability value $R \mid x$ is of importance. For a given value of x, the reliability $R \mid x = 1 - F(x; \mu, \sigma)$ can be inverted to the $(1 - R)$-quantile x_{1-R}, as was done in Section 10.18. A lower confidence limit on R is therefore equivalent to an upper confidence limit on x_{1-R}. Thus, Equation (10.76) needs to be solved for the order $q = 1 - R$ such that the $\chi^2_{1,1-\alpha}$ variate equals the given value x. See Example 10.9 for an illustration of the calculations.

APPLICATIONS

10.21 General

The Normal distribution is the appropriate postulate for a measurement variable X if it can be argued that realizations x are the result of many independent random causes that *add up* to generate x. The central limit theorem (see Section 2.4) ensures that X is distributed Normally, at least approximately.

For example, often the sum of noise signals from components produces the measurable noise signal at the output stage of an electronic circuit. Similarly, the errors of machining processes often aggregate additively, so that produced dimensions are Normally distributed. Additionally, project costs are the summations of many source costs; project completion times are the sums of completion times of critical-path tasks; and the lifetime of a system with standby components and perfect switching devices is the sum of component lifetimes. In each case, the measured variable is modeled by a Normal distribution, at least approximately.

When such reasoning is not justified, the Normal distribution may still be postulated if the following assumptions appear reasonable:

1. Most measurements can be expected to cluster closely around a central value.
2. Deviations from the central value are equally likely to occur in either direction.

3. The likelihood of deviations rapidly diminishes with the size of the deviation.

That is, a symmetrical, bell-shaped frequency pattern is postulated so that the Normal distribution is a reasonable model. The above assumptions are usually met for the *error* of measuring a *constant*.

For example, repeated measurements of a given dimension will show variation at the limit of instrument precision. The conditions causing this variation are often consonant with the above assumptions, so that a Normal error model fits the data well.

Note, however, that the above assumptions cannot *generally* be made for engineering variables, which most often are not only inherently positive but also highly skewed, so that the Normal model would be inappropriate.

In some engineering situations, decisions focus on the *central region* of the measurement model $f(x;\theta)$. The Normal distribution may then provide an adequate approximation to the true distribution of X in that central region.

For example, the cost of engineering tasks is an important factor influencing decisions. A common decision criterion for the evaluation of tasks is the *expected cost value*. The Normal model for such cost variables then often yields reasonable decisions even when the true cost distributions can be shown to deviate from the Normal model in the tails.

Take note, however, that most engineering decisions focus on the *tails* of distributions. Choosing an appropriate model then becomes highly important. Blithely assuming the Normal distribution can lead to erroneous decisions because estimated decision quantities in extreme tails could then be in error by an order of magnitude!

10.22 Life Testing

The Normal distribution sometimes finds application as a *life model* for the time-to-failure of equipment subject to wear. An important decision function is then the reliability R of the equipment at a specified operating time x or, equivalently, the operating time at which the equipment features a specified reliability value. Chapter 12 introduces some basic concepts of reliability analysis.

In Section 12.16 the *hazard function* is introduced as

$$h(x) = \frac{f(x)}{1 - F(x)},$$

which *increases* with x for the Normal model. This means that the probability of failure increases with equipment age. It can be shown that for the Normal model the hazard function approaches a straight line with slope $1/\sigma^2$ as x increases. Figure 10.4 shows a typical hazard and reliability function. The Normal life

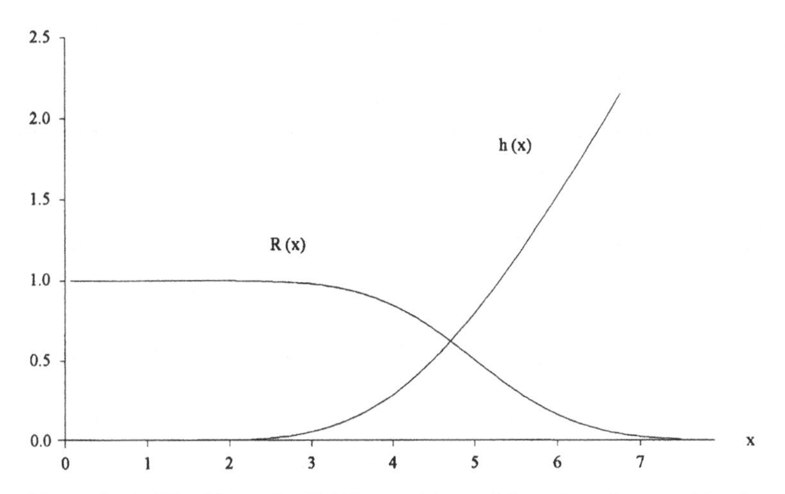

Figure 10.4. The Normal reliability and hazard functions for $\mu = 5$ and $\sigma = 1$.

model is justified when the empirical hazard function (12.64) behaves in the manner shown in Fig. 10.4. Note, however, that plotting this empirical function requires a large sample. To determine lower confidence limits on reliability, see Sections 10.18 and 10.20.

It is sometimes required to determine the replacement time of an aged unit before it fails (see also Section 17.27). If the unit failure-replacement cost C_f is larger than the unit planned age-replacement cost C_r, the decision function to minimize is the average cost per unit operating time for the equipment, at the replacement time t:

$$Ca(t) = \frac{C_f\,\Phi\left(\frac{t-\mu}{\sigma}\right) + C_r\left[1 - \Phi\left(\frac{t-\mu}{\sigma}\right)\right]}{\int_0^t \left[1 - \Phi\left(\frac{x-\mu}{\sigma}\right)\right] dx}, \tag{10.77}$$

where Φ is the standard Normal cdf. Minimizing (10.77) over t gives the solution equation for the optimal replacement time t^*:

$$\frac{\phi\left(\frac{t^*-\mu}{\sigma}\right)}{\sigma\left[1 - \Phi\left(\frac{t^*-\mu}{\sigma}\right)\right]} \cdot \int_0^{t^*} \left[1 - \Phi\left(\frac{x-\mu}{\sigma}\right)\right] dx + 1 - \Phi\left(\frac{t^* - \mu}{\sigma}\right) - \frac{C_f}{C_f - C_r} = 0, \tag{10.78}$$

where ϕ is the standard Normal pdf.

When units are only replaced upon failure (no planned replacements), the average cost per time unit for the equipment unit is

$$C = \frac{C_f}{\mu}. \tag{10.79}$$

The difference between (10.79) and (10.78) gives an estimate of the average savings per time unit of planned replacement for an equipment unit. See Example 10.10 for an illustration of the computations.

EXAMPLE 10.1

Calculate the 95-percentile of the Chi-squared and *t*-distributions at 16 degrees of freedom, of the *F*-distribution at (10, 16) degrees of freedom, and of the noncentral *t*-distribution at 16 degrees of freedom and noncentrality parameter 10.

1. χ^2-percentile

$\alpha := 0.95$

$v := 16 \qquad F(x) := \frac{1}{2 \cdot \Gamma\left(\frac{v}{2}\right)} \cdot \int_0^x \left(\frac{t}{2}\right)^{\frac{v}{2}-1} \cdot \exp\left(-\frac{t}{2}\right) dt$

$x := 25 \qquad X95 := \mathrm{root}(F(x) - \alpha, x) \qquad X95 = 26.296$

2. *t*-percentile

$\alpha := 0.95$

$v := 16 \qquad F(x) := \frac{\Gamma\left(\frac{v+1}{2}\right)}{\sqrt{v \cdot \pi} \cdot \Gamma\left(\frac{v}{2}\right)} \cdot \int_{-6}^x \left(1 + \frac{t^2}{v}\right)^{-\frac{v+1}{2}} dt$

$x := 1.5 \qquad T95 := \mathrm{root}(F(x) - 0.95, x) \qquad T95 = 1.746$

3. *F*-percentile

$\alpha := 0.95$

$v1 := 10 \qquad v2 := 16 \qquad F(x) = \frac{\Gamma\left(\frac{v1+v2}{2}\right)}{\Gamma\left(\frac{v1}{2}\right) \cdot \Gamma\left(\frac{v2}{2}\right)} \cdot v1^{\frac{v1}{2}} \cdot v2^{\frac{v2}{2}}$

$\cdot \int_0^x y^{\frac{v1}{2}-1} \cdot (v2 + v1 \cdot y)^{-\frac{v1+v2}{2}} dy$

$x := 2 \qquad F95 := \mathrm{root}(F(x) - 0.95, x) \qquad F95 = 2.493$

4. Noncentral *t*-percentile

$\alpha := 0.95 \qquad z95 := 1.645$

$v := 16 \qquad \delta := 10$

approximate value: $t := \frac{\delta + z95 \cdot \sqrt{1 + \frac{\delta^2 - z95^2}{2 \cdot v}}}{1 - \frac{z95^2}{2 \cdot v}} \qquad t = 14.536$

exact value:

x integrand: $I(x, t) := x^{v-1} \cdot \exp\left(-\frac{x^2}{2}\right) \cdot \int_0^{\frac{x \cdot t}{\sqrt{v}}} \exp\left[-\frac{(u-\delta)^2}{2}\right] du$

x integration limits: $a := 0 \qquad b := 10$

$$F(t) := \frac{2^{1-\frac{\nu}{2}}}{\Gamma\left(\frac{\nu}{2}\right) \cdot \sqrt{2 \cdot \pi}} \cdot \int_a^b I(x, t)\, dx$$

Plot the integrand $I(x)$ to check the integration limits a and b:

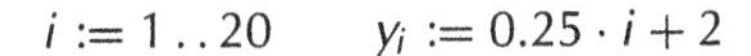

$i := 1 \,..\, 20 \qquad y_i := 0.25 \cdot i + 2$

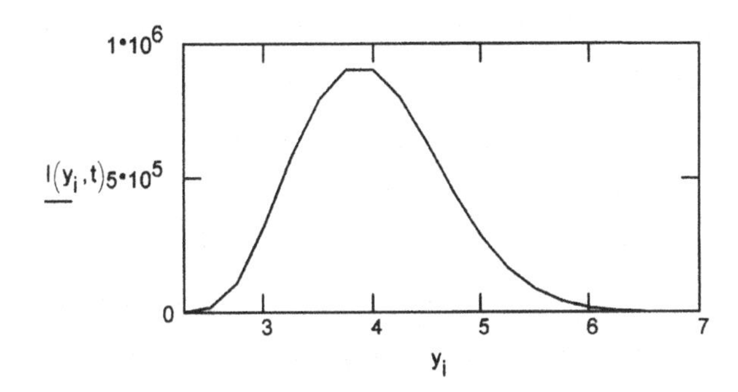

$T := \text{root}(F(t) - 0.95, t)$ $\qquad\qquad$ $T = 14.640$

EXAMPLE 10.2

A sample of 17 specimens was drawn from a machining process, and a critical dimension (in millimeters) was measured on each:

12.065, 11.992, 11.992, 11.921, 11.954, 11.945, 12.029, 11.948, 11.885,
11.997, 11.982, 12.109, 11.966, 12.081, 11.846, 12.007, 12.011.

Check graphically if a Normal distribution could represent this machining dimension.

$n := 17 \qquad i := 1 \,..\, n$

$x_i :=$ $\qquad$ $s := \text{sort}(x)$

12.065
11.992
11.992
11.921
11.954
11.945
12.029
11.948
11.885

median plotting position: $p_i := \dfrac{i - 0.3}{n + 0.4}$

standard Normal quantile: $t := 0 \quad z_i := \text{root}(\text{cnorm}(t) - p_i, t)$

ordinate of data plot: $y_i := 0.5 + 0.21 \cdot z_i$

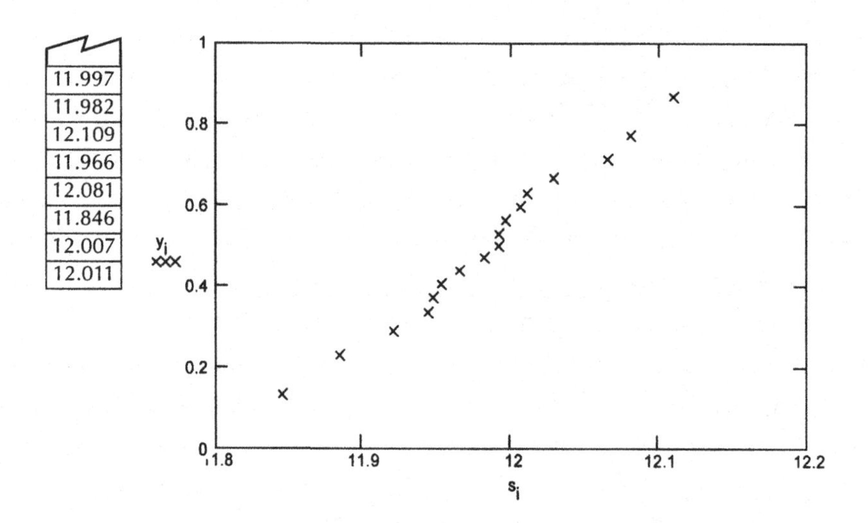

The data plot looks reasonably linear; thus a Normal process appears plausible.

EXAMPLE 10.3

A sample of 22 truck tires was placed on an accelerated wear test. The coded values of duration are:

17.8, 22.7, 27, 32.5, 34.3, 34.6, 37, 38.4, 40*, 42.3, 43.6,
45.5, 48.6, 50*, 55.2, 56.4, 57.8, 60*, 60*, 62.6, 66.9, 72.3,

where asterisks denote the times when tires where removed from test for a destructive materials test. Check graphically if a Normal model reasonably represents this wear phenomenon.

$n := 22 \qquad r := 18 \qquad i := 1 \,..\, r$

(The x_i are the 18 actual wear data, and q_i gives the order of these wear data in the sample.)

$x_i :=$	$q_i :=$
17.8	1
22.7	2
27	3
32.5	4
34.3	5
34.6	6
37	7
38.4	8
42.3	10
43.6	11
45.5	12
48.6	13

median plotting position: $p_i = 1 - \dfrac{n+0.7}{n+0.4} \displaystyle\prod_{j=1}^{i} \dfrac{n-q_j+0.7}{n-q_j+1.7}$

$t := 0 \qquad z_i := \mathrm{root}(\mathrm{cnorm}(t) - p_i, t)$

ordinates for the data plot: $y_i := 0.5 + 0.21 \cdot z_i$

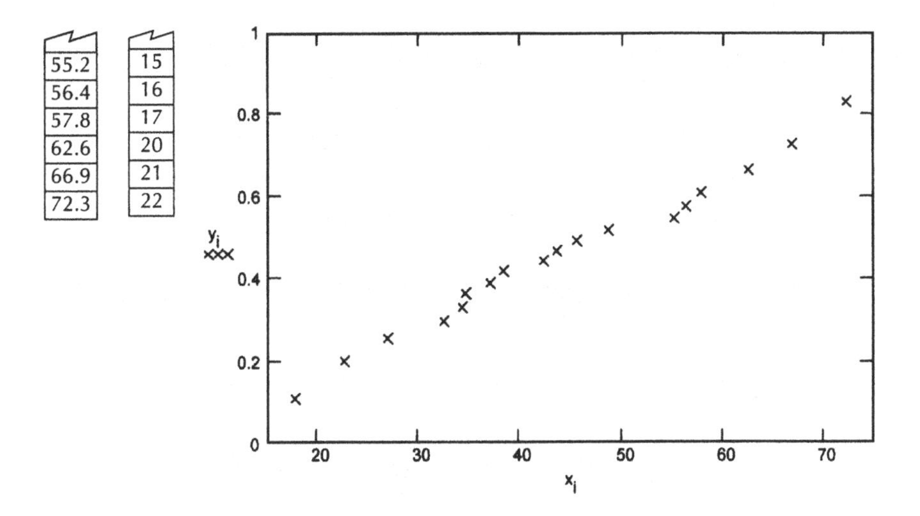

The data plot looks fairly linear; thus a Normal model appears justified for the wear process.

EXAMPLE 10.4

For the data of Example 10.2, estimate the Normal model, and test the fit at the 5% level of significance. Estimate the 5-percentile and its standard error.

$n := 17 \qquad i := 1..n$

$x_i := \qquad s := \text{sort}(x)$

x_i
12.065
11.992
11.992
11.921
11.954
11.945
12.029
11.948
11.885
11.997
11.982
12.109
11.966
12.081
11.846
12.007
12.011

1. Estimation

$$\mu := \frac{1}{n} \cdot \sum_{i=1}^{n} x_i \qquad \mu = 11.984$$

$$\sigma := \sqrt{\frac{1}{n-1} \cdot \sum_{i=1}^{n} (x_i - \mu)^2} \qquad \sigma = 0.067$$

2. Standard errors

$$\text{SE}\mu := \frac{\sigma}{\sqrt{n}} \qquad \text{SE}\mu = 0.016$$

$$\text{SE}\sigma := \frac{\sigma}{\sqrt{2 \cdot (n-1)}} \qquad \text{SE}\sigma = 0.012$$

3. Test-of-fit

Use the biased estimate of σ:

$$\sigma b := \sigma \cdot \sqrt{\frac{n-1}{n}} \qquad w_i := \text{cnorm}\left(\frac{s_i - \mu}{\sigma b}\right)$$

$$A := \left[-n - \frac{1}{n} \cdot \sum_{i=1}^{n} (2 \cdot i - 1) \cdot (\ln(w_i) + \ln(1 - w_{n-i+1}))\right] \cdot \left(1 + \frac{0.75}{n} + \frac{2.25}{n^2}\right)$$

$A = 0.194$

The 5% critical value is 0.752. Since the sample value A is smaller, there is no reason to reject the estimated model at the 5% significance level.

4. Probability plot

$$p_i := \frac{i - 0.3}{n + 0.4} \qquad t := 0 \qquad z_i := \text{root}(\text{cnorm}(t) - p_i, t)$$

ordinate of data plot: $\quad y_i := 0.5 + 0.21 \cdot z_i$

ordinate for the model plot: $\quad r_i := 0.5 + 0.21 \cdot \dfrac{s_i - \mu}{\sigma}$

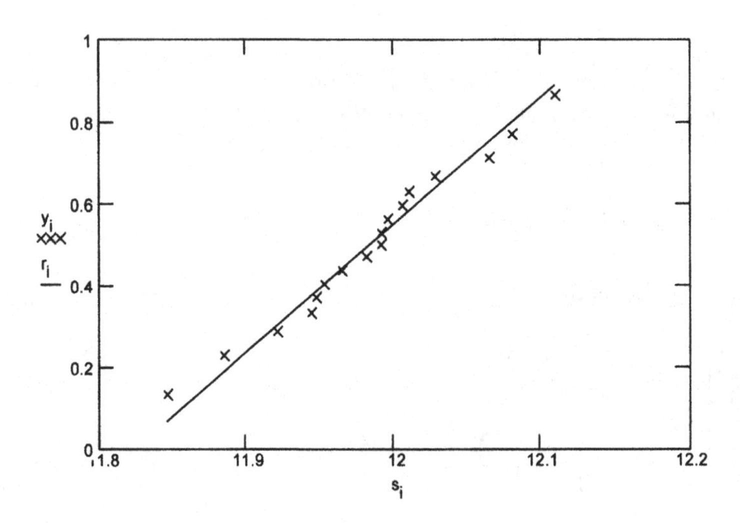

5. 5-percentile

$z05 := -1.645$

$x05 := \mu + z05 \cdot \sigma \qquad x05 = 11.874$

standard error:

$SEx := \sqrt{SE\mu^2 + (z05 \cdot SE\sigma)^2} \qquad SEx = 0.025$

EXAMPLE 10.5

For the data of Example 10.3, estimate the Normal parameters and their standard errors. Estimate the 80-percentile of wear duration and its standard error.

$n := 22 \quad r := 18 \quad i := 1..r$

$m := n - r \quad j := 1..m$

$$\phi(t) := \frac{1}{\sqrt{2 \cdot \pi}} \cdot \exp\left(-\frac{t^2}{2}\right) \qquad \Phi(t) := \text{cnorm}(t)$$

$x_i :=$ $c_j :=$

x_i
17.8
22.7
27
32.5
34.3
34.6
37
38.4
42.3
43.6
45.5
48.6
55.2
56.4
57.8
62.6
66.9
72.3

c_j
40
50
60
60

1. Estimation

starting values:

$$a := \frac{1}{r} \cdot \sum_{i=1}^{r} x_i \qquad b := \sqrt{\frac{1}{r} \cdot \sum_{i=1}^{r} (x_i - a)^2} \qquad a = 44.2 \quad b = 14.9$$

GIVEN

$$\sum_{i=1}^{r} \frac{x_i - a}{b} + \sum_{j=1}^{m} \frac{\phi\left(\frac{c_j - a}{b}\right)}{1 - \Phi\left(\frac{c_j - a}{b}\right)} = 0$$

$$\sum_{i=1}^{r} \left(\frac{x_i - a}{b}\right)^2 + \sum_{j=1}^{m} \frac{\frac{c_j - a}{b} \cdot \phi\left(\frac{c_j - a}{b}\right)}{1 - \Phi\left(\frac{c_j - a}{b}\right)} = r$$

$$\begin{pmatrix} \mu \\ \sigma \end{pmatrix} := \text{FIND}(a, b) \qquad \mu = 47.8 \quad \sigma = 16.2$$

2. Standard errors

$$h(t) := \frac{\phi(t)}{1 - \Phi(t)}$$

$$z_i := \frac{x_i - \mu}{\sigma} \qquad y_j := \frac{c_j - \mu}{\sigma} \qquad s1 := \sum_{i=1}^{r} z_i \qquad s2 := \sum_{i=1}^{r} (z_i)^2$$

$$s3 := \sum_{j=1}^{m} h(y_j) \cdot (h(y_j) - y_j) \qquad s4 := \sum_{j=1}^{m} h(y_j) \cdot [1 + y_j \cdot (h(y_j) - y_j)]$$

$$s5 := \sum_{j=1}^{m} y_j \cdot h(y_j) \qquad s6 := \sum_{j=1}^{m} (y_j)^2 \cdot h(y_j) \cdot (h(y_j) - y_j)$$

$$I := \frac{1}{\sigma^2} \cdot \begin{pmatrix} r + s3 & 2 \cdot s1 + s4 \\ 2 \cdot s1 + s4 & 3 \cdot s2 + 2 \cdot s5 + s6 - r \end{pmatrix} \qquad V := I^{-1}$$

$$V = \begin{pmatrix} 12.819 & 1.119 \\ 1.119 & 7.559 \end{pmatrix} \qquad \begin{matrix} SE\mu := \sqrt{V_{0,0}} \\ SE\sigma := \sqrt{V_{1,1}} \end{matrix} \qquad \begin{matrix} SE\mu = 3.6 \\ SE\sigma = 2.7 \end{matrix}$$

3. Probability plot

$q_i :=$ (q_i is the order of the actual wear data in the total sample.)

q_i
1
2
3
4
5
6
7
8
10
11
12
13
15
16
17
20
21
22

median plotting position: $p_i := 1 - \dfrac{n+0.7}{n+0.4} \cdot \prod_{j=1}^{i} \dfrac{n-q_j+0.7}{n-q_j+1.7}$

$t := 0 \qquad zp_i := \mathrm{root}(\mathrm{cnorm}(t) - p_i, t)$

ordinates for the data plot: $y_i := 0.5 + 0.21 \cdot zp_i$

ordinates for the model plot: $r_i := 0.5 + 0.21 \cdot \dfrac{x_i - \mu}{\sigma}$

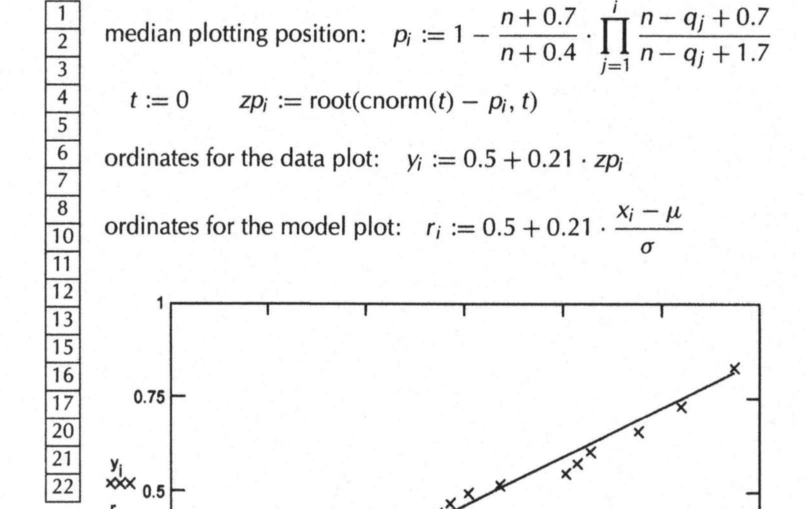

4. 80-percentile

$z80 := \mathrm{root}(\mathrm{cnorm}(t) - 0.8, t) \qquad z80 = 0.842$

$x80 := \mu + z80 \cdot \sigma \qquad x80 = 61.5$

standard error: $d_0 := 1 \qquad d_1 := z80$ (derivatives of $x80$)

$\mathrm{varX} := \sum_{i=0}^{1} \sum_{j=0}^{1} d_i \cdot d_j \cdot V_{i,j} \qquad \mathrm{SE}x := \sqrt{\mathrm{var}X} \qquad \mathrm{SE}x = 4.5$

EXAMPLE 10.6

For the data of Example 10.2, test the following hypotheses at the 5% level of significance:

a. $Ho: \mu = 12.000$ against $Ha: \mu = 12.050$
b. $Ho: \sigma = 0.060$ against $Ha: \sigma = 0.100$

From the data: $n := 17 \qquad xb := 11.984 \qquad s := 0.067$

a. $\mu o := 12.000 \qquad \mu a := 12.050$

95-percentile of t, at $v = 16$, is $\quad t95 := 1.746$

Critical value of Xb is $\quad tc := \mu o + t95 \cdot \frac{s}{\sqrt{n}} \qquad tc = 12.028$

Since $xb < tc$, accept H_o.

power of test:

$$Td := \frac{tc - \mu o}{\frac{s}{\sqrt{n}}} \qquad \delta := \frac{\mu a - \mu o}{\frac{s}{\sqrt{n}}} \qquad v := n - 1$$

$$P := 1 - \frac{2^{1-\frac{v}{2}}}{\Gamma\left(\frac{v}{2}\right)\cdot\sqrt{2\cdot\pi}} \cdot \int_2^8 x^{v-1}\cdot\exp\left(-\frac{x^2}{2}\right)\cdot\int_0^{\frac{x\cdot Td}{\sqrt{v}}} \exp\left[-\frac{(u-\delta)^2}{2}\right] du\, dx$$

$P = 0.904$

b. $\sigma o := 0.060 \qquad \sigma a := 0.100$

95-percentile of χ^2 at $v = 16$ degrees of freedom is $\quad K := 26.3$

Critical value of S^2 is

$$tc := \frac{\sigma o^2}{n-1} \cdot K \qquad tc = 0.0059 \qquad s^2 = 0.0045$$

Since $s^2 < tc$, accept H_o.

power of test:

$$L := \left(\frac{\sigma o}{\sigma a}\right)^2 \cdot K \qquad L = 9.468$$

$$P := 1 - \frac{1}{2\cdot\Gamma\left(\frac{v}{2}\right)} \cdot \int_0^L \left(\frac{t}{2}\right)^{\frac{v}{2}-1} \cdot \exp\left(-\frac{t}{2}\right) dt \qquad P = 0.893$$

EXAMPLE 10.7

Suppose that the process, from which the data of Example 10.2 were drawn, was adjusted to increase the process mean. Suppose that a sample of 15 specimens from the changed process gave a sample mean of 12.042 and a sample standard deviation of 0.074. At the 5% level of significance, test the null hypotheses that:

a. the difference in means is zero, against the alternative that this difference is 0.060 (new mean is larger);
b. the variance ratio is 1, against the alternative that it is 3 (new variance is larger).

From the data:

$nx := 17 \quad xbx := 11.984 \qquad ny := 15 \quad xby := 12.042$

$sx := 0.067 \qquad sy := 0.074$

a. Difference in means

$\Delta M := 0.060$ 95-percentile of t, at $v = 30$, is $t95 := 1.697$

pooled standard deviation: $Sp := \sqrt{\dfrac{(nx-1)\cdot sx^2 + (ny-1)\cdot sy^2}{nx+ny-2}}$

The test statistic Δxb is $xby - xbx = 0.058$

Critical value of t is

$$tc := t95 \cdot Sp \cdot \sqrt{\frac{1}{nx} + \frac{1}{ny}} \qquad tc = 0.042$$

Since $\Delta xb > tc$, reject H_0 (the mean *has* increased).

Power of the test:

$$Td := \frac{tc}{Sp \cdot \sqrt{\frac{1}{nx} + \frac{1}{ny}}} \qquad \delta := \frac{\Delta M}{Sp \cdot \sqrt{\left(\frac{1}{nx} + \frac{1}{ny}\right)}} \qquad v := nx + ny - 2$$

$$P := 1 - \frac{2^{1-\frac{v}{2}}}{\Gamma\left(\frac{v}{2}\right)\cdot\sqrt{2\cdot\pi}} \cdot \int_2^8 x^{v-1} \cdot \exp\left(-\frac{x^2}{2}\right) \cdot \int_0^{\frac{x\cdot Td}{\sqrt{v}}} \exp\left[-\frac{(u-\delta)^2}{2}\right] du\, dx$$

$P = 0.769$

b. Variance ratio

$R := 3$

The test statistic is

$$\left(\frac{sy}{sx}\right)^2 = 1.22$$

The critical test value is the 95-percentile F, at (14,16) degrees of freedom:

$F95 := 2.373.$

Since the test statistic is less than the critical value, accept H_0 (the variance has *not* changed).

Power of the test:

$v1 := 14 \qquad v2 := 16$

integration limit:

$$L := \frac{F95}{R}$$

$$P := 1 - \frac{\Gamma\left(\frac{v1+v2}{2}\right)}{\Gamma\left(\frac{v1}{2}\right)\cdot\Gamma\left(\frac{v2}{2}\right)} \cdot v1^{\frac{v1}{2}} \cdot v2^{\frac{v2}{2}} \cdot \int_0^L y^{\frac{v1}{2}-1} \cdot (v2 + v1\cdot y)^{-\frac{v1+v2}{2}} dy$$

$P = 0.667$

EXAMPLE 10.8

For the data of Example 10.2, determine a) the 95% confidence interval on the 5-percentile of the machined dimension, and b) calculate the 90% lower confidence limit on the probability that the machined dimension will be greater than 11.900.

a. From example 10.4:

$n := 17 \quad xb := 11.984 \quad s := 0.067 \quad x05 := 11.874$

Given:

$q := 0.05 \quad zq := -1.645 \quad v := n - 1 \quad \delta := zq \cdot \sqrt{n} \qquad \delta = -6.783$

1. We want the 2.5-percentile of t' at $d = -6.783$; from (10.26), this equals the negative of the 97.5-percentile of t' at $\delta = +6.783$; following Example 10.1:

   $z975 := 1.96$

   $v := 16 \qquad \delta := -\delta$

   approximate value: $t := \dfrac{\delta + z975 \cdot \sqrt{1 + \frac{\delta^2 - z975^2}{2 \cdot v}}}{1 - \frac{z975^2}{2 \cdot v}} \qquad t = 11.099$

   exact value:

   x integrand: $\quad I(x, t) := x^{v-1} \cdot \exp\left(-\dfrac{x^2}{2}\right) \cdot \displaystyle\int_0^{\frac{x \cdot t}{\sqrt{v}}} \exp\left[-\dfrac{(u - \delta)^2}{2}\right] du$

   x integration limits: $\quad a := 0 \qquad b := 10$

   $F(t) := \dfrac{2^{1-\frac{v}{2}}}{\Gamma\left(\frac{v}{2}\right) \cdot \sqrt{2 \cdot \pi}} \cdot \displaystyle\int_a^b I(x, t)\, dx \qquad T := \mathrm{root}(F(t) - 0.975, t)$

   $T = 11.105$

   Thus $\quad t025 := -11.105$

2. Similarly, we obtain $\quad t975 := -4.359$

   The required exact confidence interval is therefore

   $xL := xb + \dfrac{s}{\sqrt{n}} \cdot t025 \qquad xL := 11.804$

   $xU := xb + \dfrac{s}{\sqrt{n}} \cdot t975 \qquad xU := 11.913$

b. Given: $\quad x := 11.900$

the t value is

$td := \dfrac{x - xb}{s} \cdot \sqrt{n} \qquad td = -5.169 \qquad t := -td$

approximate determination of δ from (10.25): $\qquad z90 := 1.28$

$d := 5 \qquad g(d) := \dfrac{d + z90 \cdot \sqrt{1 + \frac{d^2 - z90^2}{2 \cdot v}}}{1 - \frac{z90^2}{2 \cdot v}}$

$da := \mathrm{root}(g(d) - t, d), \qquad da = 3.435$

exact determination of δ:

x integrand: $\quad I(x,\delta) := x^{\nu-1}\cdot\exp\left(-\frac{x^2}{2}\right)\cdot\int_0^{\frac{x\cdot t}{\sqrt{\nu}}}\exp\left[-\frac{(u-\delta)^2}{2}\right]du$

x integration limits: $\quad a := 0 \qquad b := 10$

$$d := da \qquad F(\delta) := \frac{2^{1-\frac{\nu}{2}}}{\Gamma\left(\frac{\nu}{2}\right)\cdot\sqrt{2\cdot\pi}}\cdot\int_a^b I(x,\delta)\,dx$$

$de := \text{root}(F(d) - 0.90, d) \qquad de = 3.365$

From (10.65):

$$zq := \frac{de}{\sqrt{n}} \qquad zq = 0.816$$

The required confidence limit is therefore

$L := \text{cnorm}(zq) \qquad L = 0.793$

EXAMPLE 10.9

Consider the data of Example 10.3. Calculate a) 90% confidence intervals on the distribution parameters, b) a 90% confidence interval on the 80-percentile wear duration, and c) a 95% lower confidence limit on the reliability value at the duration $x = 20$.

From Example 10.5: $n := 22 \qquad r := 18 \qquad i := 1..r$

$m := n - r \qquad j := 1..m$

$x_i :=$

17.8
22.7
27
32.5
34.3
34.6
37
38.4
42.3
43.6
45.5
48.6
55.2
56.4
57.8
62.6
66.9
72.3

$c_j :=$

40
50
60
60

$\mu := 47.847 \qquad \text{SE}\mu := 3.580$

$\sigma := 16.192 \qquad \text{SE}\sigma := 2.749$

$x80 := 61.474 \qquad \text{SE}x := 4.479$

a. Parameters

i) Normal approximation:
Standard Normal 95-percentile: $z95 := 1.645$

$\mu L := \mu - z95\cdot\text{SE}\mu \qquad \mu L = 42.0$

$\mu U := \mu + z95\cdot\text{SE}\mu \qquad \mu U = 53.7$

$\sigma L := \sigma - z95\cdot\text{SE}\sigma \qquad \sigma L = 11.7$

$\sigma U := \sigma + z95\cdot\text{SE}\sigma \qquad \sigma U = 20.7$

ii) Likelihood ratio method:
Chi-sqared 90-precentile at $\nu = 1$: $\quad K90 := 2.71$

Standard Normal functions: $\quad \phi(t) := \frac{1}{\sqrt{2\cdot\pi}}\exp\left(-\frac{t^2}{2}\right)$

$\Phi(t) := \text{cnorm}(t)$

log-likelihood function:

$$LL(\mu,\sigma) := -r \cdot \ln(\sigma) - \frac{1}{2} \cdot \sum_{i=1}^{r} \left(\frac{x_i - \mu}{\sigma}\right)^2 + \sum_{j=1}^{m} \ln\left(1 - \Phi\left(\frac{c_j - \mu}{\sigma}\right)\right)$$

location parameter μ:

$$(10.45):\quad ML(a,b) := \sum_{i=1}^{r} \left(\frac{x_i - a}{b}\right)^2 + \sum_{j=1}^{m} \frac{\left(\frac{c_j-a}{b}\right) \cdot \phi\left(\frac{c_j-a}{b}\right)}{1 - \Phi\left(\frac{c_j-a}{b}\right)} - r$$

$b := \sigma \qquad s(a) := \mathrm{root}(ML(a,b), b)$

$\qquad\qquad LR(a) := 2 \cdot LL(\mu,\sigma) - 2 \cdot LL(a, s(a))$

$a := \mu - 6 \qquad \mu L := \mathrm{root}(LR(a) - K90, a) \qquad \mu L = 41.9$

$a := \mu + 6 \qquad \mu U := \mathrm{root}(LR(a) - K90, a) \qquad \mu U = 54.2$

scale parameter σ:

$$(10.44):\quad ML(a,b) := \sum_{i=1}^{r} \frac{x_i - a}{b} + \sum_{j=1}^{m} \frac{\phi\left(\frac{x_j-a}{b}\right)}{1 - \Phi\left(\frac{x_j-a}{b}\right)}$$

$a := \mu \qquad g(b) := \mathrm{root}(ML(a,b), a)$

$\qquad\qquad LR(b) := 2 \cdot LL(\mu,\sigma) - 2 \cdot LL(g(b), b)$

$b := \sigma - 4 \qquad \sigma L := \mathrm{root}(LR(b) - K90, b) \qquad \sigma L = 13.1$

$b := \sigma + 4 \qquad \sigma U := \mathrm{root}(LR(b) - K90, b) \qquad \sigma U = 21.5$

b. 80-percentile

i) Normal approximation:

$XL := x80 - z95 \cdot SEx \qquad XL = 54.1$

$XU := x80 + z95 \cdot SEx \qquad XU = 68.8$

ii) Likelihood ratio method:

$t := 1 \qquad z80 := \mathrm{root}(\mathrm{cnorm}(t) - 0.8, t) \qquad z80 = 0.842$

$$(10.75):\quad ML(xq,b) := \sum_{i=1}^{r} \frac{x_i - xq}{b} \cdot \left(\frac{x_i - xq}{b} + z80\right) + \sum_{j=1}^{m} \frac{\frac{c_j-xq}{b} \cdot \phi\left(\frac{c_j-xq}{b} + z80\right)}{1 - \Phi\left(\frac{c_j-xq}{b} + z80\right)} - r$$

$b := \sigma \qquad s(xq) := \mathrm{root}(ML(xq,b), b)$

$\qquad\qquad g(xq) := xq - z80 \cdot s(xq)$

$\qquad\qquad LR(xq) := 2 \cdot LL(\mu,\sigma) - 2 \cdot LL(g(xq), s(xq))$

$xq := x80 - 15 \qquad XL := \text{root}(LR(xq) - K90, xq) \qquad XL = 55.0$

$xq := x80 + 10 \qquad XU := \text{root}(LR(xq) - K90, xq) \qquad XU = 70.4$

c. Reliability

reliability value at $xq = 20$:

$$R := 1 - \text{cnorm}\left(\frac{20-\mu}{\sigma}\right) \qquad R = 0.96$$

lower confidence limit:

$$(10.75):\quad ML(zq, b) := \sum_{i=1}^{r} \frac{x_i - 20}{b} \cdot \left(\frac{x_i - 20}{b} + zq\right) + \sum_{j=1}^{m} \frac{\frac{c_j-20}{b} \cdot \phi\left(\frac{c_j-20}{b} + zq\right)}{1 - \Phi\left(\frac{c_j-20}{b} + zq\right)} - r$$

$b := \sigma \qquad s(zq) := \text{root}(ML(zq, b), b)$

$g(zq) := 20 - zq \cdot s(zq)$

$LR(zq) := 2 \cdot LL(\mu, \sigma) - 2 \cdot LL(g(zq), s(zq))$

$zq := -1.2 \qquad zU := \text{root}(LR(zq) - K90, zq) \qquad zU = -1.164$

$RL := 1 - \text{cnorm}(zU) \qquad RL = 0.88$

EXAMPLE 10.10

Suppose that the accelerated wear conditions of Example 10.3 apply to a particular tire application. Calculate the optimal tire replacement time if the planned replacement cost per tire is \$300 while the cost is \$800 for replacing a tire that failed in operation.

$Cr := 300 \qquad$ From example 10.5: $\mu := 47.847 \qquad V := \begin{pmatrix} 12.819 & 1.119 \\ 1.119 & 7.559 \end{pmatrix}$

$Cf := 800 \qquad \sigma := 16.192$

1. Optimum replacement time

$$\phi(a) := \frac{1}{\sqrt{2 \cdot \pi}} \cdot \exp\left(-\frac{a^2}{2}\right) \qquad \Phi(a) := \text{cnorm}(a)$$

$$G(ts, \mu, \sigma) := \frac{\phi\left(\frac{ts-\mu}{\sigma}\right)}{\sigma \cdot \left(1 - \text{cnorm}\left(\frac{ts-\mu}{\sigma}\right)\right)} \cdot \int_0^{ts} \left(1 - \text{cnorm}\left(\frac{x-\mu}{\sigma}\right)\right) dx + 1 - \text{cnorm}\left(\frac{ts-\mu}{\sigma}\right) - \frac{Cf}{Cf - Cr}$$

$ts := 40 \qquad T(\mu, \sigma) := \text{root}(G(ts, \mu, \sigma), ts) \qquad T(\mu, \sigma) = 36.0$

2. Standard error

$i := 0..1 \qquad j := 0..1 \qquad d_0 := \frac{d}{d\mu} T(\mu, \sigma) \qquad d_1 := \frac{d}{d\sigma} T(\mu, \sigma)$

$\mathrm{var}T := \sum_i \sum_j d_i \cdot d_j \cdot V_{i,j} \qquad \mathrm{se}T := \sqrt{\mathrm{var}T} \qquad \mathrm{se}T = 2.6$

3. Age-replacement cost/time unit

$$Ca(t) := \frac{Cf \cdot \mathrm{cnorm}\left(\frac{t-\mu}{\sigma}\right) + Cr \cdot \left(1 - \mathrm{cnorm}\left(\frac{t-\mu}{\sigma}\right)\right)}{\int_0^t \left(1 - \mathrm{cnorm}\left(\frac{x-\mu}{\sigma}\right)\right) dx} \qquad Ca(T(\mu, \sigma)) = 12.3$$

4. Failure-replacement cost/time unit

$C := \frac{Cf}{\mu} \qquad C = 16.7$

5. Cost function

$$i := 1..60 \qquad t_i := i \qquad c_i := \frac{Cf \cdot \mathrm{cnorm}\left(\frac{t_i-\mu}{\sigma}\right) + Cr \cdot \left(1 - \mathrm{cnorm}\left(\frac{t_i-\mu}{\sigma}\right)\right)}{\int_0^{t_i} \left(1 - \mathrm{cnorm}\left(\frac{x-\mu}{\sigma}\right)\right) dx}$$

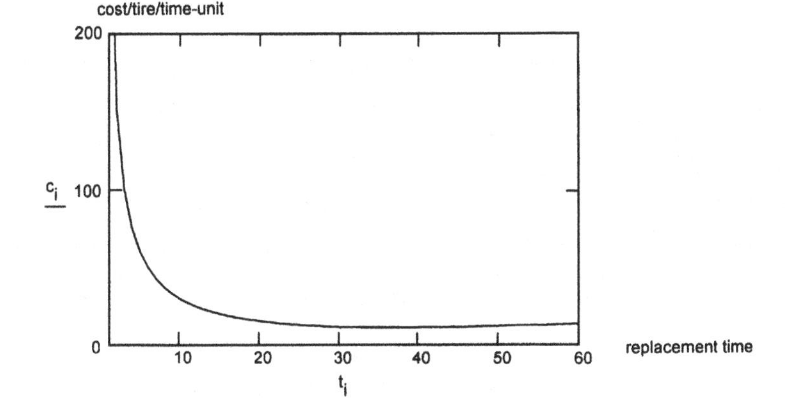

CHAPTER ELEVEN

Log-Normal Distributions

INTRODUCTION

11.1 Definition

A continuous random variable X has a *Log-Normal distribution* if its pdf has the form

$$f(x; \mu, \sigma) = \frac{1}{x\sigma\sqrt{2\pi}} \exp\left\{-\frac{1}{2}\left(\frac{\ln(x) - \mu}{\sigma}\right)^2\right\};$$

$$x > 0, \quad \sigma > 0, \quad -\infty < \mu < \infty. \qquad (11.1)$$

The similarity of this pdf with the Normal pdf (10.1) is apparent. Indeed, the two distribution families are closely related: If the logarithm of a variable X is Normally distributed, then the variable X itself has a Log-Normal distribution. Because of this close relationship, the Normal parameters μ and σ are customarily carried over to the Log-Normal pdf. Note, however, that their character has changed: The parameter $\eta = \exp(\mu\}$ is now a *scale* parameter, whereas σ is a *shape* parameter. Reducing the variable X to $Z = X/\exp(\mu\}$ eliminates μ from the pdf:

$$f(z; \sigma) = \frac{1}{z\sigma\sqrt{2\pi}} \exp\left\{-\frac{1}{2}\left(\frac{\ln(z)}{\sigma}\right)^2\right\}. \qquad (11.2)$$

Figure 11.1 shows this reduced pdf for several values of σ, indicating the *shape flexibility* of this distribution.

The Log-Normal distribution was introduced more than 100 years ago by Galton, who pointed out that the logarithm of the product of certain naturally occurring random variables has a Normal distribution. This distribution has become a prominent measurement model in engineering as an alternative to the Normal model, since its sample space is positive and its shape more naturally fits many data frequency patterns in engineering.

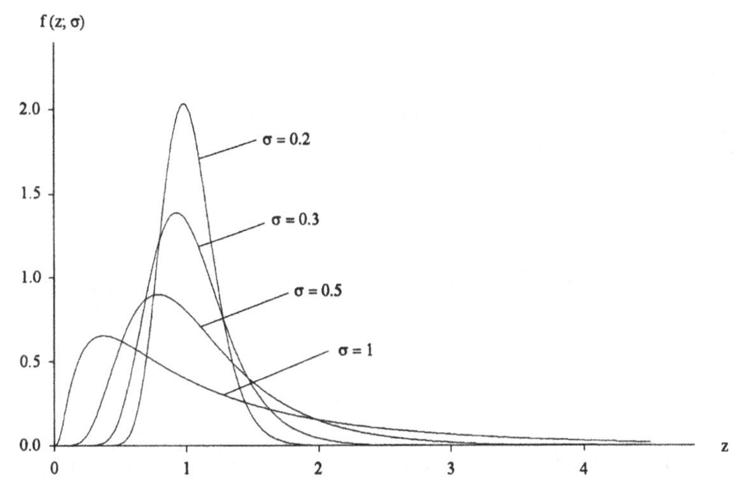

Figure 11.1. Log-Normal pdfs for the reduced variable z.

PROPERTIES: TWO-PARAMETER MODEL

11.2 Relations to the Standard Normal Functions

The Log-Normal distribution function is

$$F(x; \mu, \sigma) = \frac{1}{\sigma\sqrt{2\pi}} \int_0^x \frac{1}{t} \exp\left\{-\frac{1}{2}\left(\frac{\ln(t) - \mu}{\sigma}\right)^2\right\} dt \tag{11.3}$$

and is related to the tabulated standard Normal cdf $\Phi(z)$ (see Section 10.2) as

$$F(x; \mu, \sigma) = \Phi\left(\frac{\ln(x) - \mu}{\sigma}\right). \tag{11.4}$$

Similarly, the Log-Normal pdf is related to the standard Normal pdf $\phi(z)$ as

$$f(x; \mu, \sigma) = \frac{1}{x\sigma}\phi\left(\frac{\ln(x) - \mu}{\sigma}\right). \tag{11.5}$$

11.3 Log-Normal Variable

The rth moment of X about the origin is

$$\mu_r'(X) = \exp\left\{\mu r + \frac{r^2\sigma^2}{2}\right\}, \tag{11.6}$$

so that the expected value of X is

$$\mu_1'(X) = \exp\left\{\mu + \frac{\sigma^2}{2}\right\}. \tag{11.7}$$

The variance of X is

$$\mu_2(X) = \exp\{2\mu + \sigma^2\}[\exp\{\sigma^2\} - 1]. \tag{11.8}$$

Thus, the coefficient of variation of X is

$$cv(X) = \sqrt{\exp\{\sigma^2\} - 1}. \tag{11.9}$$

The quantile of order q is given by

$$x_q = \exp\{\mu + z_q\sigma\}, \tag{11.10}$$

where z_q is the corresponding standard Normal quantile, obtained from standard Normal tables, or numerically as the inverse of the standard Normal cdf: $\Phi(z_q) = q$. Note that Mathcad 6+ has a built-in function that gives Log-Normal quantiles.

The mode value is

$$x_{\mathrm{m}} = \exp\{\mu - \sigma^2\}. \tag{11.11}$$

Writing $c^2 = \exp\{\sigma^2\} - 1$, the first shape factor (see Section 1.10) becomes

$$\gamma_1 = c^3 + 3c > 0, \tag{11.12}$$

showing that all Log-Normal models are positively skewed. The second shape factor is

$$\gamma_2 = c^8 + 6c^6 + 15c^4 + 16c^2 + 3 > 3. \tag{11.13}$$

As $\sigma \to 0$, we have $c \to 0$, $\gamma_1 \to 0$, and $\gamma_2 \to 3$. These are the Normal shape factors; see Section 10.3. Thus, the Log-Normal model approaches the Normal model as $\sigma \to 0$.

The matrix of minimum-variance-bounds (see section 3.2) for estimators of μ and σ is the same as for the Normal distribution:

$$\begin{bmatrix} V_{\mu\mu} & V_{\mu\sigma} \\ V_{\mu\sigma} & V_{\sigma\sigma} \end{bmatrix} = \begin{bmatrix} \frac{\sigma^2}{n} & 0 \\ 0 & \frac{\sigma^2}{2n} \end{bmatrix}, \tag{11.14}$$

giving the lower bound on the variances of estimators for μ and σ.

11.4 Reproductive Property

The Log-Normal distribution features a reproductive property: If k independent Log-Normal variables X_i, with parameters μ_i and σ_i, are multiplied as

$$T = \prod_{i=1}^{k} b_i X_i^{a_i},$$

where the a_i and b_i are constants, then T is again Log-Normal with parameters

$$\mu = \sum_{i=1}^{k} \ln(b_i) + \sum_{i=1}^{k} a_i \mu_i \quad \text{and} \quad \sigma^2 = \sum_{i=1}^{k} a_i^2 \sigma_i^2. \tag{11.15}$$

Thus, for example, if $k = 2$, $b_1 = b_2 = 1$, and $a_1 = a_2 = 1$, then the product $T = X_1 X_2$ of two Log-Normal variables is also Log-Normal with parameters $\mu = \mu_1 + \mu_2$ and $\sigma^2 = \sigma_1^2 + \sigma_2^2$. If $k = 2$, $b_1 = b_2 = 1$, and $a_1 = 1$, $a_2 = -1$, then the quotient $T = X_1/X_2$ of two Log-Normal variables is Log-Normal with parameters $\mu = \mu_1 - \mu_2$ and $\sigma^2 = \sigma_1^2 + \sigma_2^2$. If $a_i = 1/n$ and $b_i = 1$, then $T = \prod_i x_i^{1/n}$ is the *geometric* mean of n Log-Normal variables; this mean is also Log-Normal with parameters $\mu = \frac{1}{n}\sum_i \mu_i$ and $\sigma^2 = \frac{1}{n^2}\sum_i \sigma_i^2$.

11.5 Simulation

Random observations x_i from a Log-Normal process with known parameters μ and σ are obtained from *standard* Normal random variates z_i as

$$x_i = \exp\{\mu + \sigma z_i\}. \tag{11.16}$$

See Section 10.5 for the simulation of standard Normal observations z_i. Note that Mathcad 6+ has a built-in function that generates random Log-Normal observations x_i.

PROPERTIES: THREE-PARAMETER MODEL

11.6 Definition and Properties

Some engineering measurement variables do not produce values below some threshold δ, which thus limits the sample space.

For example, steels feature an *endurance stress level* δ below which any number of stress reversals will not fail a specimen in fatigue. It is clearly of importance to estimate δ for such a material.

The value δ represents a *location* parameter for the distribution, which then becomes a three-parameter model. The pdf is now

$$f(x; \delta, \mu, \sigma) = \frac{1}{(x-\delta)\sigma\sqrt{2\pi}} \exp\left\{ -\frac{1}{2}\left(\frac{\ln(x-\delta)-\mu}{\sigma}\right)^2 \right\};$$
$$x > \delta, \quad \sigma > 0, \quad -\infty < u < \infty. \tag{11.17}$$

The cdf relates to the standard Normal cdf $\Phi(z)$ as

$$F(x; \delta, \mu, \sigma) = \Phi\left(\frac{\ln(x-\delta)-\mu}{\sigma}\right). \tag{11.18}$$

All location measures of the two-parameter model shift to the right by the amount δ, while central moments remain the same. The expected value is therefore

$$\mu_1'(X) = \delta + \exp\left\{\mu + \frac{\sigma^2}{2}\right\}. \tag{11.19}$$

The mode is

$$x_m = \delta + \exp\{\mu - \sigma^2\}. \tag{11.20}$$

The quantile is

$$x_q = \delta + \exp\{\mu + \sigma z_q\}. \tag{11.21}$$

The variance and the shape factors remain unchanged, but the coefficient of variation now becomes

$$cv(X) = \frac{\exp\{\mu + \frac{\sigma^2}{2}\}\sqrt{\exp\{\sigma^2\} - 1}}{\delta + \exp\{\mu + \frac{\sigma^2}{2}\}}. \tag{11.22}$$

For the three-parameter Log-Normal distribution the minimum-variance-bounds for estimators of δ, μ, and σ can be obtained from the inverse of the *expected* information matrix (3.4) of Section 3.1. This information matrix can be inverted directly. The resulting symmetrical variance–covariance matrix is

$$\begin{bmatrix} V_{\delta\delta} = \frac{\sigma^2}{nw} & V_{\delta\mu} = \frac{-\sigma^2 \exp\left\{-\mu+\frac{\sigma^2}{2}\right\}}{nw} & V_{\delta\sigma} = \frac{\sigma^3 \exp\left\{-\mu+\frac{\sigma^2}{2}\right\}}{nw} \\ V_{\delta\mu} & V_{\mu\mu} = \frac{\sigma^2\left[w+\exp\{-2\mu+\sigma^2\}\right]}{nw} & V_{\mu\sigma} = \frac{-\sigma^3 \exp\{-2\mu+\sigma^2\}}{nw} \\ V_{\delta\sigma} & V_{\mu\sigma} & V_{\sigma\sigma} = \frac{\sigma^2\left[w+2\sigma^2 \exp\{-2\mu+\sigma^2\}\right]}{2nw} \end{bmatrix}, \tag{11.23}$$

where

$$w = \exp\{-2\mu + \sigma^2\}[\exp\{\sigma^2\}(\sigma^2 + 1) - 2\sigma^2 - 1].$$

The *local* information matrix (3.20) of Section 3.7 is given by

$$\begin{bmatrix} I_{\delta\delta} = \frac{(\mu+1)s_2 - s_3}{\sigma^2} - s_2 & I_{\delta\mu} = \frac{s_1}{\sigma^2} & I_{\delta\sigma} = -\frac{2s_1}{\sigma} \\ I_{\delta\mu} & I_{\mu\mu} = \frac{n}{\sigma^2} & I_{\mu\sigma} = 0 \\ I_{\delta\sigma} & I_{\mu\sigma} & I_{\sigma\sigma} = \frac{2n}{\sigma^2} \end{bmatrix}, \tag{11.24}$$

where

$$s_1 = \sum_i (x_i - \delta)^{-1}, \qquad s_2 \sum_i (x_i - \delta)^{-2}, \qquad s_3 = \sum_i \ln(x_i - \delta)(x_i - \delta)^{-2}.$$

PROBABILITY PLOT

11.7 Plotting Procedure

A probability plot of the data (see Section 3.13) should always be used to check the Log-Normal distributional assumption. The technique of plotting follows that for the Normal model (see Section 10.10) except that the *logarithms* of the ordered data $x_{(i)}$ are plotted instead of the data $x_{(i)}$. Thus, Normal probability paper can be used for the plot.

It is recommended to plot the approximate median position

$$p_i = \frac{i - 0.3}{n + 0.4} \tag{11.25}$$

on the probability scale versus $\ln(x_{(i)})$. The plot should follow a linear trend if the data came from a two-parameter Log-Normal process (see Section 3.14). If the low end of the plot curves downward, the presence of a threshold parameter δ can be suspected. Replotting with $\ln(x_{(i)} - \delta)$ for different values of δ until the plot straightens yields a *preliminary* estimate of δ.

A computer plot is usually more convenient than plotting with probability paper. The probability scale then needs to be linearized numerically. This is accomplished by first finding the standard Normal quantiles z_i from $\Phi(z_i) = p_i$, where p_i is obtained from (11.25). The linear probability scale is then given for the endpoints $z_{0.01}$ and $z_{0.09}$ by

$$y_i = 0.500 + 0.210 z_i, \tag{11.26}$$

which is plotted versus $\ln(x_{(i)})$, or versus $\ln(x_{(i)} - \delta)$ if a threshold parameter is suspected. See Example 11.1 for an illustration.

The Log-Normal distribution often models strength phenomena, in particular fatigue life of engineering materials. In the context of fatigue testing, samples are sometimes *censored* in the right tail. That is, some specimens do not fail within a reasonable test duration, and the test is terminated before a complete sample is on hand. Probability plotting proceeds as described above for the first r available failure data.

In engineering, *multiply* censored samples typically occur in the context of life testing of equipment. However, the Log-Normal distribution is not recommended as a life model (see the comments in Section 11.16). Should a multiply censored sample require Log-Normal probability plotting, Section 10.11 on Normal probability plotting can be consulted, replacing $x_{(i)}$ by $\ln(x_{(i)})$.

POINT ESTIMATES: TWO-PARAMETER MODEL

11.8 Maximum Likelihood Estimates

Inference procedures follow directly those for the Normal model, with $\ln(x_i)$ replacing x_i. Thus, the likelihood function of a complete sample of n independent Log-Normal observations is

$$L(\mu, \sigma) = \sigma^{-n}(2\pi)^{-\frac{n}{2}} \prod_{i=1}^{n} x_i^{-1} \exp\left\{-\frac{1}{2}\sum_{i=1}^{n}\left(\frac{\ln(x_i) - \mu}{\sigma}\right)^2\right\}, \tag{11.27}$$

and the maximum likelihood estimates (see Section 3.6) reduce to

$$\widehat{\mu} = \frac{1}{n}\sum_{i=1}^{n} \ln(x_i) \tag{11.28}$$

and

$$\widehat{\sigma} = \sqrt{\frac{1}{n}\sum_{i=1}^{n}[\ln(x_i) - \widehat{\mu}]^2}. \tag{11.29}$$

As for the Normal model, the estimator $\widehat{\mu}$ is unbiased, with minimum variance. Its standard error is

$$se(\widehat{\mu}) = \frac{\sigma}{\sqrt{n}} \tag{11.30}$$

for all n. The unbiased version of $\widehat{\sigma}$ is

$$\widehat{\sigma}_{\mathrm{u}} = \sqrt{\frac{1}{n-1}\sum_{i=1}^{n}[\ln(x_i) - \widehat{\mu}]^2} \tag{11.31}$$

with standard error

$$se(\widehat{\sigma}_{\mathrm{u}}) = \frac{\sigma}{\sqrt{2(n-1)}}. \tag{11.32}$$

The standard errors of parameter functions $g(\mu, \sigma)$ are estimated from the error propagation formula (see Section 3.3), which is evaluated at the ML estimates $\widehat{\mu}$ and $\widehat{\sigma}_{\mathrm{u}}$. Using the above standard errors for a complete sample we get:

$$\mathrm{Var}(g) \doteq \left(\frac{\partial g}{\partial \mu}\right)^2 \frac{\sigma^2}{n} + \left(\frac{\partial g}{\partial \sigma}\right)^2 \frac{\sigma^2}{2(n-1)}. \tag{11.33}$$

For a statistical test-of-fit the *Anderson–Darling* statistic is recommended:

$$A = -n - \frac{1}{n}\sum_{i=1}^{n}(2i - 1)[\ln(w_i) + \ln(1 - w_{n-i+1})]. \tag{11.34}$$

Here w_i is the standard Normal cdf $\Phi([\ln(x_{(i)}) - \widehat{\mu}]/\widehat{\sigma})$; note that the *biased* estimate $\widehat{\sigma}$ is used here. The value A is then modified to account for sample size:

$$A_m = A\left(1 + \frac{0.75}{n} + \frac{2.25}{n^2}\right). \tag{11.35}$$

Critical values[1] for A_m are:

α	0.100	0.050	0.025	0.010
A_c	0.631	0.752	0.873	1.035

Example 11.2 demonstrates the estimation process.

11.9 Moment Estimates

When observations on X are *not* available, but estimates of the mean μ_1' and the variance μ_2 are on hand, a *moment estimate* of σ can be obtained from the coefficient of variation (11.9):

$$\widetilde{\sigma} = \sqrt{\ln\left[\mu_2 + (\mu_1')^2\right] - 2\ln(\mu_1')}. \tag{11.36}$$

A moment estimate of μ then follows from the expected value (11.7):

$$\widetilde{\mu} = \ln(\mu_1') - \frac{\widetilde{\sigma}^2}{2}. \tag{11.37}$$

Because the efficiencies of these estimates are low, unless σ is very small, they should not be used when sample information is on hand.

11.10 Censored Samples

Consider a sample that is *type I censored on the right* (see Section 12.2). This type of sample typically occurs in the context of fatigue testing. That is, from a sample of size n the first r observations x_i are on hand; the remaining $m = n - r$ items are censored at $x = c$. The likelihood function of such a sample is

$$L(\mu, \sigma) = \sigma^{-r}(2\pi)^{-\frac{r}{2}} \prod_{i=1}^{r} x_i^{-1} \exp\left\{-\frac{1}{2}\sum_{i=1}^{r}\left(\frac{\ln(x_i) - \mu}{\sigma}\right)^2\right\} \cdot (1 - F(c; \mu, \sigma))^m, \tag{11.38}$$

where F is the Log-Normal cdf (11.3). The maximum likelihood (ML) equations are

$$\sum_{i=1}^{r} z_i + m h(z) = 0 \tag{11.39}$$

[1] Extracted from R. B. D'Agostino, M. A. Stephens, *Goodness-of-Fit Techniques*, Marcel Dekker Inc., New York, 1986, by courtesy of the publisher.

and

$$\sum_{i=1}^{r} z_i^2 + mzh(z) = r, \tag{11.40}$$

where $z_i = [\ln(x_i) - \mu]/\sigma$, $z = [\ln(c) - \mu]/\sigma$, and

$$h(z) = \frac{\phi(z)}{1 - \Phi(z)} \tag{11.41}$$

is the standard Normal hazard function (10.46). The functions ϕ and Φ are the standard Normal pdf (10.2) and cdf (10.3), respectively. The simultaneous solution of these two equations is straightforward with an equation solver. Good starting values for the solution process can usually be obtained from the mean and standard deviation of the log data, ignoring the censoring information.

The standard errors of the estimates are obtained from the inverse of the *local* information matrix (see Section 3.7), which is given by

$$\begin{bmatrix} I_{\mu\mu} = \frac{1}{\sigma^2}\{r + mh(z)[h(z) - z]\} & I_{\mu\sigma} = \frac{1}{\sigma^2}\{2\sum_{i=1}^{r} z_i + mh(z)(1 + z[h(z) - z])\} \\ I_{\mu\sigma} & I_{\sigma\sigma} = \frac{1}{\sigma^2}\{3\sum_{i=1}^{r} z_i^2 + mzh(z)[2 + zh(z) - z^2] - r\} \end{bmatrix}, \tag{11.42}$$

where $h(z)$ is defined by (11.41). See Example 11.3 for an illustration.

INTERVAL ESTIMATES: TWO-PARAMETER MODEL

11.11 Complete Samples

One of the attractive features of the Log-Normal distribution is that exact sampling distributions of useful statistics are available, following the pattern for Normal statistics. Thus, the formulas of Sections 10.14 to 10.18 for confidence intervals and tests are used directly, but on the *logarithms* of the data x_i. That is, if a Normal formula requires the sample mean, then

$$\bar{x} = \frac{1}{n}\sum_{i=1}^{n} \ln(x_i)$$

is used. If a sample variance is required, then

$$s^2 = \frac{1}{n-1}\sum_{i=1}^{n}\left[\ln(x_i) - \frac{1}{n}\sum_{i=1}^{n}\ln(x_i)\right]^2$$

is used.

In addition, confidence intervals can be constructed for any monotonic function of a single parameter. For instance, the Log-Normal median value is $x_m = \exp\{\mu\}$, and a complete-sample $(1-\alpha)$-level confidence interval on x_m is obtained from (10.48) as

$$\left(\exp\left\{\bar{x} + t_{n-1,\frac{\alpha}{2}} \cdot \frac{s}{\sqrt{n}}\right\}, \exp\left\{\bar{x} - t_{n-1,\frac{\alpha}{2}} \cdot \frac{s}{\sqrt{n}}\right\}\right). \tag{11.43}$$

Similarly, a confidence interval on the coefficient of variation $cv(X) = \sqrt{\exp\{\sigma^2\} - 1}$ is obtained from (10.52) as

$$\left(\sqrt{\exp\left\{\frac{s^2(n-1)}{\chi^2_{n-1,1-\frac{\alpha}{2}}}\right\} - 1}, \sqrt{\exp\left\{\frac{s^2(n-1)}{\chi^2_{n-1,\frac{\alpha}{2}}}\right\} - 1}\right). \tag{11.44}$$

Confidence intervals on a quantile use the Normal results of Section 10.18. Example 11.4 provides an illustration.

11.12 Censored Samples

For censored samples, the asymptotic sampling distributions of ML estimators can be used to obtain approximate confidence intervals on parameters and parameter functions, provided the sample is large. Recall from Section 3.7 that the sampling pdf of the ML estimator $\widehat{\theta}$ is asymptotically Normal, with mean value θ and variance MVB_θ, where θ stands for μ or σ:

$$f_N(\widehat{\theta}; \theta, \sqrt{MVB_\theta}). \tag{11.45}$$

Thus, the $(1 - \alpha)$-level confidence interval on θ is obtained as

$$(l_1, l_2) = \theta \pm z_{\frac{\alpha}{2}} \cdot \sqrt{MVB_\theta} \tag{11.46}$$

such that

$$Pr(l_1 \leq \theta \leq l_2) = 1 - \alpha.$$

For type I right-censored data, the MVBs are obtained from the inverse of the information matrix (11.42).

For parameter functions $g(\theta)$, the invariance property of ML estimators is used (see Section 3.7). That is, the sampling pdf of $\widehat{g} = g(\widehat{\theta})$ is asymptotically Normal, with mean value g and variance approximated by the error propagation formula (3.11) of Section 3.3.

For small censored samples, the preceding Normal approximation produces confidence limits that are somewhat in error, particularly for highly censored samples. The likelihood-ratio method (see Section 3.9) tends to give superior results that are more accurate for small samples.

Recall that the statistic

$$LR(\theta) = 2\ln[L(\widehat{\theta})] - 2\ln[L(\theta)] \tag{11.47}$$

is approximately Chi-squared distributed with $\nu = 1$ degree of freedom. Here, L is the *time-censored* likelihood function (11.38). Thus, a $(1-\alpha)$-level confidence interval on θ comprises those values θ for which $LR(\theta) \leq \chi^2_{1,1-\alpha}$. Since there are two parameters, one parameter in (11.47) needs to be expressed in terms of the other by its ML equation. Thus, a confidence interval on μ is obtained from

$$LR(\mu) = 2\ln[L(\widehat{\mu}, \widehat{\sigma})] - 2\ln[L(\mu, \sigma\{\mu\})], \tag{11.48}$$

where $\sigma\{\mu\}$ is defined by the ML equation (11.40).

Similarly, a confidence interval on σ is obtained from

$$LR(\sigma) = 2\ln[L(\widehat{\mu}, \widehat{\sigma})] - 2\ln[L(\mu\{\sigma\}, \sigma)], \tag{11.49}$$

where $\mu\{\sigma\}$ is defined by the ML equation (11.39). See Example 11.5 for an illustration.

A confidence interval on a parameter function $g(\theta)$ can be constructed, but with more difficulty, since the relation among parameters, implied by $g(\theta)$, constrains the solution process. As an illustration, consider the quantile $x_q = \exp\{\mu + \sigma z_q\}$, so that

$$\mu\{x_q, \sigma\} = \ln(x_q) - \sigma z_q. \tag{11.50}$$

Substituting (11.50) into the likelihood function (11.38) and differentiating with respect to σ gives the constraint relation where c is the censoring time:

$$\frac{1}{\sigma^2}\sum_{i=1}^{r}\left\{\ln\left(\frac{x_i}{x_q}\right) + \sigma z_q\right\}\ln\left(\frac{x_i}{x_q}\right) + \frac{m}{\sigma}\cdot\frac{\phi\left(\frac{\ln(c/x_q)}{\sigma} + z_q\right)\ln(c/x_q)}{1 - \Phi\left(\frac{\ln(c/x_q)}{\sigma} + z_q\right)} = r. \tag{11.51}$$

A confidence interval on x_q is then obtained from

$$LR(x_q) = 2\ln[L(\widehat{\mu}, \widehat{\sigma})] - 2\ln[L(\mu\{x_q, \sigma\{x_q\}\}, \sigma\{x_q\})], \tag{11.52}$$

where $\mu\{x_q, \sigma\}$ is given by (11.50) and $\sigma\{x_q\}$ is the solution of (11.51). See Example 11.5 for an illustration of the computations.

ESTIMATES: THREE-PARAMETER MODEL

11.13 Point Estimates

The likelihood function of a sample of n independent observations from a three-parameter Log-Normal process is

$$L(\delta, \mu, \sigma) = \sigma^{-n}(2\pi)^{-\frac{n}{2}}\prod_{i=1}^{n}(x_i - \delta)^{-1}\exp\left\{-\frac{1}{2}\sum_{i=1}^{n}\left(\frac{\ln(x_i - \delta) - \mu}{\sigma}\right)^2\right\}. \tag{11.53}$$

The maximum likelihood equations (see Section 3.6) are

$$\widehat{\mu} = \frac{1}{n}\sum_{i=1}^{n} \ln(x_i - \widehat{\delta}), \tag{11.54}$$

$$\widehat{\sigma}^2 = \frac{1}{n}\sum_{i=1}^{n}[\ln(x_i - \widehat{\delta}) - \widehat{\mu}]^2, \tag{11.55}$$

and

$$(\widehat{\sigma}^2 - \widehat{\mu})\sum_{i=1}^{n}(x_i - \widehat{\delta})^{-1} + \sum_{i=1}^{n}\frac{\ln(x_i - \widehat{\delta})}{x_i - \widehat{\delta}} = 0. \tag{11.56}$$

Eliminating $\widehat{\mu}$ and $\widehat{\sigma}^2$ from these equations gives a solution function in $\widehat{\delta}$ only:

$$\varphi(\widehat{\delta}) = [s_2(\widehat{\delta}) - s_1(\widehat{\delta})]\sum_{i=1}^{n}(x_1 - \widehat{\delta})^{-1} + s_3(\widehat{\delta}) = 0, \tag{11.57}$$

where

$$s_1(\delta) = \frac{1}{n}\sum_{i}\ln(x_i - \delta),$$

$$s_2(\delta) = \frac{1}{n}\sum_{i}[\ln(x_i - \delta) - s_1(\delta)]^2,$$

and

$$s_3(\delta) = \sum_{i}(x_i - \delta)^{-1}\ln(x_i - \delta).$$

It is prudent to graph $\varphi(\delta)$ to see if a solution exists and to obtain a good starting value for the equation solver. If a solution exists, it corresponds to a local maximum of the likelihood function. The resulting ML estimates are usually reasonable in that they produce a good model fit to the data. If a solution does not exist, then either the three-parameter Log-Normal assumption is inappropriate or $\widehat{\delta} = x_{(1)}$. In the latter case one can censor $x_{(1)}$ after translating all other data as $x_i - x_{(1)}$. A two-parameter model can then be estimated on the remaining $(n - 1)$ translated data.

Approximate standard errors of the estimates are obtained from the *expected* variance–covariance matrix (11.23) or from the inverse of the *local* information matrix (11.24). A satistical test-of-fit for the three-parameter Log-Normal model is not available. Probability plotting is recommended as a reliable check on the fit of the estimated model to the data. See Example 11.6.

11.14 Interval Estimates

For three-parameter Log-Normal models, convenient small-sample methods for *confidence intervals* are not available. The asymptotic formula (11.46) can be used for this case. However, caution is advised, because the sampling distributions deviate from Normality unless the sample is large.

When the threshold value δ is not of primary interest, one may construct confidence intervals on μ and σ by conditioning on the estimated value $\widehat{\delta}$, transforming the data to $y_i = x_i - \widehat{\delta}$, and applying two-parameter methods. Often such conditional inferences are insensitive to $\widehat{\delta}$, since the likelihood function does not hold much information on δ.

APPLICATIONS

11.15 General Measurement Model

The Log-Normal distribution is a prominent general model for random measurement variables in the applied sciences, because it features shape flexibility and a positive sample space. Furthermore, its close relation to the Normal distribution allows exact inference procedures for many Log-Normal statistics. Thus, for fitting a distribution to positively skewed frequency plots of positive data, the Log-Normal model is highly attractive, competing with the Gamma and Weibull models; see Chapters 13 and 17.

There is, however, an *a priori* basis for postulating the Log-Normal model, based on its reproductive property (see Section 11.4). If a measurement variable X is the *multiplicative* effect of several independent Log-Normal causes, then X itself has a Log-Normal distribution. If there are many such causes, of unknown distributions, then a central limit theorem (see Section 2.4) supports the same postulate, because the logarithm of the effect X is the sum of the logarithms of the underlying causes.

In engineering, the "proportional effect" model of a strength-degradation phenomenon may be used with the above line of reasoning to postulate a Log-Normal distribution for the measured strength. That is, the incremental damage sustained by a load-carrying component during a load cycle is assumed to be proportional to the damage present at the beginning of that cycle. The incremental damages thus multiply up to the total damage at failure. In fact, the excellent fit of the Log-Normal distribution to a variety of fatigue-strength data suggested the proportional-effect model in the first place. Thus, the Log-Normal distribution has become a prominent model for material strength phenomena.

In addition to applications of a central limit theorem, a Log-Normal model can, of course, be postulated when the measured variable X relates logarithmically to a Normal variable Y.

For example, in certain acoustical problems the sound intensity Y of a source may be Normally distributed. If X is measured in decibels, then X is Log-Normally distributed, since a decibel is a logarithmic measure of sound intensity Y.

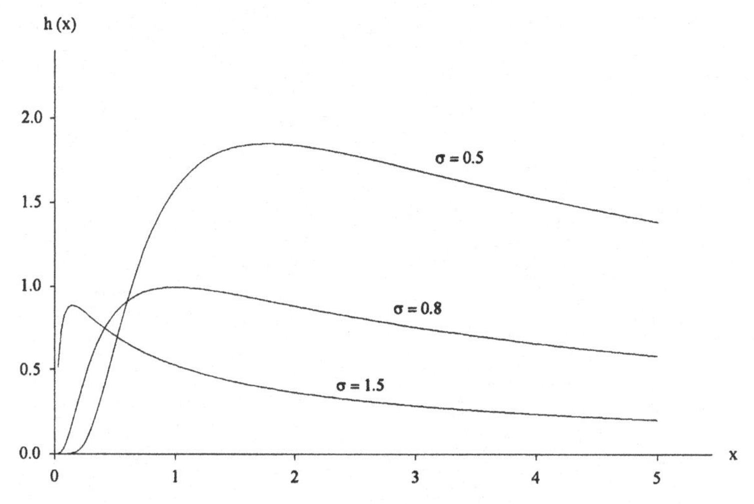

Figure 11.2. Log-Normal hazard functions for $\mu = 0$.

11.16 Life Testing

The Log-Normal distribution is sometimes advocated as a *life model* for the operating time-to-failure of equipment. In this context, the *hazard function* (see Section 12.16) becomes a unique and significant characteristic of a life model. It categorizes the change over operating time of the equipment's failure risk:

$$h(x) = \frac{f(x)}{1 - F(x)}.$$

The so-called "bathtub curve" (see Figure 12.2 of Section 12.16) indicates the possible general shape of this function for equipments subject to wear. Figure 11.2 graphs the hazard function for the Log-Normal distribution. This function rises from the origin to a peak somewhere near the mean value and *then decreases to zero!* It is difficult to rationalize this shape of the hazard function for real-life equipment, as it is opposite to the shape of the bathtub curve. The application of the Log-Normal model to life phenomena in engineering reliability work is therefore not recommended.

EXAMPLE 11.1

A sample of 19 specimens of a newly designed composite material was tested for rupture strength in tension with the following results (1,000 psi):

137, 122, 116, 116, 135, 117, 120, 120, 139, 125,
122, 131, 109, 130, 126, 147, 105, 113, 116.

Check graphically if a Log-Normal distribution could represent the rupture strength of this material.

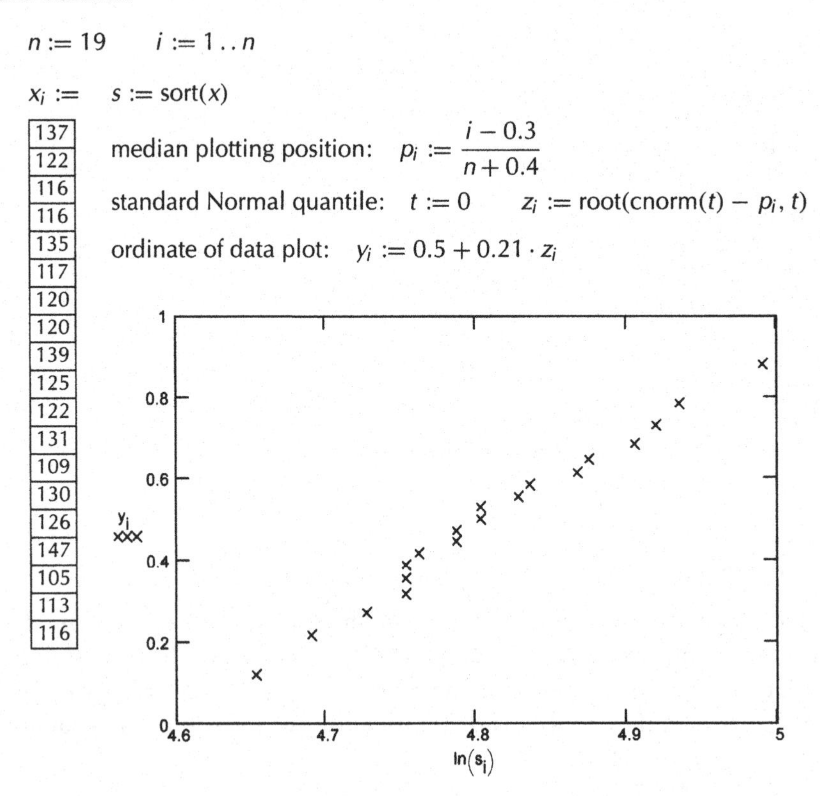

The data plot looks linear; thus, a Log-Normal model appears reasonable.

EXAMPLE 11.2

For the data of Example 11.1, estimate the Log-Normal model, and test the fit at the 5% level of significance. Estimate the 5-percentile and its standard error.

$n := 19 \qquad i := 1 \,..\, n$

$x_i := \qquad s := \mathrm{sort}(x)$

x_i
137
122
116
116
135
117
120
120
139

1. Estimation

$$\mu := \frac{1}{n} \cdot \sum_{i=1}^{n} \ln(x_i) \qquad \mu = 4.812$$

$$\sigma := \sqrt{\frac{1}{n-1} \cdot \sum_{i=1}^{n} (\ln(x_i) - \mu)^2} \qquad \sigma = 0.087$$

125
122
131
109
130
126
147
105
113
116

2. Standard errors

$$\mathrm{SE}\mu := \frac{\sigma}{\sqrt{n}} \qquad \mathrm{SE}\mu = 0.020$$

$$\mathrm{SE}\sigma := \frac{\sigma}{\sqrt{2\cdot(n-1)}} \qquad \mathrm{SE}\sigma = 0.014$$

3. Test-of-fit

$$w_i := \mathrm{cnorm}\left(\frac{\ln(s_i)-\mu}{\sigma}\right)$$

$$A := \left[-n - \frac{1}{n}\cdot\sum_{i=1}^{n}(2\cdot i-1)\cdot(\ln(w_i) + \ln(1-w_{n-i+1}))\right]\cdot\left(1+\frac{0.75}{n}+\frac{2.25}{n^2}\right) \qquad A = 0.210$$

The 5% critical value is 0.752. The sample value is smaller, so there is no reason to reject the estimated model.

4. Probability plot

median plotting position: $p_i := \dfrac{i-0.3}{n+0.4}$

$t := 0 \quad z_i := \mathrm{root}(\mathrm{cnom}(t) - p_i, t)$

ordinate of data plot: $y_i := 0.5 + 0.21\cdot z_i$

ordinate of model plot: $r_i := 0.5 + 0.21\cdot\dfrac{\ln(s_i)-\mu}{\sigma}$

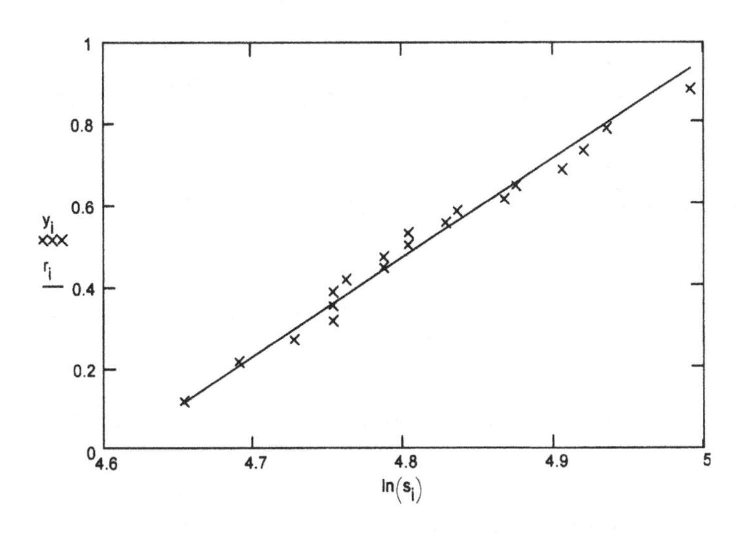

5. Five-percentile

$z05 := -1.645$

$x05(\mu, \sigma) := \exp(\mu + z05 \cdot \sigma)$ $\qquad x05(\mu, \sigma) = 106.6$

standard error:

$$dm := \frac{d}{d\mu} x05(\mu, \sigma) \qquad ds := \frac{d}{d\sigma} x05(\mu, \sigma)$$

$$seX := \sqrt{(dm \cdot SE\mu)^2 + (ds \cdot SE\sigma)^2} \qquad seX = 3.3$$

EXAMPLE 11.3

It is postulated that the fatigue life, in stress cycles ×1000, of a composite material is a Log-Normal process. Of the 18 specimens put on test, 15 failed prior to the test termination at 180,000 cycles:

32.4, 48.8, 49.8, 52.7, 56.8, 58.9, 59.0, 66.3,
72.1, 77.9, 86.4, 115.0, 120.8, 164.8, 169.7.

Check the Log-Normal assumption graphically. Estimate the parameters and their standard errors. Estimate the 5-percentile fatigue life.

$n := 18 \qquad r := 15 \qquad i := 1..r \qquad m := n - r \qquad c := 180$

$x_i :=$

32.4
42.8
49.8
52.7
56.8
58.9
59
66.3
72.1
77.9
86.4
115
120.8
154.8
169.7

1. Probability plot

median plotting position:

$$p_i := \frac{i - 0.3}{n + 0.4} \qquad t := 0 \qquad z_i := \text{root}(\text{cnorm}(t) - p_i, t)$$

ordinate of data plot: $y_i := 0.5 + 0.21 \cdot z_i$

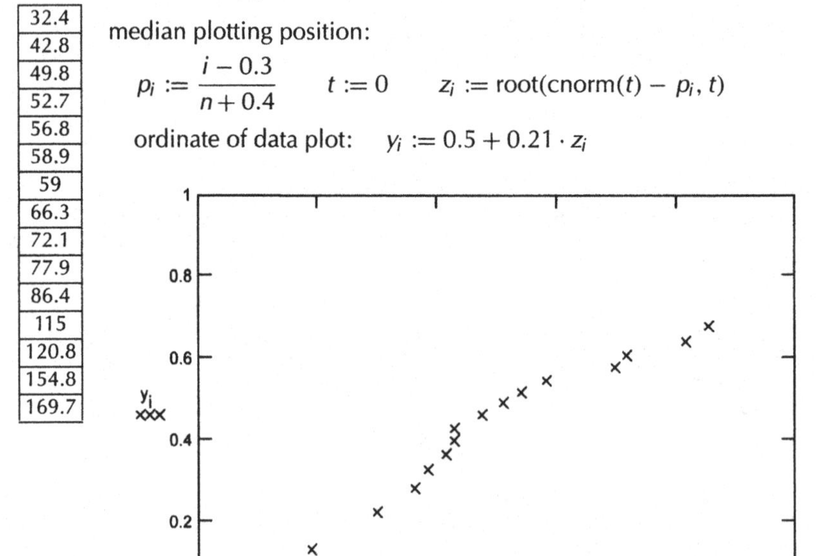

The Log-Normal postulate looks reasonable.

2. Estimation

starting values: $a := \frac{1}{r} \cdot \sum_i \ln(x_i)$ $\quad a = 4.287$

$$b := \sqrt{\frac{1}{r} \cdot \sum_i (\ln(x_i) - a)^2} \qquad b = 0.457$$

ML estimates:

$$\phi(t) := \frac{1}{\sqrt{2 \cdot \pi}} \cdot \exp\left(-\frac{t^2}{2}\right) \qquad zc(a, b) := \frac{\ln(c) - a}{b}$$

$$h(a, b) := \frac{\phi(zc(a, b))}{(1 - \text{cnorm}(zc(a, b)))}$$

GIVEN:

$$\sum_i \frac{\ln(x_i) - a}{b} + m \cdot h(a, b) = 0 \qquad \sum_i \left(\frac{\ln(x_i) - a}{b}\right)^2 + m \cdot zc(a, b) \cdot h(a, b) = r$$

$$\begin{pmatrix} \mu \\ \sigma \end{pmatrix} := \text{FIND}(a, b) \qquad \mu = 4.490$$

$$\sigma = 0.627$$

standard errors:

$$z_i := \frac{\ln(x_i) - \mu}{\sigma} \qquad \text{zc} := \frac{\ln(c) - \mu}{\sigma} \qquad h(zc) := \frac{\phi(zc)}{(1 - \text{cnorm}(zc))}$$

$$I := \frac{1}{\sigma^2} \cdot \begin{bmatrix} r + m \cdot h(zc) \cdot (h(zc) - zc) & 2 \cdot \sum_i z_i + m \cdot h(zc) \cdot (1 + \text{zc} \cdot (h(zc) - zc)) \\ 2 \cdot \sum_i z_i + m \cdot h(zc) \cdot (1 + zc \cdot (h(zc) - zc)) & 3 \cdot \sum_i (z_i)^2 + m \cdot zc \cdot h(zc) \cdot (2 + zc \cdot h(zc) - zc^2) - r \end{bmatrix}$$

$$V := I^{-1} \qquad V = \begin{pmatrix} 0.0227 & 0.0018 \\ 0.0018 & 0.0144 \end{pmatrix} \qquad \begin{array}{l} SE\mu := \sqrt{V_{0,0}} \\ SE\sigma := \sqrt{V_{1,1}} \end{array} \qquad \begin{array}{l} SE\mu = 0.151 \\ SE\sigma = 0.120 \end{array}$$

Model check:

ordinate for model plot: $r_i := 0.5 + 0.21 \cdot \frac{\ln(x_i) - \mu}{\sigma}$

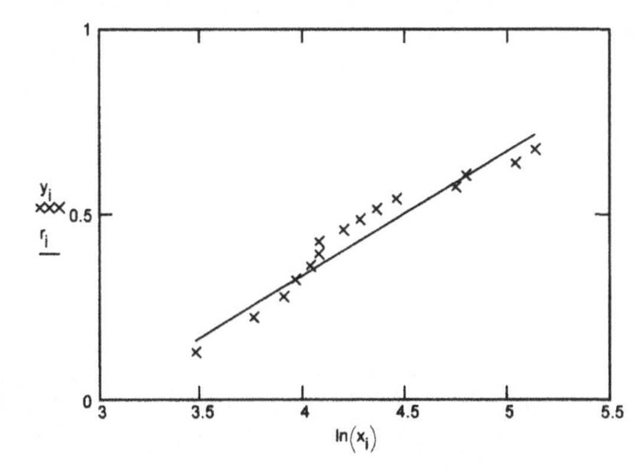

The model fit looks reasonable.

3. Five-percentile

$zq := 1.645$ $\qquad$ $x(\mu, \sigma) := \exp(\mu + zq \cdot \sigma)$ $\qquad$ $x(\mu, \sigma) = 31.8$

standard error: $j := 0..1$ $\qquad$ $k := 0..1$

$d_0 := \frac{d}{d\mu} x(\mu, \sigma)$ $\qquad$ $d_1 := \frac{d}{d\sigma} x(\mu, \sigma)$

$\mathrm{var}X := \sum_j \sum_k d_j \cdot d_k \cdot V_{j,k}$ $\qquad$ $\mathrm{SE}x := \sqrt{\mathrm{var}X}$ $\qquad$ $\mathrm{SE}x = 7.5$

EXAMPLE 11.4

For the data of Example 11.1, calculate the 90% confidence intervals on the parameters, on the 5-percentile, and on the coefficient of variation.

From Example 11.2:

$n := 19$ $\qquad$ $xb := 4.812$ $\qquad$ $s := 0.087$

1. **confidence interval on μ:** 5% t-quantile at $v = 18$ is $\qquad$ $t05 := -1.734$

   from(10.48):

   $\mu L := xb + t05 \cdot \frac{s}{\sqrt{n}}$ $\qquad$ $\mu = 4.777$

   $\mu U := xb - t05 \cdot \frac{s}{\sqrt{n}}$ $\qquad$ $\mu U = 4.847$

2. **confidence interval on σ:**

   $5\%\chi^2$ – quantile at $v = 18$ is $\qquad$ $K05 := 9.39$
   $95\%\chi^2$ – quantile at $v = 18$ is $\qquad$ $K95 := 28.9$

   from (10.52):

   $\sigma L := s \cdot \sqrt{\frac{n-1}{K95}}$ $\qquad$ $\sigma L = 0.069$

   $\sigma U := s \cdot \sqrt{\frac{n-1}{K05}}$ $\qquad$ $\sigma U = 0.120$

3. **5-percentile**:

   given

   $q := 0.05$ $\qquad$ $zq := -1.645$ $\qquad$ $v := 18$

   from (10.65):

   $\delta := zq \cdot \sqrt{n}$ $\qquad$ $\delta = -7.170$

   a) We require the 5-percentile of t' at $\delta = -7.17$; from (10.26) this equals the negative of the 95-percentile of t' at $\delta = 7.17$; following Example 10.1:

      $z95 := 1.645$ $\qquad$ $\delta := -\delta$

approximate t' value:

$$t := \frac{\delta + z95 \cdot \sqrt{1 + \frac{\delta^2 - z95^2}{2 \cdot v}}}{1 - \frac{z95^2}{2 \cdot v}} \qquad t = 10.482$$

exact t' value:

x integrand: $I(x, t) := x^{v-1} \cdot \exp\left(-\frac{x^2}{2}\right) \cdot \int_0^{\frac{x \cdot t}{\sqrt{v}}} \exp\left[-\frac{(u-\delta)^2}{2}\right] du$

X integration limits: $a := 0 \qquad b := 10$

$$F(t) := \frac{2^{1-\frac{v}{2}}}{\Gamma\left(\frac{v}{2}\right) \cdot \sqrt{2 \cdot \pi}} \cdot \int_a^b I(x, t)\, dx \qquad T := \mathrm{root}(F(t) - 0.95, t)$$

$$T = 10.563$$

Thus $t05 := -10.563$

b) Similarly, we obtain the 95-percentile of t' at $\delta = -7.17$ as $t95 := -5.075$
From (11.44), the required confidence interval is

$$xL := \exp\left(xb + \frac{s}{\sqrt{n}} \cdot t05\right) \qquad xL = 99.6$$

$$xU := \exp\left(xb + \frac{s}{\sqrt{n}} \cdot t95\right) \qquad xU = 111.1$$

4. coefficient of variation:

estimate:

$$\mathrm{cv} := \sqrt{\exp(s^2) - 1} \qquad \mathrm{cv} = 0.087$$

We can use the calculated confidence limits for σ directly in (11.9):

$$\mathrm{cv}L := \sqrt{\exp(\sigma L^2) - 1} \qquad \mathrm{cv}L = 0.069$$

$$\mathrm{cv}U := \sqrt{\exp(\sigma U^2) - 1} \qquad \mathrm{cv}U = 0.121$$

EXAMPLE 11.5

For the data of Example 11.3 obtain the 90% confidence intervals on the parameters, on the 5-percentile, and on the coefficient of variation.

From Example 11.3:

$n := 18 \qquad r := 15 \qquad m := n - r$

$\mu := 4.49 \qquad x05 := 31.8 \qquad V := \begin{pmatrix} 0.0227 & 0.0018 \\ 0.0018 & 0.0114 \end{pmatrix}$

$\sigma := 0.627 \qquad \mathrm{SE}x := 7.5$

1. Normal approximations

a) parameters:

$z05 := 1.645$

$\mu L := \mu + z05 \cdot \sqrt{V_{0,0}}$ $\qquad \mu L = 4.242$

$\mu U := \mu - z05 \cdot \sqrt{V_{0,0}}$ $\qquad \mu U = 4.738$

$\sigma L := \sigma + z05 \cdot \sqrt{V_{1,1}}$ $\qquad \sigma L = 0.451$

$\sigma U := \sigma - z05 \cdot \sqrt{V_{1,1}}$ $\qquad \sigma U = 0.803$

b) 5-percentile:

$xL := x05 + z05 \cdot SEx$ $\qquad xL = 19.5$

$xU := x05 - z05 \cdot SEx$ $\qquad xU = 44.1$

c) coefficient of variation:

$cvL := \sqrt{\exp(\sigma L^2) - 1}$ $\qquad cvL = 0.475$

$cvU := \sqrt{\exp(\sigma U^2) - 1}$ $\qquad cvU = 0.951$

2. Likelihood ratio method

a) parameters:
chi squared 90-percentile at $v = 1$: $K90 := 2.71$
standard Normal functions:

$$\phi(t) := \frac{1}{\sqrt{2 \cdot \pi}} \cdot \exp\left(-\frac{t^2}{2}\right) \qquad \Phi(t) := \text{cnorm}(t)$$

From (11.38), the log-likelihood function is $i := 1 .. r$ $\qquad c := 180$

$x_i :=$

32.4
42.8
49.8
52.7
56.8
58.9
59
66.3
72.1
77.9
86.4
115
120.8
154.8
169.7

$$LL(\mu, \sigma) := -r \cdot \ln(\sigma) - \sum_i \ln(x_i) - \frac{1}{2} \cdot \sum_i \left(\frac{\ln(x_i) - \mu}{\sigma}\right)^2 + m \cdot \ln\left(1 - \Phi\left(\frac{\ln(c) - \mu}{\sigma}\right)\right)$$

parameter μ:

$$ML(a, b) := \sum_i \left(\frac{\ln(x_i) - a}{b}\right)^2 + m \cdot \frac{\ln(c) - a}{b} \cdot \frac{\phi\left(\frac{\ln(c) - a}{b}\right)}{1 - \phi\left(\frac{\ln(c) - a}{b}\right)} - r$$

$b := \sigma$ $\qquad s(a) := \text{root}(ML(a, b), b)$

$LR(a) := 2 \cdot LL(\mu, \sigma) - 2 \cdot LL(a, s(a))$

$a := \mu - 0.3$ $\qquad \mu L := \text{root}(LR(a) - K90, a)$ $\qquad \mu L = 4.237$

$a := \mu + 0.2$ $\qquad \mu U := \text{root}(LR(a) - K90, a)$ $\qquad \mu U = 4.760$

parameter σ:

$$ML(a,b) := \sum_i \left(\frac{\ln(x_i)-a}{b}\right) + m\cdot\frac{\phi\left(\frac{\ln(c)-a}{b}\right)}{1-\Phi\left(\frac{\ln(c)-a}{b}\right)}$$

$a := \mu$ $\quad g(b) := \text{root}(ML(a,b), a)$

$LR(b) := 2\cdot LL(\mu,\sigma) - 2\cdot LL(g(b), b)$

$b := \sigma - 0.2$ $\quad \sigma L := \text{root}(LR(b) - K90, b)$ $\quad \sigma L = 0.471$

$b := \sigma + 0.2$ $\quad \sigma U := \text{root}(LR(b) - K90, b)$ $\quad \sigma U = 0.890$

b) 5-percentile:

from (11.51):

$$ML(xq,b) := \frac{1}{b^2}\cdot\sum_i\left(\ln\left(\frac{x_i}{x05}\right) + b\cdot z05\right)\cdot\ln\left(\frac{x_i}{x05}\right) + \frac{\frac{m}{b}\cdot\phi\left(\frac{\ln\left(\frac{c}{xq}\right)}{b} + z05\right)\cdot\ln\left(\frac{c}{x05}\right)}{1-\Phi\left(\frac{\ln\left(\frac{c}{xq}\right)}{b} + z05\right)} - r$$

$b := \sigma$ $\quad s(xq) := \text{root}(ML(xq,b), b)$

$g(xq) := \ln(xq) - z05\cdot s(xq)$

$LR(xq) := 2\cdot LL(\mu,\sigma) - 2\cdot LL(g(xq), s(xq))$

$xq := x05 - 15$ $\quad xL := \text{root}(LR(xq) - K90, xq)$ $\quad xL = 23.5$

$xq := x05 + 15$ $\quad xU := \text{root}(LR(xq) - K90, xq)$ $\quad xU = 41.9$

c) coefficient of variation:

$cvL := \sqrt{\exp(\sigma L^2) - 1}$ $\quad cvL = 0.498$

$cvU := \sqrt{\exp(\sigma U^2) - 1}$ $\quad cvU = 1.098$

EXAMPLE 11.6

The fatigue life (stress cycles $\times 1000$) of a steel is postulated to be a Log-Normal process. A sample of 21 specimens was fatigued to failure at a given mean stress and stress amplitude. The ordered test results are:

566.9, 574.8, 599.6, 627.1, 628.3, 638.8, 640.4, 642.4, 677.5, 680.8, 684.4, 695.7, 703.4, 726.5, 779.3, 785.6, 792.5, 817.5, 838.6, 884.3, 969.4.

Check the Log-Normal assumption graphically. If warranted, estimate the parameters and their standard errors. Estimate the 10-percentile fatigue life.

$n := 21 \qquad i := 1..n$

$x_i :=$

566.9
574.8
599.6
627.1
628.3
638.8
640.4
642.4
677.5
680.8
684.4
695.7
703.4
726.5
779.3
785.6
792.5
817.5
838.6
884.3
969.4

1. Probability plot

median plotting position: $p_i = \dfrac{i - 0.3}{n + 0.4}$

Standard Normal quantile: $t := 0 \qquad z_i := \mathrm{root}(\mathrm{cnorm}(t) - p_i, t)$

ordinate of data plot: $y_i := 0.5 + 0.21 \cdot z_i$

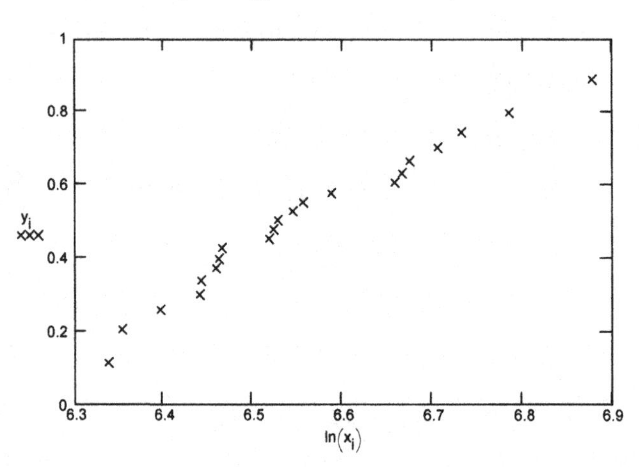

The Log-Normal postulate looks reasonable. The downward sloping trend at the left end of the plot suggests the presence of a threshold parameter.

2. Estimation (3-parameter model)

$$s1(\delta) := \frac{1}{n} \cdot \sum_i \ln(x_i - \delta) \qquad s2(\delta) := \frac{1}{n} \cdot \sum_i (\ln(x_i - \delta) - s1(\delta))^2$$

$$s3(\delta) := \sum_i \frac{\ln(x_i - \delta)}{x_i - \delta}$$

solution function: $\varphi(\delta) := (s2(\delta) - s1(\delta)) \cdot \sum_i \dfrac{1}{x_i - \delta} + s3(\delta)$

$j := 1..20 \quad t_j := 450 + 2 \cdot j$

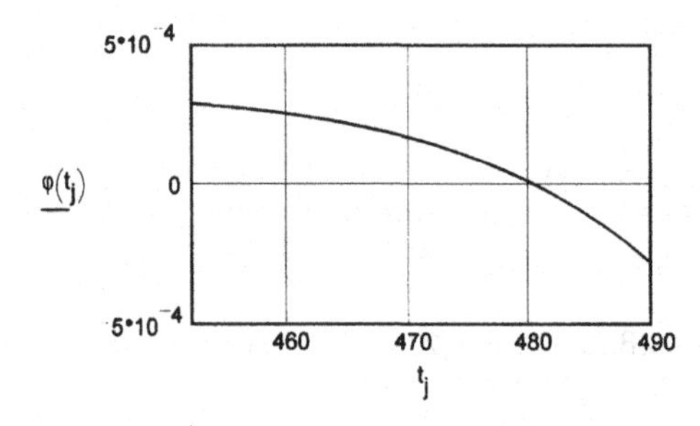

$d := 480$ $\qquad$ $\delta := \text{root}(\varphi(d), d)$ $\qquad$ $\delta = 480.4$

$\mu := s1(\delta)$ $\qquad$ $\mu = 5.345$

$\sigma := \sqrt{s2(\delta)}$ $\qquad$ $\sigma = 0.454$

standard errors:

$w := \exp(-2 \cdot \mu + \sigma^2) \cdot [\exp(\sigma^2) \cdot (\sigma^2 + 1) - 2 \cdot \sigma^2 - 1]$

from (11.23):

$$V := \frac{\sigma^2}{n \cdot w} \cdot \begin{bmatrix} 1 & -\exp\left(-\mu + \frac{\sigma^2}{2}\right) & \sigma \cdot \exp\left(-\mu + \frac{\sigma^2}{2}\right) \\ -\exp\left(-\mu + \frac{\sigma^2}{2}\right) & w + \exp(-2 \cdot \mu + \sigma^2) & -\sigma \cdot \exp(-2 \cdot \mu + \sigma^2) \\ \sigma \cdot \exp\left(-\mu + \frac{\sigma^2}{2}\right) & -\sigma \cdot \exp(-2 \cdot \mu + \sigma^2) & \frac{W + 2 \cdot \sigma^2 \cdot \exp(-2 \cdot \mu + \sigma^2)}{2} \end{bmatrix}$$

$$V = \begin{pmatrix} 5013.468 & -26.52 & 12.041 \\ -26.52 & 0.15 & -0.064 \\ 12.041 & -0.064 & 0.034 \end{pmatrix}$$

$SE\delta := \sqrt{V_{0,0}}$ $\qquad$ $SE\delta = 70.8$

$SE\mu := \sqrt{V_{1,1}}$ $\qquad$ $SE\mu = 0.387$

$SE\sigma := \sqrt{V_{2,2}}$ $\qquad$ $SE\sigma = 0.184$

model check:

ordinate for model plot: $\quad r_i := 0.5 + 0.21 \cdot \dfrac{\ln(x_i - \delta) - \mu}{\sigma}$

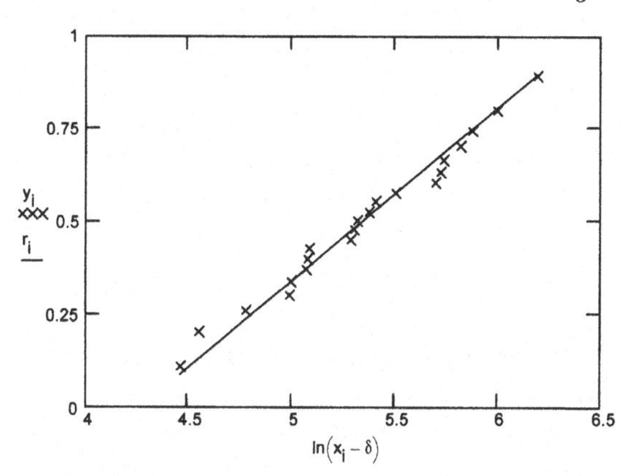

The model fit looks reasonable.

3. Ten-percentile

$zq := -1.28$ $\qquad$ $x(\delta, m, \sigma) := \delta + \exp(\mu + zq \cdot \sigma)$ $\qquad$ $x(\delta, \mu, \sigma) = 598$

standard error:

$j := 0..2$ $\qquad$ $k := 0..2$

$d_0 := \dfrac{d}{d\delta} x(\delta, \mu, \sigma)$ $\qquad$ $d_1 = \dfrac{d}{d\mu} x(\delta, \mu, \sigma)$ $\qquad$ $d_2 := \dfrac{d}{d\sigma} x(\delta, \mu, \sigma)$

$\text{var}X := \sum_j \sum_k d_j \cdot d_k \cdot V_{j,k}$ $\qquad$ $SEx := \sqrt{\text{var}X}$ $\qquad$ $SEx = 16$

CHAPTER TWELVE

Exponential Distributions

INTRODUCTION

12.1 Definition

A continuous random variable X has an *Exponential distribution* if its pdf has the form

$$f(x;\mu,\sigma) = \frac{1}{\sigma}\exp\left\{-\frac{x-\mu}{\sigma}\right\}; \quad x \geq \mu, \quad \mu \geq 0, \quad \sigma > 0 \tag{12.1}$$

(see Fig. 12.1). This pdf has location-scale structure (see Section 9.2). The pdf of the reduced variable $z = (x - \mu)/\sigma$ is therefore parameter free:

$$f(z) = \exp\{-z\}; \qquad z \geq 0. \tag{12.2}$$

In engineering applications of this model, μ is often zero, resulting in the one-parameter Exponential pdf:

$$f(x;\sigma) = \frac{1}{\sigma}\exp\left\{-\frac{x}{\sigma}\right\}; \quad x \geq 0, \quad \sigma > 0. \tag{12.3}$$

The Exponential distribution is widely used to describe phenomena involving events that occur randomly in space or time, at a *constant rate*. This distribution is therefore intimately connected with the Poisson distribution (see Chapter 8). In Section 13.16 it is shown that these two distributions provide equivalent descriptions of such phenomena.

For example, consider the time-to-failure (TTF) of an electronic component. One can argue that such a component does not "wear out" under normal operating conditions, and that it may therefore only fail when a randomly occurring *excessive* load (voltage, shock, temperature, etc.) is placed on the component. Under these assumptions, the Exponential distribution models component TTF closely.

The Exponential model also arises as a special case of both the Gamma and Weibull distributions, when their shape parameter takes the value $\lambda = 1$.

Historically, the mathematically simple Exponential model was the focus for much of the early statistical development of reliability analysis, of which it is

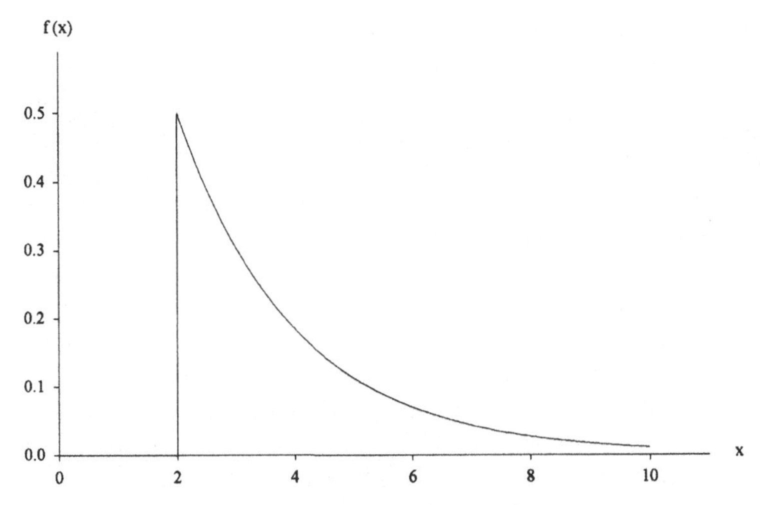

Figure 12.1. The Exponential pdf with $\mu = 2$ and $\sigma = 2$.

still the core. Numerous authors have contributed to this development over the past forty years, notably Epstein, Sobel, and Sarhan.

CENSORING

12.2 Data Structure

Engineering applications of the Exponential distribution are largely associated with the life modeling of components that are not subject to wear. Since it is difficult to make reliable components fail in a reasonable length of time, the data base resulting from life tests is often *censored*. That is, a decision is made to limit the test duration for economic reasons. Some of the components under test may then not have failed at the test termination. Sometimes units have to be removed from testing for a variety of reasons, and thus their eventual failure times are not known. In other cases, data are collected from the field when, at the time of the analysis, only some failures have occurred while other units are still functioning at their individual operating times. In all of these cases, the failure times of the survivors are *right-censored*, meaning that the unknown failure times are larger than the known censoring times. The data base then comprises r actual failure times, x_i, from a sample of size n and $m = n - r$ censoring times c_j. Probability plotting and parameter estimation procedures need to use all of this information to produce a valid model.

A sample may be *singly* censored, usually on the right: The first r failure times are on hand, and the $m = n - r$ survivors are censored. In that case the correct order of failed items is known. However, a sample may also be *multiply* censored, where operating items were removed from test at any point. In that case the correct order of failed items, beyond the first censored item, is not known, and probability plotting methods need to be adjusted.

When it is decided to remove items from test at fixed censoring times c_j, or when field items have accumulated different operating times c_j, the sample is said to be *time-censored* or *type-I censored.* When it is decided beforehand to remove items from test at the ith failure time $x_{(i)}$, the sample is said to be *failure-censored* or *type-II censored.* Once the data are on hand, plotting and parameter estimation procedures are similar for the two types of censored samples. However, the *sampling distributions* of estimators are distinct for the two types of censoring. For *time*-censored samples the *number r* of failures before censoring is a random variable, whereas for *failure*-censored samples the censoring-*time* $x_{(r)}$ is random. Generally, it is more difficult to determine the sampling distributions of estimators for the time-censored case since the censoring times c_j are often not known beforehand and may depend on some unknown random process (see the illustration in Section 17.11). For the Exponential distribution, results are available for both types of censoring. The more commonly useful methods are presented in this chapter.

PROPERTIES

12.3 Exponential Distribution

The cdf of the Exponential variable X, as defined by (12.1), is

$$F(x; \mu, \sigma) = 1 - \exp\left\{-\frac{x-\mu}{\sigma}\right\}. \tag{12.4}$$

The mean value of X is

$$\mu_1'(X) = \mu + \sigma, \tag{12.5}$$

and the variance of X is

$$\mu_2(X) = \sigma^2. \tag{12.6}$$

The coefficient of variation of X is a function of μ and σ:

$$cv(X) = \frac{\sigma}{\mu + \sigma}. \tag{12.7}$$

The mode value is simply

$$x_m = \mu. \tag{12.8}$$

The quantile of order q is

$$x_q = \mu - \sigma \ln(1 - q). \tag{12.9}$$

The first and second shape factors (see Section 1.10) are fixed: $\gamma_1 = 2$ and $\gamma_2 = 9$. Hence there is no shape flexibility.

The expected value and variance of the rth order statistic $x_{(r)}$ are

$$E\{x_{(r)}\} = \sigma \sum_{i=1}^{r} \frac{1}{n-i+1} \tag{12.10}$$

and

$$\mathrm{Var}(x_{(r)}) = \sigma^2 \sum_{i=1}^{r} \frac{1}{(n-i+1)^2}. \tag{12.11}$$

For the one-parameter model (12.3) the properties are obtained from the above formulas with $\mu = 0$. For this case the minimum-variance-bound (see Section 3.2) for estimators of the scale parameter σ is

$$MVB_\sigma = \frac{\sigma^2}{n}. \tag{12.12}$$

12.4 Reproductive Properties

The Exponential distribution inherits from its relation to the Weibull distribution the reproductive property that its sample minimum $X_{(1)}$ is again Exponential, but with the scale parameter decreased to σ/n (see Section 17.4). This feature is easily verified by substituting the cdf (12.4) into expression (2.5) of Section 2.5.

From its relation to the Gamma distribution, the Exponential model inherits the reproductive property that the sum of n observations from an Exponential process is Gamma distributed with parameters σ and n (see Section 13.3). Thus, the average of n such observations is Gamma distributed with parameters σ/n and n. Equivalently, the statistic $2n\overline{X}/\sigma$ is χ^2 distributed with $\nu = 2n$ degrees of freedom (see Section 13.6).

Most applications of the Exponential distribution are based on its *memoryless property*, which implies that the conditional probability of an event, given some related information, equals the unconditional probability of that event:

$$Pr(X \geq a + b \,|\, X \geq b) = Pr(X \geq a).$$

Thus, past events do not influence the probability of future events. In the context of reliability engineering, this means that an Exponentially distributed component with an operating history is as good as a new component, implying that there is *no wear out* through use.

12.5 Simulation

Random observations from an Exponential process with known parameter values are simulated as

$$x_i = \mu - \sigma \ln(u_i), \tag{12.13}$$

where u_i is a Uniform random number on (0, 1) (see Section 2.8). Note that Mathcad 6+ has a built-infunction that generates random Exponential observations x_i for $\mu = 0$. It is prudent to check the adequacy of a simulated sample by comparing at least its first two moments with those of the given model.

PROBABILITY PLOT

12.6 Plotting Procedure

A probability plot of ordered data (see Section 3.13) is a good way to check if the Exponential assumption is justified for the data on hand. Taking natural logs of the cdf (12.4) gives a linear relation in $x_{(i)}$:

$$-\ln[1 - F(x_{(i)})] = -\frac{\mu}{\sigma} + \frac{x_{(i)}}{\sigma}.$$

Replacing $F(x_{(i)})$ by a *plotting position* p_i such as the approximate median value of $F(x_{(i)})$,

$$p_i = \frac{i - 0.3}{n + 0.4}, \tag{12.14}$$

gives

$$-\ln[1 - p_i] = a + bx_{(i)}.$$

When the sample is *complete* or *right-tail censored*, $-\ln[1 - p_i]$ can be plotted versus the ordered data $x_{(i)}$. The result approximates a straight line, if the data came from an Exponential process (see Section 3.14). A noticeable departure of the plot from a straight line would put the Exponential postulate in question. The parameters μ and σ could be estimated from the intercept a and the slope b of the plot. However, because the statistical properties of the resulting estimates are not known, this procedure is not recommended.

When the sample is multiply censored, the correct order of the failure data in the total sample is not known. The order must be established by taking account of the information implicit in the censored data. This is done by using the distribution-free *product limit* estimate of the reliability function (see Section 3.15), which results in a modified plotting position.

Consider a multiply right-censored sample of size n, with r failure times, x_i, and $m = n - r$ survival times c_j. Label the combined and *ordered* set of failure and survival data as $\{t_{(k)}\}_n$. Thus, some of the $t_{(k)}$ are failure data x_i and some are survival data c_j. The revised median plotting position is then

$$p_k = 1 - \frac{n + 0.7}{n + 0.4} \prod_{q}^{k} \frac{n - q + 0.7}{n - q + 1.7}, \tag{12.15}$$

where $q \leq k$ is the subscript of a *failure* time in the combined, ordered set $\{t_{(k)}\}_n$. When there are no censored data, (12.15) reduces to the complete-sample median plotting position (12.14). As in the preceding section, $-\ln(1-p_i)$ is plotted versus the ordered failure data $x_{(i)}$. See Example 12.1 for an illustration.

POINT ESTIMATES: ONE-PARAMETER MODEL

12.7 Complete Samples

The likelihood function of a complete sample of n independent observations on an Exponential variable X is

$$L(\sigma) = \sigma^{-n} \exp\left\{-\frac{n\bar{x}}{\sigma}\right\}. \tag{12.16}$$

The maximum likelihood estimator (see Section 3.6) is therefore

$$\hat{\sigma} = \bar{x}. \tag{12.17}$$

It is seen that the statistic $\bar{x}$ is *sufficient*, that is, it holds all sample information relevant to the estimation of σ. The sampling variance of this estimator equals the MVB (12.12), so that the standard error of the estimate is

$$\text{se}(\hat{\sigma}) = \frac{\sigma}{\sqrt{n}}. \tag{12.18}$$

The standard errors of estimated parameter functions $g(\sigma)$ are approximated from the error propagation formula (see Section 3.3), which is evaluated at $\hat{\sigma}$. Using the above standard error for a complete sample, we get

$$\text{Var}(g) \doteq \left(\frac{\partial g}{\partial \sigma}\right)^2 \frac{\sigma^2}{n}. \tag{12.19}$$

See Example 12.2 where a percentile function is estimated.

For a statistical test-of-fit the *Anderson–Darling* statistic is recommended:

$$A = -n - \frac{1}{n}\sum_{i=1}^{n}(2i-1)[\ln(w_i) + \ln(1 - w_{n-i+1})], \tag{12.20}$$

where w_i is the Exponential cdf (12.4), with $\mu = 0$ and evaluated at the order statistic $x_{(i)}$. The value A is then modified to generalize it for all sample sizes n:

$$A_{\text{m}} = A\left(1 + \frac{0.6}{n}\right). \tag{12.21}$$

Critical values[1] for A_m are:

α	0.100	0.050	0.025	0.010
A_c	1.062	1.321	1.591	1.959

See Example 12.2 for an illustration.

12.8 Censored Samples

For a multiply censored sample, as defined in Section 12.6, the likelihood function becomes

$$L(\sigma) = \sigma^{-r} \exp\left\{ -\frac{1}{\sigma} \sum_{k=1}^{n} t_k \right\}. \tag{12.22}$$

Here $r < n$ is the number of observed values x_i in a sample of size n, and $\{t_k\}_n$ is the combined data set comprising failure and censoring data.

When the sample is *time*-censored, the maximum likelihood estimator is

$$\widehat{\sigma}_{\mathrm{I}} = \frac{1}{r} \sum_{k=1}^{n} t_k. \tag{12.23}$$

The asymptotic sampling variance of $\widehat{\sigma}_{\mathrm{I}}$ is

$$\mathrm{Var}(\widehat{\sigma}_{\mathrm{I}}) = \frac{\sigma^2}{r}. \tag{12.24}$$

The standard error of $\widehat{\sigma}_{\mathrm{I}}$ is therefore $\sigma/\sqrt{r}$. See Example 12.3 for an illustration. When the sample is *singly* censored on the right at the time $c > x_{(r)}$, as is often the case in practice, the above estimator specializes to

$$\widehat{\sigma}_{\mathrm{I}'} = \frac{1}{r} \left(\sum_{i=1}^{r} x_i + (n-r)c \right). \tag{12.25}$$

When the sample is *failure*-censored, with b_i items censored at the failure time x_i, the ML estimator is

$$\widehat{\sigma}_{\mathrm{II}} = \frac{1}{r} \sum_{i=1}^{r} (1+b_i) x_i, \tag{12.26}$$

where $n - r = \sum_i b_i$ is the total number of censored items. The standard error of $\widehat{\sigma}_{\mathrm{II}}$ is

$$\mathrm{se}(\widehat{\sigma}_{\mathrm{II}}) = \frac{\sigma}{\sqrt{r}}. \tag{12.27}$$

[1] Extracted from R. B. D'Agostino, M. A. Stephens, *Goodness-of-Fit Techniques*, Marcel Dekker Inc., New York, 1986, by courtesy of the publisher.

When there is *single* censoring on the right, of $(n-r)$ items at the rth failure time $x_{(r)}$, the estimator (12.26) specializes to

$$\widehat{\sigma}_{\mathrm{II}'} = \frac{1}{r}\left(\sum_{i=1}^{r} x_{(i)} + (n-r)x_{(r)}\right). \tag{12.28}$$

The quantities $T = r\widehat{\sigma}$ from the above expressions are often termed the *total time on test*. These are decision quantities of economic importance when the test cost is a function of the test times.

Statistical tests-of-fit for a general multiply censored Exponential sample are not available. However, for the practically important case of *single* censoring on the right, of either type, the complete-sample test-of-fit of Section 12.7 can be applied to the *spacings* between the observations $x_{(i)}$. It follows from the memoryless property of the Exponential model (see Section 12.4) that these spacings are themselves Exponential variables. Suppose the spacings are scaled in proportion to their expected values (12.10),

$$E\{x_{(i)} - x_{(i-1)}\} = \frac{\sigma}{n-i+1},$$

to

$$s_i = (n-i+1)\left(x_{(i)} - x_{(i-1)}\right); \quad i = 1, 2, \ldots, r; \quad x_{(0)} = 0. \tag{12.29}$$

The r variables s_i are then identically and independently distributed as Exponentials with scale parameter σ. A test-of-fit can be applied, treating these transformed data as a complete sample of size r. See Example 12.4 for an illustration.

INTERVAL ESTIMATES: ONE-PARAMETER MODEL

12.9 Complete or Failure-Censored Samples

For samples that are singly failure-censored on the right, confidence intervals on σ are obtained from the exact sampling distribution of the following quantity related to the total time on test $T = r\widehat{\sigma}$ from (12.28):

$$\frac{2r\widehat{\sigma}_{\mathrm{II}'}}{\sigma} \sim \chi^2_{2r}. \tag{12.30}$$

That is, the quantity $2r\widehat{\sigma}_{\mathrm{II}'}/\sigma$ is exactly distributed as a Chi-squared variable with $2r$ degrees of freedom. A two-sided, equal-tail, $(1-\alpha)$-level confidence interval can therefore be constructed (see Section 3.8) from the probability statement

$$Pr\left(\chi^2_{2r,\frac{\alpha}{2}} \leq \frac{2r\widehat{\sigma}_{\mathrm{II}'}}{\sigma} \leq \chi^2_{2r,1-\frac{\alpha}{2}}\right) = 1-\alpha.$$

Here $\chi^2_{2r,q}$ is the q-quantile of the χ^2 distribution, at $\nu = 2r$ degrees of freedom. The confidence interval on σ is then

$$\left(\frac{2r\widehat{\sigma}_{\mathrm{II}'}}{\chi^2_{2r,1-\frac{\alpha}{2}}}, \frac{2r\widehat{\sigma}_{\mathrm{II}'}}{\chi^2_{2r,\frac{\alpha}{2}}} \right). \tag{12.31}$$

For complete samples, r is replaced by n and $\widehat{\sigma}_{\mathrm{II}'}$ by $\widehat{\sigma}$. Confidence intervals for functions of σ, such as quantiles and reliability values, follow directly from the results on σ. See Example 12.2 for an illustration.

When the null hypothesis $H_0 : \sigma = \sigma_0$ is to be tested against the alternative hypothesis $H_a : \sigma = \sigma_a\sigma_o$, the critical test value t_c for the statistic (12.28) is obtained from (12.30) as

$$t_c = \frac{\sigma_0 \chi^2_{2r,1-\alpha}}{2r}, \tag{12.32}$$

where α is the significance level of the test (type I error probability; see Section 3.10). The operating characteristic β (type II error probability) of the test is similarly obtained as a function of the alternative value σ_a from

$$\frac{\chi^2_{2r,1-\alpha}}{\chi^2_{2r,\beta}} = \frac{\sigma_a}{\sigma_0}. \tag{12.33}$$

Thus, for given values of r and σ_a, β is determined from (12.33). Conversely, for a specified β, (12.33) gives the smallest integer r such that the left side of (12.33) is equal to, or greater than, σ_a/σ_0, implying that the power $(1 - \beta)$ of the test is at least as specified. See Section 12.21 for more details.

12.10 Multiply Censored Samples

For large multiply censored samples, the asymptotically Normal sampling distribution of the estimator $\widehat{\sigma}_{\mathrm{I}}$ or $\widehat{\sigma}_{\mathrm{II}}$ can be used to construct approximate confidence intervals on σ. This sampling pdf is

$$f_{\mathrm{N}}(\widehat{\sigma}; \sigma, \sqrt{\mathrm{Var}(\widehat{\sigma})}), \tag{12.34}$$

where the unknown parameter σ is replaced by the appropriate estimate $\widehat{\sigma}_{\mathrm{I}}$ or $\widehat{\sigma}_{\mathrm{II}}$. Thus, a $(1 - \sigma)$-level confidence interval on σ is obtained as

$$(l_1, l_2) = \widehat{\sigma} \pm z_{\frac{\alpha}{2}} \sqrt{\mathrm{Var}(\widehat{\sigma})}, \tag{12.35}$$

such that

$$Pr(l_1 \le \sigma \le l_2) = 1 - \alpha.$$

For small to moderate-sized samples the likelihood ratio method tends to give more accurate results than the preceding Normal method. Recall from Section 3.9 that the statistic

$$LR(\sigma) = 2\ln[L(\widehat{\sigma})] - 2\ln[L(\sigma)] \tag{12.36}$$

is approximately Chi-squared distributed with $\nu = 1$ degree of freedom. A $(1-\alpha)$-level confidence interval on σ therefore comprises the values σ for which $LR(\sigma) \leq \chi^2_{1,1-\alpha}$. The likelihood function $L(\sigma)$ is given by (12.22). Example 12.3 illustrates the computations.

POINT ESTIMATES: TWO-PARAMETER MODEL

12.11 Complete Samples

For a complete sample from a two-parameter Exponential distribution, the likelihood function is

$$L(\mu, \sigma) = \sigma^{-n} \exp\left\{-\frac{1}{\sigma}\sum_{i=1}^{n}(x_i - \mu)\right\}. \tag{12.37}$$

Since L increases with μ, while $\mu \leq x_{(1)}$, the ML estimator of μ is $\widehat{\mu} \leq x_{(1)}$. L is then maximized at $\widehat{\sigma} = \bar{x} - x_{(1)}$. These estimators are independent. However, they are biased. The unbiased versions[2] are

$$\widehat{\mu} = \frac{nx_{(1)} - \bar{x}}{n-1}, \tag{12.38}$$

with standard error

$$\mathrm{se}(\widehat{\mu}) = \frac{\sigma}{\sqrt{n(n-1)}}, \tag{12.39}$$

and

$$\widehat{\sigma} = \frac{n\left(\bar{x} - x_{(1)}\right)}{n-1}, \tag{12.40}$$

with standard error

$$\mathrm{se}(\widehat{\sigma}) = \frac{\sigma}{\sqrt{n-1}}. \tag{12.41}$$

[2] A. E. Sarhan, B. G. Greenberg, *Contributions to Order Statistics*, Wiley, New York, 1962.

For a statistical test-of-fit the *Anderson–Darling* statistic (12.20) is recommended. The value A is modified[3] to account for the sample size n:

$$A_{\mathrm{m}} = A\left(1 + \frac{5.4}{n} - \frac{11}{n^2}\right). \tag{12.42}$$

Critical values of A_{m} are as given in Section 12.7.

12.12 Multiply Censored Samples

When the sample is multiply censored as defined in Section 12.6, the likelihood function is

$$L(\mu, \sigma) = \sigma^{-r} \exp\left\{-\frac{1}{\sigma}\sum_{k=1}^{n}(t_k - \mu)\right\}, \quad \text{for all} \quad t_k > \mu. \tag{12.43}$$

Here $r \leq n$ is the number of observed values x_i in a sample of size n, and $\{t_k\}_n$ is the combined data set comprising failure and censoring data.

For a *time*-censored sample the ML estimator of the location parameter is

$$\widehat{\mu}_{\mathrm{I}} = x_{(1)}. \tag{12.44}$$

Simple results for the bias and standard error of this estimator are not available. The ML estimator of the scale parameter is

$$\widehat{\sigma}_{\mathrm{I}} = \frac{1}{r}\sum_{k=1}^{n}\left(t_k - x_{(1)}\right). \tag{12.45}$$

A bias correction is also not available for this estimator. Conditional on the observed value $x_{(1)}$, the standard error of $\widehat{\sigma}_{\mathrm{I}}$ is approximately

$$\mathrm{se}(\widehat{\sigma}_{\mathrm{I}}) \doteq \frac{\sigma}{\sqrt{r}}. \tag{12.46}$$

When the sample is *failure*-censored on the right at $x_{(r)}$, the unbiased minimum-variance estimate of μ is

$$\widehat{\mu}_{\mathrm{II}} = \frac{1}{n(r-1)}\left(nr\,x_{(1)} - \sum_{i=1}^{r} x_i - (n-r)x_{(r)}\right), \tag{12.47}$$

with standard error

$$\mathrm{se}(\widehat{\mu}_{\mathrm{II}}) = \frac{\sigma}{n}\sqrt{\frac{r}{r-1}}. \tag{12.48}$$

[3] Extracted from R. B. D'Agostino, M. A. Stephens, *Goodness-of-Fit Techniques*, Marcel Dekker Inc., New York, 1986, by courtesy of the publisher.

The unbiased minimum-variance estimator of σ is

$$\widehat{\sigma}_{\mathrm{II}} = \frac{1}{r-1}\left(\sum_{i=1}^{r} x_i - n x_{(1)} + (n-r)x_{(r)}\right), \tag{12.49}$$

with standard error

$$\mathrm{se}(\widehat{\sigma}_{\mathrm{II}}) = \frac{\sigma}{\sqrt{r-1}}. \tag{12.50}$$

These estimators are independent. A statistical test-of-fit can be constructed as in Section 12.8 for *single* censoring.

INTERVAL ESTIMATES: TWO-PARAMETER MODEL

12.13 Complete or Singly Failure-Censored Samples

Consider a sample that is complete or singly failure-censored on the right. The following quantities are independent, with exact sampling distributions:

$$\frac{2n(x_{(1)} - \mu)}{\sigma} \sim \chi_2^2 \tag{12.51}$$

and

$$\frac{2r\widehat{\sigma}}{\sigma} \sim \chi_{2r-2}^2, \tag{12.52}$$

where χ_ν^2 is the Chi-squared distribution with ν degrees of freedom. It follows that the ratio of these quantities divided by the ratio of their degrees of freedom is an F variable (see Section 10.9):

$$\frac{2n(x_{(1)} - \mu)}{2\sigma} \Big/ \frac{2r\widehat{\sigma}}{\sigma(2r-2)} = \frac{n(r-1)(x_{(1)} - \mu)}{r\widehat{\sigma}} \sim F_{2,2r-2} \tag{12.53}$$

with $(2, 2r-2)$ degrees of freedom.

A two-sided, equal-tail, $(1-\alpha)$-level confidence interval on μ is constructed from the probability statement

$$Pr\left(F_{2,2r-2,\frac{\alpha}{2}} \le \frac{n(r-1)(x_{(1)} - \mu)}{r\widehat{\sigma}} \le F_{2,2r-2,1-\frac{\alpha}{2}}\right) = 1 - \alpha.$$

However, the qth quantile $F_{2,2r-2,q}$ can be expressed as

$$F_{2,2r-2,q} = (r-1)\{(1-q)^{\frac{1}{1-r}} - 1\}. \tag{12.54}$$

Hence the confidence interval on μ reduces to

$$\left(x_{(1)} - \frac{\widehat{\sigma}r}{n}\left\{\left(\frac{\alpha}{2}\right)^{\frac{1}{1-r}} - 1\right\}, x_{(1)} - \frac{\widehat{\sigma}r}{n}\left\{\left(1 - \frac{\alpha}{2}\right)^{\frac{1}{1-r}} - 1\right\}\right). \tag{12.55}$$

When the lower bound from (12.55) turns out to be negative, its appropriate value is zero.

A confidence interval on σ is similarly constructed from the probability statement

$$Pr\left(\chi^2_{2r-2,\frac{\alpha}{2}} \leq \frac{2r\widehat{\sigma}}{\sigma} \leq \chi^2_{2r-2,1-\frac{\alpha}{2}}\right) = 1 - \alpha$$

as

$$\left(\frac{2r\widehat{\sigma}}{\chi^2_{2r-2,1-\frac{\alpha}{2}}}, \frac{2r\widehat{\sigma}}{\chi^2_{2r-2,\frac{\alpha}{2}}}\right), \tag{12.56}$$

where $\chi^2_{2r-2,q}$ is the q-quantile of the χ^2 distribution at $\nu = (2r - 2)$ degrees of freedom. When the sample is complete, r is replaced by n in the above formulas. See Example 12.5 for an illustration.

12.14 Multiply Censored Samples

When the sample is multiply *time-censored*, a confidence interval on μ can be obtained from the approximate sampling distribution of

$$\frac{2(2r-1)(x_{(1)} - \mu)}{2r\widehat{\sigma}_{\mathrm{I}}} \sim F_{2,2r-1},$$

where $\widehat{\sigma}_{\mathrm{I}}$ is given by (12.45). The approximate two-sided, equal-tail, $(1-\alpha)$-level confidence interval on μ reduces to

$$\left(x_{(1)} - \frac{\widehat{\sigma}_{\mathrm{I}} r}{n}\left\{\left(\frac{\alpha}{2}\right)^{\frac{2}{1-2r}} - 1\right\}, x_{(1)} - \frac{\widehat{\sigma}_{\mathrm{I}} r}{n}\left\{\left(1 - \frac{\alpha}{2}\right)^{\frac{2}{1-2r}} - 1\right\}\right). \tag{12.57}$$

To construct a confidence interval on σ, one can treat the quantity $2r\widehat{\sigma}_{\mathrm{I}}/\sigma$ as approximately Chi-squared with $(2r - 1)$ degrees of freedom, conditional on the value $x_{(1)}$. An approximate conditional confidence interval on σ is thus

$$\left(\frac{2r\widehat{\sigma}_{\mathrm{I}}}{\chi^2_{2r-1,1-\frac{\alpha}{2}}}, \frac{2r\widehat{\sigma}_{\mathrm{I}}}{\chi^2_{2r-1,\frac{\alpha}{2}}}\right). \tag{12.58}$$

A lower, $(1 - \alpha)$-level confidence limit on the quantile $x_q = \mu - \sigma \ln(1 - q)$ can be obtained[4] for *complete* or *failure-censored* samples as

$$L = x_{(1)} + kS, \tag{12.59}$$

[4] Engelhardt, M., Bain, L. J., "Tolerance Limits and Confidence Limits on Reliability for the Two-Parameter Exponential Distribution," *Technometrics*, Vol. 20, No. 1, pp. 37–39, 1978.

where

$$k = \frac{1}{n}\left\{1 - \left(\frac{(1-q)^n}{\alpha}\right)^{\frac{1}{r-1}}\right\} \tag{12.60a}$$

and

$$S = \sum_{i=1}^{r} x_i - nx_{(1)} + (n-r)x_{(r)}. \tag{12.60b}$$

The above formula is exact for

$$n < \ln(\alpha)/\ln(1-q). \tag{12.61}$$

For larger n, Equation (12.59) still provides an excellent approximation, particularly for the small quantiles and high confidence levels that are often required for engineering decisions. See Example 12.6 for an illustration. For *time-censored* samples, the approximation (12.59) is less accurate but can be improved by substituting $(r + 0.5)$ for r in (12.60). See also Section 12.18, where the equivalent lower confidence bound on reliability is given by expression (12.84).

APPLICATIONS

The Exponential distribution cannot be recommended as a general model of measurement phenomena, because its shape is not only fixed but also of a peculiar spiked form that does not comfortably fit the usual more gradual rise and fall of data frequency functions. However, there are two engineering applications where the Exponential model may provide the appropriate distributional postulate for specific measurement variables: inter-arrival time (IAT) and time-to-failure (TTF).

12.15 Inter-Arrival Times

The inter-arrival time (IAT) variable arises from counting processes where it measures the time (or space) between adjoining events. The prominent application is to *waiting lines*, which model, for example, communication networks and industrial processes featuring congested segments. When the underlying counting process is Poisson (see Chapter 8), the IAT is exactly Exponential, so that the two models are equivalent descriptions of that process. See Section 13.16, where this equivalence is detailed for waiting-line phenomena.

12.16 Time-to-Failure

The time-to-failure variable arises in the reliability analysis of components (or systems of components). When these components are subject to the *constant hazard* of randomly occurring failure-causing loads, the Exponential distribution

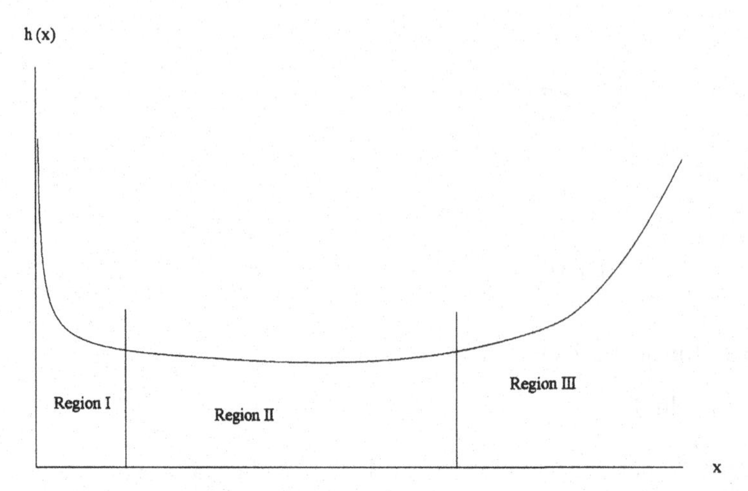

Figure 12.2. An idealized hazard function for the life cycle of a mechanical device.

models the TTF well. To elucidate this connection, consider the *hazard function*, which is an important quantity characterizing life phenomena.

The hazard function (or hazard rate, mortality function, intensity function) is defined as

$$h(x) = \frac{f(x)}{1 - F(x)}. \tag{12.62}$$

Note that the hazard function cannot be interpreted as a conditional probability of failure, given survival to the time x, since its integral is unbounded. In fact, the hazard functions of many distributions diverge themselves with increasing x (see, for example, Fig. 17.6 in Section 17.22). For the Exponential model, the hazard function is constant:

$$h(x) = \frac{1}{\sigma}. \tag{12.63}$$

The implication is that the failure hazard is independent of the age of the component. The connection to the memoryless property (see Section 12.4) is immediate.

Most engineering components (e.g., mechanical devices) are subject to wear, implying that their life history affects the failure hazard. Thus, the hazard function is *not* constant over the life cycle of a more general component. See Fig. 12.2 for the appearance of a general hazard function. In engineering circles, such a graph is termed a "bathtub curve." When a sufficiently large failure sample is on hand, the empirical version of the hazard function can be plotted, much like a relative frequency histogram:

$$h(x_{(i)}) = \frac{1}{(x_{(i)} - x_{(i-1)})(n - i + 1)}, \tag{12.64}$$

where $x_{(0)} = \mu$ (or zero). By revealing the hazard trend over the sample range, such a plot may lead to the identification of an appropriate life model.

Three distinct hazard regimes can be discerned from Fig. 12.2. Region I, where the hazard function decreases, is sometimes termed the region of *infant mortality*. Decreasing hazard functions are associated with manufactured products that may contain assembly faults, or other serious defects, precipitating early failures. As these weakling units are removed from the manufactured lot, the robust units remain, so that the failure hazard of the lot decreases. Quality control procedures of a manufacturing process may take advantage of this hazard regime by instituting a *run-in* or *burn-in* operation, to improve the quality of the surviving product lot.

In region II, often termed the *random failure region*, the hazard function does not change rapidly. This region covers the useful product life period. Region III is the *wear-out region*, where the hazard function increases with product age, due to deterioration processes. Several statistical distributions allow for increasing or decreasing hazard functions. See, in particular, Chapter 17 on Weibull distributions.

It is clear that the Exponential distribution can only serve as an approximate general life model for period II. However, for some types of product the hazard function is, in fact, constant. Most electronic components fall in this class. The remainder of this chapter outlines some basic concepts of *reliability analysis*, as it applies to constant-hazard life processes.

It should be noted that inference procedures for the Exponential distribution are sensitive to departures from it. Hence, unless a strong case can be made for a constant hazard function, a better strategy is to assume a more general model for which the Exponential distribution is a special case (Gamma or Weibull).

LIFE TESTING

12.17 Reliability Function

A key quantity in assessing the reliability of a device is the statistical *reliability value*. It is defined as the probability of a device to survive an operating period x (often termed the "mission time"), under given operating conditions:

$$R \mid x = 1 - F(x; \theta). \tag{12.65}$$

This measure is conditional on the chosen model F and its estimated parameters θ. When considered as a function of x, $R(x)$ is the *reliability function*. For the Exponential model, the reliability function is

$$R(x) = \exp\left\{-\frac{x-\mu}{\sigma}\right\}. \tag{12.66}$$

The Exponential reliability value $R \mid x$ is estimated by maximum likelihood as

$$\widehat{R} \mid x = \exp\left\{-\frac{x-\widehat{\mu}}{\widehat{\sigma}}\right\}. \tag{12.67}$$

The variance of this statistic is approximated from the error propagation formula (see Section 3.3) as

$$\mathrm{Var}(\widehat{R}\,|\,x) \doteq \frac{R^2}{\sigma^2}\,(\mathrm{Var}(\mu) + \ln^2(R)\cdot \mathrm{Var}(\sigma)). \tag{12.68}$$

Here the parameter variances correspond to the particular sample on hand. For instance, the *complete*-sample standard errors (12.39) and (12.41) reduce (12.68) to

$$\mathrm{Var}(\widehat{R}\,|\,x) \doteq \frac{R^2}{n-1}\left(\frac{1}{n} + \ln^2(R)\right). \tag{12.69}$$

When the sample comes from a one-parameter Exponential process, the μ term drops out of (12.68), and (12.69) simplifies further to

$$\mathrm{Var}(\widehat{R}\,|\,x) \doteq \frac{[R\ln(R)]^2}{n-1}. \tag{12.70}$$

In the high-reliability region ($R > 0.95$, say), where most engineering reliability work is focused, the ML estimate (12.67) is reasonably accurate. However, because economically consequential product decisions often hinge on a fraction of a percent in the reliability estimate, its value is preferably estimated without bias. The following unbiased, minimum-variance estimators are available for the one-parameter model, where T is the *total time on test*:

1. Complete sample:

$$R\,|\,x = \left(1 - \frac{x}{T}\right)^{n-1}, \qquad x < T \tag{12.71}$$

with

$$T = \sum_{i=1}^{n} x_i. \tag{12.72}$$

2. Censored on right at time c:

$$R\,|\,x = \left(1 - \frac{x}{T}\right)^{r}, \qquad x < T \tag{12.73}$$

with

$$T = \sum_{i=1}^{n} x_i + (n-r)c. \tag{12.74}$$

3. Censored on right at rth failure time:

$$R\,|\,x = \left(1 - \frac{x}{T}\right)^{r-1}, \qquad x < T \tag{12.75}$$

with

$$T = \sum_{i=1}^{n} x_i + (n - r)x_{(r)}. \tag{12.76}$$

In each case, $R \mid x = 0$ when $x \geq T$.

For the two-parameter Exponential model, corresponding results are:

1. Complete sample:

$$R \mid x = \frac{n-1}{n}\left(1 - \frac{x - x_{(1)}}{T - x_{(1)}}\right)^{n-1}, \qquad x < T \tag{12.77}$$

with

$$T = \sum_{i=1}^{n} \left(x_i - x_{(1)}\right) + x_{(1)}. \tag{12.78}$$

2. Censored on right at time c:

$$R \mid x = \frac{n-1}{n}\left(1 - \frac{x - x_{(1)}}{T - x_{(1)}}\right)^{r-1}, \qquad x < T \tag{12.79}$$

with

$$T = \sum_{i=1}^{n} \left(x_i - x_{(1)}\right) + (n - r)\left(c - x_{(1)}\right) + x_{(1)}. \tag{12.80}$$

3. Censored on right at rth failure time:

$$R \mid x = \frac{n-1}{n}\left(1 - \frac{x - x_{(1)}}{T - x_{(1)}}\right)^{r-2}, \qquad x < T \tag{12.81}$$

with

$$T = \sum_{i=1}^{n} \left(x_i - x_{(1)}\right) + (n - r)\left(x_{(r)} - x_{(1)}\right) + x_{(1)}. \tag{12.82}$$

Again, when $x \geq T$, $R \mid x = 0$.

12.18 Confidence Limit on Reliability

To reduce the risk of reliability-based decisions, the engineer often prefers to work with a *lower* confidence limit $R_L \mid x$ on the reliability value, rather than the point estimate of $R \mid x$ and its standard error.

For the *one-parameter* Exponential model, a lower $(1 - \alpha)$-level confidence limit on $R \mid x$ is obtained directly from a similar confidence limit l on the scale parameter σ (see Section 12.10) by the relation

$$R_L \mid x = \exp\left\{-\frac{x}{l}\right\}. \tag{12.83}$$

That is, if l is a lower $(1-\alpha)$-level confidence limit on σ, then (12.83) is the corresponding lower confidence limit on R.

To obtain a lower $(1-\alpha)$-level confidence limit on $R \mid x$ for the *two-parameter* Exponential model, one can start with a probability statement on a quantile:

$$Pr(X_q \geq x) = Pr(R \geq 1-q) = 1-\alpha.$$

One then needs to find the order q such that the $(1-\alpha)$-level, lower confidence limit on x_q equals the specified mission time x. Equating the lower confidence limit (12.59, 12.60) to x, and solving for $1-q$, gives the required lower confidence limit on $R \mid x$ as

$$R_L \mid x = \alpha^{\frac{1}{n}}\left(1 - n\frac{x - x_{(1)}}{S}\right)^{\frac{r-1}{n}}, \tag{12.84}$$

where S is given by (12.60b) and the sample-size condition (12.61) applies. See Example 12.7 for an illustration.

12.19 Expected Number of Failures

When managing a large system that comprises many similar components (e.g., a communications network with many digital relays), a reliability problem is the prediction of the number of failures over some future period T, so that spare-part levels can be set. For Exponentially distributed components, the expected total number of failures, before replacement, is

$$\mathrm{E}\{N\} = \sum_{i=1}^{n} \frac{F(t_i + u_i) - F(t_i)}{1 - F(t_i)} = n - \sum_{i=1}^{n} \exp\left\{-\frac{u_i}{\sigma}\right\}, \tag{12.85}$$

where u_i is the anticipated usage time of the ith unit during the decision period T. See Section 17.26 for further discussion. The above estimate is useful when failed components are replaced by a modified design with a much longer life characteristic.

When components are replaced upon failure by units from the same population, the system comprises n Poisson renewal processes, each with Exponentially distributed times-between-failures (see also Section 13.16). The expected number of failures for the ith component, during its usage time u_i, is therefore the Poisson average, which is u_i/σ. The expected total number of failures is thus

$$\mathrm{E}\{N\} = \sum_{i=1}^{n} \frac{u_i}{\sigma}. \tag{12.86}$$

12.20 Acceptance Tests

One of the problems encountered in engineering reliability work is the need to *test* product samples to provide information on product life characteristics for decision making. For example, a decision may be required whether product development should continue or production should commence. The objective of setting up statistical *acceptance tests* is to maximize the information obtained from the test for a given test budget or, equivalently, to minimize the test cost for a required level of information from the test. This objective typically translates to determining the sample size n of test specimens and the test termination rule. The test then results in a *decision*, at a specified confidence level, whether a desired life criterion is met. For the Exponential life model several statistical test plans are available. Some useful plans are presented in the following sections.

For the one-parameter Exponential model, any life characteristic of interest can be expressed as a function of the parameter σ. Hence, a decision would concern two hypothesized states of σ: the null hypothesis $H_0 : \sigma = \sigma_0$ and an alternative hypothesis such as $H_a : \sigma = \sigma_a > \sigma_0$. Here $\sigma = \sigma_0$ means that the (unknown) parameter σ equals the hypothesized value σ_0 (see Chapter 3 on *Tests*). The test parameters chosen by the decision maker are the type I error probability α that H_0 is rejected although H_0 is true and the type II error probability β that H_0 is accepted although H_a is true. That is,

$$Pr(\text{reject } H_0 \mid H_0 \text{ is true}) = \alpha \tag{12.87}$$

and

$$Pr(\text{accept } H_0 \mid H_a \text{ is true}) = \beta. \tag{12.88}$$

12.21 Failure-Censored Test

Consider placing n Exponentially distributed specimens on test until r failures have occurred. That is, the test sample will be failure-censored on the right. Since we know the sampling distribution of the ML estimate $\widehat{\sigma}_{II'}$ from (12.30), expression (12.87) yields the critical value $\widehat{\sigma}_c$ from

$$Pr\left(\frac{2r\widehat{\sigma}}{\sigma_0} > \chi^2_{2r,1-\alpha}\right) = \alpha$$

as

$$\widehat{\sigma}_c = \frac{\sigma_0 \chi^2_{2r,1-\alpha}}{2r}, \tag{12.89}$$

where $\chi^2_{2r,1-\alpha}$ is the Chi-squared quantile of order $1-\alpha$, at $2r$ degrees of freedom. The critical value $\widehat{\sigma}_c$ defines the *rejection region*: If the test statistic $\widehat{\sigma}_{II'}$, from (12.28), is greater than the critical value, then H_0 is rejected. Thus for any given value of r, an α-level critical value $\widehat{\sigma}_c$ can be obtained.

If the power of the test, $1-\beta$, is specified for a given alternative $H_a:\sigma = \sigma_a > \sigma_0$, Equation (12.88) becomes

$$Pr\left(\frac{2r\widehat{\sigma}}{\sigma_a} < \chi^2_{2r,\beta}\right) = \beta,$$

so that

$$\frac{2r\widehat{\sigma}_c}{\sigma_a} = \chi^2_{2r,\beta}.$$

Substituting (12.89) into the above gives

$$\frac{\chi^2_{2r,1-\alpha}}{\chi^2_{2r,\beta}} = \frac{\sigma_a}{\sigma_0}. \tag{12.90}$$

Now the quantity r needs to be determined so that (12.90) is satisfied. Since r is integer valued, an exact solution for (12.90) is rarely possible. The usual practice is to find the smallest value of r such that the probability (12.88) is just less than β. This means that the left-hand side of (12.90) is just less than the constant σ_a/σ_0. For the value of r so obtained, the critical value $\widehat{\sigma}_c$ is then given by (12.89).

To compute solutions to (12.90), it is convenient to express the χ^2-quantiles in terms of their Poisson equivalents (see Section 13.16) as

$$F_\chi(a;\nu=2r) = 1 - F_p\left(r-1;\frac{\alpha}{2}\right) = 1 - \exp\left\{-\frac{a}{2}\right\}\sum_{i=0}^{r-1}\frac{\left(\frac{a}{2}\right)^i}{i!}.$$

Thus, the χ^2-quantile of order $1-\alpha$, at $2r$ degrees of freedom, is the value a satisfying

$$1 - \exp\left\{-\frac{a}{2}\right\}\sum_{i=0}^{r-1}\frac{\left(\frac{a}{2}\right)^i}{i!} = 1-\alpha. \tag{12.91}$$

The other quantile in (12.90) is treated similarly. See Example 12.8 for an illustration.

Note that the above test plan is independent of the sample size n and is therefore the same whether or not failed items are replaced. What depends on n, however, is the *test duration* $x_{(r)}$. Its expected value is given by (12.10), if failed items are not replaced:

$$\mathrm{E}\{x_{(r)}\} = \sigma\sum_{i=1}^{r}\frac{1}{n-i+1}. \tag{12.92}$$

If failed specimens are, however, replaced immediately upon failure, then there are n items on test during the entire test period $x'_{(r)}$ and the expected test duration is

$$\mathrm{E}\{x'_{(r)}\} = \frac{r}{n}\mathrm{E}\{X\} = \frac{r}{n}\sigma, \tag{12.93}$$

which is shorter than (12.92). See Example 12.8.

12.22 Time-Censored Test

There are sometimes economic reasons for limiting the test duration at a chosen time T. One would then consider a *time-censored, replacement test* plan, where the test statistic is the observed number r of failures. That is, if r is smaller than a critical value r_c, then the null hypothesis $H_0 : \sigma = \sigma_0$ is rejected in favor of the alternative $H_a : \sigma = \sigma_a > \sigma_0$. The test sample is thus time-censored on the right at T. From (12.87):

$$Pr(r < r_c \mid \sigma_0) = \alpha.$$

From the connection of Exponential and Poisson variables (see Section 13.16) the distribution of r is Poisson with mean value nT/σ. Expressed equivalently in terms of the waiting time to the $(r-1)$th event, the above Poisson probability becomes

$$Pr\left(\chi^2_{2r_c} > \frac{2nT}{\sigma_0}\right) = \alpha,$$

so that the total time on test is obtained as

$$nT = \frac{\sigma_0}{2}\chi^2_{2r_c,1-\alpha}. \tag{12.94}$$

Similarly, from (12.88),

$$Pr(r > r_c \mid \sigma_a) = Pr\left(\chi^2_{2r} < \frac{2nT}{\sigma_a}\right) = \beta,$$

giving

$$nT = \frac{\sigma_a}{2}\chi^2_{2r_c,\beta}. \tag{12.95}$$

Combining the two expressions yields

$$\frac{\chi^2_{2r_c,1-\alpha}}{\chi^2_{2r_c,\beta}} = \frac{\sigma_a}{\sigma_0}, \tag{12.96}$$

which is the same as (12.90) for the failure-censored case, and the same solution process applies. Thus, the critical failure number r_c is determined by α and β, independent of T, the value of which may then be estimated by (12.95). See Example 12.9 for an illustration.

EXAMPLE 12.1

Five prototypes of an experimental engine control were available for a life test under accelerated conditions. The observed failure mode was a malfunctioning of a particular integrated circuit chip. Every time a chip failed it was replaced and the test resumed.

The test was terminated after 378 hours. At that time 16 chips had been installed of which 11 had failed. The five survivor chips are marked by an asterisk:

16, 27*, 35*, 36, 41, 51, 76*, 79, 84, 121, 125, 166, 179, 212*, 264, 378*.

Check graphically if the Exponential distribution is a plausible life model.

$n := 16 \qquad r := 11$

$k := 1..n \qquad i := 1..r$

$t_k :=$ $\quad x_i :=$ $\quad q_i :=$ $\quad$ (q_i is the order of the failure data in the total sample.)

t_k	x_i	q_i
16	16	1
27	36	4
35	41	5
36	51	6
41	79	8
51	84	9
76	121	10
79	125	11
84	166	12
121	179	13
125	264	15
166		
179		
212		
264		
378		

$$p_i := 1 - \frac{n+0.7}{n+0.4} \cdot \prod_{j=1}^{i} \frac{n-q_j+0.7}{n-q_j+1.7}$$

data ordinates : $\quad r_j := -\ln(1 - p_i)$

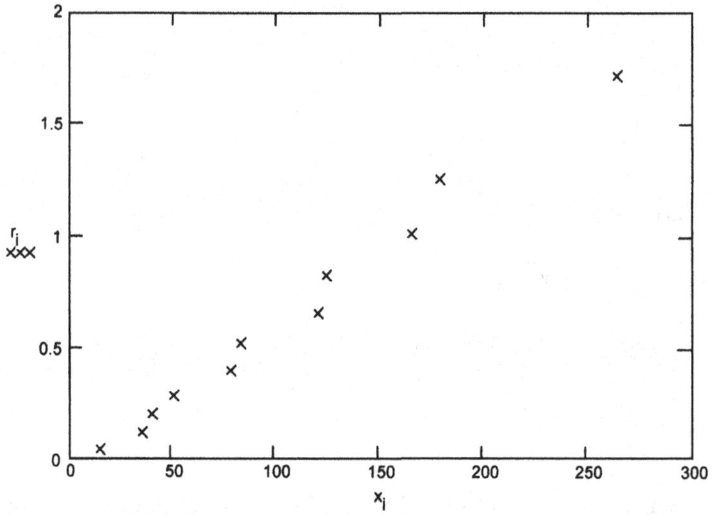

The data plot looks quite linear. The Exponential assumption is therefore reasonable.

EXAMPLE 12.2

Sixteen units of a newly designed inverter were tested to failure under magnified vibratory loads, with the following results (in hours):

0.2, 0.5, 2.0, 2.3, 3.1, 4.5, 4.8, 7.0, 10.1, 11.4, 12.0, 13.9, 18.0, 20.1, 21.5, 33.7.

Estimate the Exponential life model, verify it graphically, and test the fit at the 5% level of significance. Estimate the 20-percentile life and its 80% confidence interval.

$n := 16 \qquad i := 1..n$

$x_i :=$

0.2
0.5
2.0
2.3
3.1
4.5
4.8
7.0
10.1
11.4
12.0
13.9
18.0
20.1
21.5
33.7

1. ML parameter estimate

$$\sigma := \frac{1}{n} \cdot \sum_i x_i \qquad \sigma = 10.3$$

standard error: $SE\sigma := \dfrac{\sigma}{\sqrt{n}}$ $\qquad SE\sigma = 2.6$

2. Probability plot

$$p_i := \frac{i - 0.3}{n + 0.4}$$

data ordinates: $r_i := -\ln(1 - p_i)$

model ordinates: $z_i := \dfrac{x_i}{\sigma}$

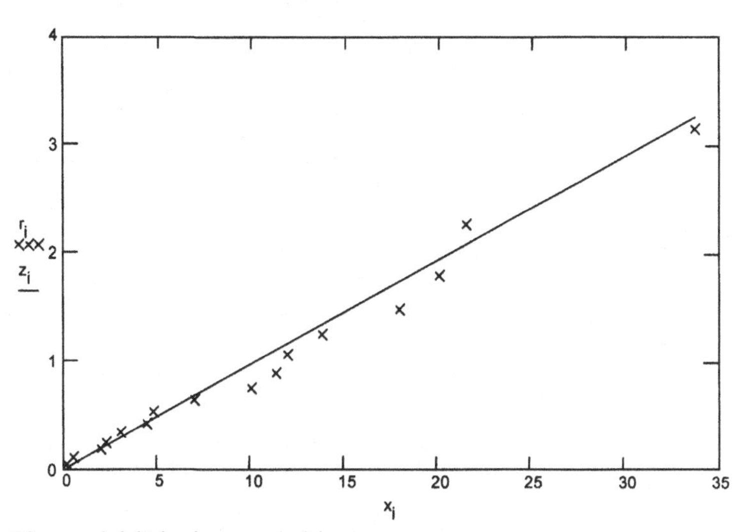

The model fit looks reasonable.

3. Test-of-fit

$$w_i := 1 - \exp\left(-\frac{x_i}{\sigma}\right)$$

$$A := \left[-n - \frac{1}{n} \cdot \sum_i (2 \cdot i - 1) \cdot (\ln(w_i) + \ln(1 - w_{n-i+1}))\right] \cdot \left(1 + \frac{0.6}{n}\right)$$

$A = 0.212$

The 5% critical value is 1.321. Because the sample value is smaller, there is no reason to reject the estimated model at the 5% significance level.

4. Percentile

$x_{20} := -\sigma \cdot \ln(1 - 0.2)$ $\qquad x_{20} = 2.3$

standard error: $SEx := -\ln(1 - 0.2) \cdot SE\sigma$ $\qquad SEx = 0.58$

80% confidence interval:

The 10- and 90-percentiles of the χ^2-variable are obtained by direct evaluation of the χ^2 cdf at 32 degrees of freedom:

$\chi_{10} := 22.271$

$\chi_{90} := 42.585$

From (12.31) the confidence limits on σ are

$s_l := \dfrac{2 \cdot n \cdot \sigma}{\chi_{90}}$ $\qquad s_l = 7.8$

$s_u := \dfrac{2 \cdot n \cdot \sigma}{\chi_{10}}$ $\qquad s_u = 14.8$

The 80% confidence limits on the percentile are therefore

$x_l := (-\ln(1 - 0.2)) \cdot s_l$ $\qquad x_l = 1.7$

$x_u := (-\ln(1 - 0.2)) \cdot s_u$ $\qquad x_u = 3.3$

EXAMPLE 12.3

For the data of Example 12.1, estimate the parameter and its 90% confidence interval.

$n := 16 \qquad k := 1..n \qquad r := 11 \qquad i := 1..r$

$t_k :=$ $\quad x_i :=$ $\quad q_i :=$ $\quad$ (q_i is the order of the failure data in the total sample.)

t_k	x_i	q_i
16	16	1
27	36	4
35	41	5
36	51	6
41	79	8
51	84	9
76	121	10
79	125	11
84	166	12
121	179	13
125	264	15
166		
179		
212		
264		
378		

1. Maximum likelihood estimate

$\sigma := \dfrac{1}{r} \cdot \sum_k t_k$ $\qquad \sigma = 172$

$SE\sigma := \dfrac{\sigma}{\sqrt{r}}$ $\qquad SE\sigma = 52$

2. Probability plot

median plotting position:

$$p_i := 1 - \frac{n + 0.7}{n + 0.4} \cdot \prod_{j=1}^{i} \frac{n - q_j + 0.7}{n - q_j + 1.7}$$

data ordinates: $q_i := -\ln(1 - p_i)$

model ordinates: $z_i := \frac{x_i}{\sigma}$

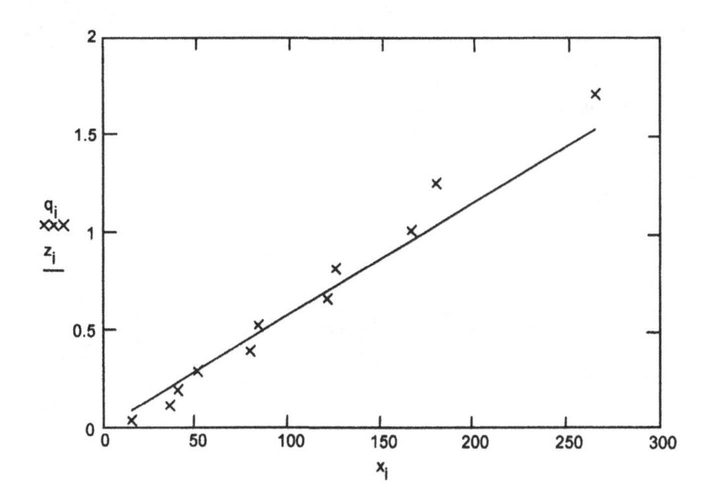

3. Confidence intervals

i) Normal approximation: Standard Normal 95-percentile is $z_{95} := 1.645$

$\sigma_l := \sigma - z_{95} \cdot \text{SE}\sigma$ $\qquad \sigma_l = 87$

$\sigma_u := \sigma + z_{95} \cdot \text{SE}\sigma$ $\qquad \sigma_u = 257$

ii) Likelihood ratio method: Chi-squared 90-percentile at $\nu = 1$ is $K := 2.71$

log of likelihood function: $LL(a) := -r \cdot \ln(a) - \frac{\sum_{k=1}^{n} t_k}{a}$

likelihood ratio: $LR(a) := 2 \cdot LL(\sigma) - 2 \cdot LL(a)$

$a := \sigma - 50$ $\qquad \sigma_l := \text{root}(LR(a) - K, a)$ $\qquad \sigma_l = 109$

$b := \sigma + 80$ $\qquad s_u := \text{root}(LR(b) - K, b)$ $\qquad \sigma_u = 257$

EXAMPLE 12.4

Eighteen units of a redesigned electronic component were put on a life test under a multiple of the maximum design voltage. It was decided to terminate the test at the twelfth failure. The following failure times were observed:

2, 10, 11, 29, 43, 68, 81, 82, 99, 122, 141, 169.

Estimate the Exponential life model and test the fit at the 5% level of significance.

$n := 18 \qquad r := 12 \qquad i := 1 \,..\, r$

$x_i :=$

2
10
11
29
43
68
81
82
99
122
141
169

1. ML parameter estimate

$$\sigma := \frac{1}{r} \cdot \left[\sum_i x_i + (n - r) \cdot x_r \right] \qquad \sigma = 155.9$$

standard error: $$\mathrm{SE}\sigma := \frac{\sigma}{\sqrt{r}} \qquad \mathrm{SE}\sigma = 45.0$$

2. Test-of-fit

$$x_0 := 0 \qquad s_i := (n - i + 1) \cdot (x_i - x_{i-1})$$

$$y := \mathrm{sort}(s) \qquad w_i := 1 - \exp\left(-\frac{y_i}{\sigma}\right)$$

$$A := \left[-r - \frac{1}{r} \cdot \sum_i (2 \cdot i - 1) \cdot (\ln(w_i) + \ln(1 - w_{r-i+1})) \right] \cdot \left(1 + \frac{0.6}{r}\right) \qquad A = 1.080$$

The 5% critical value is 1.321. Because the sample value is smaller, there is no reason to reject the estimated model for X, at the 5% significance level.

3. Probability plot

$$p_i := \frac{i - 0.3}{n + 0.4}, \qquad \text{data ordinates:} \quad q_i := -\ln(1 - p_i)$$

$$\text{model ordinates:} \quad z_i := \frac{x_i}{\sigma}$$

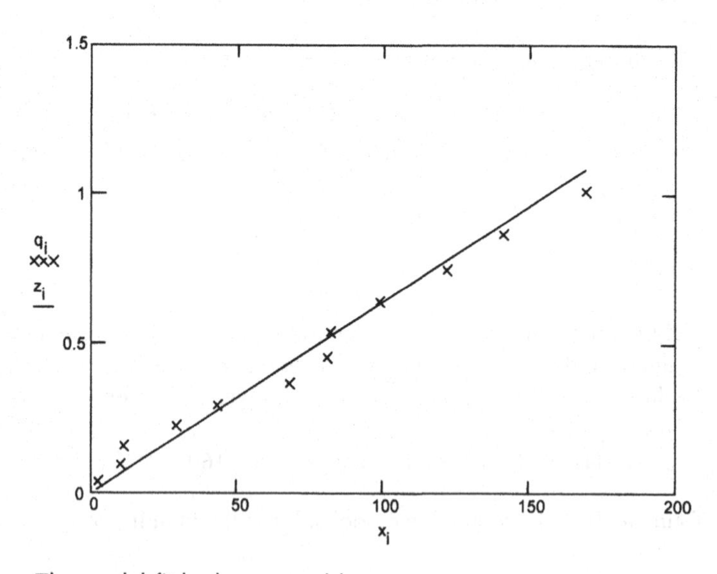

The model fit looks reasonable.

EXAMPLE 12.5

Suppose the constant $\mu = 10$ is added to the data of Example 12.2. Estimate the Exponential parameters and construct their 90% confidence intervals.

$n := 16 \qquad i := 1..n$

$x_i := \qquad y := x + 10 \qquad$ (shift the data by $\mu = 10$)

x_i
0.2
0.5
2.0
2.3
3.1
4.5
4.8
7.0
10.1
11.4
12.0
13.9
18.0
20.1
21.5
33.7

1. ML parameter estimates

$$yb := \frac{1}{n} \cdot \sum_i y_i \qquad \mu := \frac{n \cdot y_1 - yb}{n-1} \qquad \mu = 9.5$$

$$\sigma := \frac{n \cdot (yb - y_1)}{n-1} \qquad \sigma = 10.8$$

standard errors: $\mathrm{SE}\mu := \dfrac{\sigma}{\sqrt{n \cdot (n-1)}} \qquad \mathrm{SE}\mu = 0.7$

$$\mathrm{SE}\sigma := \frac{\sigma}{\sqrt{n-1}} \qquad \mathrm{SE}\sigma = 2.8$$

2. Confidence intervals

$\alpha := 0.10$

from (12.55):

$$\mu_l := y_1 - \sigma \cdot \left[\left(\frac{\alpha}{2} \right)^{\frac{1}{1-n}} - 1 \right] \qquad \mu_l = 7.8$$

$$\mu_u := y_1 - \sigma \cdot \left[\left(1 - \frac{\alpha}{2} \right)^{\frac{1}{1-n}} - 1 \right] \qquad \mu_u = 10.2$$

The 5% and 95% χ^2-quantiles at $\nu = 2n - 2 = 30$ degrees of freedom are

$\chi_{05} := 18.5 \qquad \chi_{95} := 43.8$

from (12.56): $\sigma_l := \dfrac{2 \cdot n \cdot \sigma}{\chi_{95}} \qquad \sigma_l = 7.9$

$$\sigma_u := \frac{2 \cdot n \cdot \sigma}{\chi_{05}} \qquad \sigma_u = 18.7$$

EXAMPLE 12.6

For the (artificial) data of Example 12.5, estimate the 20-percentile and its lower 90% confidence limit.

Given:

$n := 16 \qquad r := n \qquad$ sum of observations: $S_y := 325.1 \qquad$ parameters: $\mu := 9.5$

$\alpha := 0.10 \qquad q := 0.20 \qquad$ smallest observation: $Y1 := 10.2 \qquad \sigma := 10.8$

1. Estimate

From (12.9):

$y_q := \mu - \sigma \cdot \ln(1-q)$ $\qquad y_q = 11.9$

2. Confidence limit

sample size check:

$$\frac{\ln(\alpha)}{\ln(1-q)} = 10.3$$

($n > 10$; hence the confidence limit will be approximate)

from (12.60a): $k := \frac{1}{n} \cdot \left[1 - \left[\frac{(1-q)^n}{\alpha}\right]^{\frac{1}{r-1}}\right]$ $\qquad k = 5.06508 \cdot 10^{-3}$

from (12.60b): $S := Sy - n \cdot Y1$

from (12.59): $L := Y1 + k \cdot S$ $\qquad L = 11.0$

EXAMPLE 12.7

For the (artificial) data of Example 12.5, estimate the reliability value and the 90% lower confidence limit on reliability, for the mission time $y = 10$.

Given:

$n := 16$ $\quad r := n$ $\quad$ sum of observations: $Sy := 325.1$ $\quad$ parameters: $\mu := 9.5$

$\alpha := 0.10$ $\quad y := 10$ $\quad$ smallest observation: $Y1 := 10.2$ $\quad \sigma := 10.8$

1. Estimate

maximum likelihood (12.67): $R := \exp\left(-\frac{y-\mu}{\sigma}\right)$ $\qquad R = 0.95$

unbiased (12.77): $T := Sy - (n-1) \cdot Y1$ $\quad R := \frac{n-1}{n} \cdot \left(1 - \frac{y - Y1}{T - Y1}\right)$ $\qquad R = 0.94$

2. Confidence limit

sample size check: $\frac{\ln(\alpha)}{\ln(R)} = 36$ $\qquad S := Sy - n \cdot Y1$

(12.84) : $R_L := \alpha^{\frac{1}{n}} \cdot \left(1 - n \cdot \frac{y - Y1}{S}\right)^{\frac{r-1}{n}}$ $\qquad R_L = 0.88$

EXAMPLE 12.8

An electronic device has an Exponential life, with a mean life of 350 hours, estimated from tests under accelerated conditions. A design modification was made in order to improve the mean life. Set up a nonreplacement, failure-censored acceptance test to decide if the modified design has doubled the mean life, for $\alpha = 0.05$ and $\beta = 0.10$. Estimate the test duration, if 30 modified specimens are available for testing.

$\alpha := 0.05 \quad \beta := 0.10 \quad$ trial value: $r := 18 \quad i := 0..r-1$

$\sigma_o := 350 \quad \sigma_a := 700$

from (12.91):

$$a := 40 \quad K_\alpha := \text{root}\left[\exp\left(-\frac{a}{2}\right)\cdot\sum_i\left(\frac{a}{2}\right)^i\cdot\frac{1}{i!} - \alpha, a\right] \qquad K_\alpha = 51.0$$

$$b := 20 \quad K_\beta := \text{root}\left[1 - \exp\left(-\frac{b}{2}\right)\cdot\sum_i\left(\frac{b}{2}\right)^i\cdot\frac{1}{i!} - \beta, b\right] \qquad K_\beta = 25.6$$

calculated ratio: $\frac{K_\alpha}{K_\beta} = 1.99$ target ratio: $\frac{\sigma_a}{\sigma_o} = 2.00$

For $r = 17$, the calculated ratio is 2.03; hence the required failure number is $r = 18$

The critical value of the test statistic (ML estimate of sigma) is

from (12.89): $\sigma_c := \frac{\sigma_o\cdot K_\alpha}{2\cdot r}$ $\qquad \sigma_c = 496.$

Thus, if the test statistic $> \sigma_c$, reject Ho (σ has doubled),
and, if the test statistic $< \sigma_c$, accept Ho (σ has not changed).

Expected test duration (if Ho is true) for $n := 30$ available test specimens:

$j := 1..r$ from (12.92): $E := \sigma_o\cdot\sum_j\frac{1}{n-j+1}$ $\qquad E = 312$

If more test specimens were available so that a 30-unit replacement test could be run, the expected test duration (if Ho is true) would be

from (12.93): $Er := \frac{r}{n}\cdot\sigma_o$ $\qquad Er = 210$

EXAMPLE 12.9

For the decision scenario of Example 12.8, determine the fixed test duration T of a (time-censored) 30-unit replacement test plan at $\alpha = 0.05$ and $\beta = 0.10$, assuming that enough specimens are available for failure replacements.

$\sigma_o := 350 \quad K_\alpha := 51 \quad n := 30$

The critical number of failures is as found before: $\qquad r_c := 18$

From (12.94), the test is terminated at

$$T := \frac{\sigma_o\cdot K_\alpha}{2\cdot n} \qquad T = 298$$

Of course, as soon as 18 failures have occurred, the test is stopped and the decision is to accept $Ho: \sigma$ has not improved.

CHAPTER THIRTEEN

Gamma Distributions

INTRODUCTION

13.1 Definition

A continuous random variable X has a *Gamma distribution* if its pdf has the form

$$f(x;\sigma,\lambda) = \frac{1}{\sigma\Gamma(\lambda)}\left(\frac{x}{\sigma}\right)^{\lambda-1}\exp\left\{-\frac{x}{\sigma}\right\}; \quad x \geq 0; \quad \sigma,\lambda > 0. \tag{13.1}$$

This model arose during the early phases of statistical development as the *sampling* distribution of several statistics, chiefly the sum of squares of independent Normal variables and the Chi-squared statistics used for tests-of-fit. However, several features of this distribution make it attractive as a general *measurement* model in engineering, namely, its shape flexibility and its positive sample space. Applications cover a broad range of fields including waiting-line problems, reliability analysis, and hydrological and other natural processes.

The symbol Γ represents the *gamma function* and is defined as

$$\Gamma(\lambda) = \int_0^\infty x^{\lambda-1}\exp\{-x\}\,dx, \tag{13.2}$$

which equals the factorial function $(\lambda-1)!$ when λ is a positive integer. $\Gamma(\lambda)$ is tabulated[1] for $1 \leq \lambda \leq 2$. For values of λ outside that range, $\Gamma(\lambda)$ can be obtained from the recurrence relation

$$\Gamma(\lambda+1) = \lambda\Gamma(\lambda), \tag{13.3}$$

together with tabulated values. On most computational aids, Γ is a built-in function.

[1] CRC Standard Mathematical Tables, CRC Press, Cleveland, 1996. Note that $\Gamma(\lambda)$ is a built-in function in *Mathcad*.

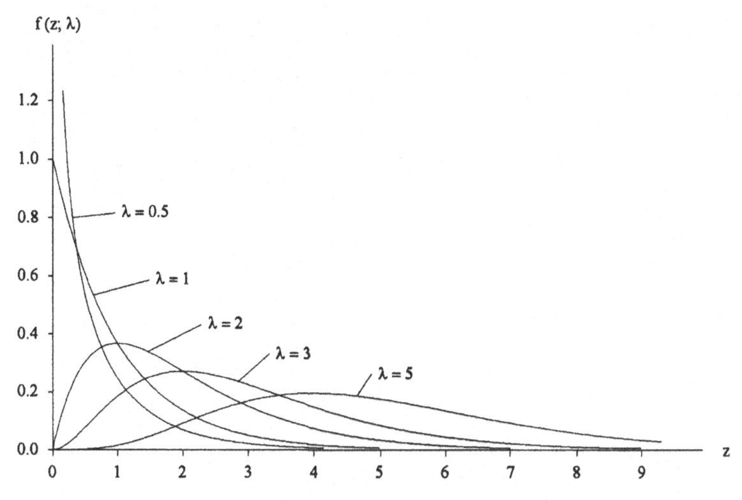

Figure 13.1. Gamma pdfs for the reduced variable z.

PROPERTIES: TWO-PARAMETER MODEL

13.2 Gamma Variable

The cdf of a Gamma variable, as defined by (13.1), cannot be expressed in closed form:

$$F(x; \sigma, \lambda) = \frac{1}{\Gamma(\lambda)} \int_0^{\frac{x}{\sigma}} z^{\lambda-1} \exp\{-z\}\, dz. \tag{13.4}$$

This cdf is often called the *gamma function ratio*, whereas the integral in (13.4) is termed the *incomplete gamma function*. Although tables are available for these functions, the cdf is more readily computed directly, using a modern aid such as Mathcad. Note that Mathcad 6+ has a built-in function that gives $F(x; \sigma = 1, \lambda)$.

This model has scale structure (see Section 9.2). The scale parameter is σ and the shape parameter is λ. The pdf of the reduced variable $Z = X/\sigma$ is therefore a function of λ only:

$$f(z; \lambda) = \frac{1}{\Gamma(\lambda)} z^{\lambda-1} \exp\{-z\}; \quad z \geq 0, \quad \lambda \geq 0. \tag{13.5}$$

Figure 13.1 shows this pdf for several values of λ, indicating the shape flexibility of the Gamma model.

The moment generating function of X (see Section 1.8) is

$$M(t) = (1 - \sigma t)^{-\lambda}. \tag{13.6}$$

Thus, the expected value of X is

$$\mu_1'(X) = \sigma\lambda. \tag{13.7}$$

The variance of X is

$$\mu_2(X) = \sigma^2\lambda. \tag{13.8}$$

The coefficient of variation of X is therefore a function of λ only:

$$cv(X) = \frac{1}{\sqrt{\lambda}}. \tag{13.9}$$

The mode value is given by

$$x_{\mathrm{m}} = \sigma(\lambda - 1). \tag{13.10}$$

The quantile of order q is defined by

$$F(x_q; \sigma, \lambda) = q \tag{13.11}$$

and is easily computed with an equation solver. Note that Mathcad 6+ has a built-in function that gives quantiles z_q for the reduced variable; the quantile x_q is then obtained as $x_q = \sigma z_q$.

The first shape factor (see Section 1.10) is

$$\gamma_1 = \frac{2}{\sqrt{\lambda}} > 0, \tag{13.12}$$

indicating the positive skew of all Gamma models. The second shape factor is

$$\gamma_2 = 3\left(1 + \frac{2}{\lambda}\right) > 3. \tag{13.13}$$

For increasing values of λ, both shape factors approach the Normal values 0 and 3, so that the Gamma pdf approaches the Normal pdf.

The matrix of minimum-variance-bounds (see Section 3.2) for estimators of σ and λ is

$$\begin{bmatrix} V_{\sigma\sigma} & V_{\sigma\lambda} \\ V_{\sigma\lambda} & V_{\lambda\lambda} \end{bmatrix} = \frac{1}{n\{\lambda\psi'(\lambda) - 1\}} \begin{bmatrix} \sigma^2\psi'(\lambda) & -\sigma \\ -\sigma & \lambda \end{bmatrix}. \tag{13.14}$$

The *digamma function*

$$\psi(\lambda) = \frac{\partial}{\partial\lambda}\ln[\Gamma(\lambda)] \tag{13.15}$$

and the *trigamma function*

$$\psi'(\lambda) = \frac{\partial}{\partial\lambda}\psi(\lambda) = \frac{\partial^2}{\partial\lambda^2}\ln[\Gamma(\lambda)] \tag{13.16}$$

are readily computed (e.g., using Mathcad) by their definitions. These functions are also tabulated[2] for $1 \leq \lambda \leq 2$; for other values of λ, they can be evaluated from the recurrence relations

$$\psi(\lambda+1) = \psi(\lambda) + \frac{1}{\lambda} \quad \text{and} \quad \psi'(\lambda+1) = \psi'(\lambda) - \frac{1}{\lambda^2}. \tag{13.17}$$

13.3 Reproductive Property

The Gamma distribution features a useful reproductive property: If k independent Gamma variables X_i with shape parameters λ_i and a common scale parameter σ are summed as

$$T = a\sum_{i=1}^{k} X_i,$$

then T is again a Gamma variable with scale parameter $a\sigma$ and shape parameter $\lambda = \sum_i \lambda_i$. For example, when $a = 1/n$ and $\lambda_i = \lambda$, then $T = \bar{X}$ is the average of n measurements from the same Gamma process, and $\bar{X}$ is Gamma distributed with parameters σ/n and $n\lambda$.

13.4 Simulation

Random observations x_i from a Gamma process with known parameters σ and λ are easily simulated if λ is integer valued, recognizing from the above reproductive property that such a Gamma variate is the sum of λ Exponentials (see Section 12.4):

$$x_i = -\sigma \sum_{j=1}^{\lambda} \ln(u_i), \tag{13.18}$$

where u_i is a Uniform random number on (0, 1).

When λ is noninteger, the following simple but slow scheme[3] can be used: Split the shape parameter into its integer portion λ_I and its decimal portion λ_D; thus, $\lambda = \lambda_I + \lambda_D$. Generate a Gamma variate for λ_D by first computing an Exponential variate $v = -\sigma \cdot ln(u)$ (see Section 12.5) and then computing a Beta variate w for the Beta shape parameters $\lambda_1 = \lambda_D$ and $\lambda_2 = 1 - \lambda_D$ (see Section 14.4). The product $x_D = vw$ is a variate from the Gamma distribution $F(x_D; \sigma, \lambda_D)$. Using (13.18), generate a Gamma variate x_I corresponding to the parameters σ and λ_I. From the reproductive property of Gamma variables it then follows that $x = x_I + x_D$ is a Gamma variate from $F(x; \sigma, \lambda)$.

[2] M. Abramowitz, I. A. Stegun, eds., *Handbook of Mathematical Functions*, Natl. Bureau of Standards, Government Printing Office, Washington, DC, 1964.

[3] Jöhnk, M. D., "Erzeugung von Betaverteilten und Gammaverteilten Zufallszahlen." *Metrika*, Vol. 8, pp. 5–15, 1964.

For moderate-sized samples it is perhaps more convenient to invert the given cdf directly. That is, a simulated observation x_i from the Gamma cdf (13.4) is simply the u_i-quantile defined by $F(x_i; \sigma, \lambda) = u_i$ (see Section 2.8). Note that Mathcad 6+ has a built-in function that generates random observations z_i from $F(z; \sigma = 1, \lambda), \lambda > 0$. A general Gamma variate is then $x_i = \sigma \cdot z_i$.

One should check the adequacy of a simulated sample by comparing at least its first two moments with those of the given model. See Example 13.5, where a small sample from a given three-parameter model is simulated directly and then processed.

PROPERTIES: THREE-PARAMETER MODEL

13.5 Definition and Properties

Measurement processes that are modeled by a Gamma distribution sometimes feature a threshold value μ, which enters the model as a location parameter. The pdf (13.1) becomes

$$f(x; \mu, \sigma, \lambda) = \frac{1}{\sigma\Gamma(\lambda)}\left(\frac{x-\mu}{\sigma}\right)^{\lambda-1} \exp\left\{-\frac{x-\mu}{\sigma}\right\}; \quad x \geq \mu; \quad \sigma, \lambda > 0. \tag{13.19}$$

Provided that $\lambda > 2$, the *expected* information matrix (see Section 3.1) for the three-parameter Gamma model is

$$\begin{bmatrix} I_{\mu\mu} = \frac{n}{\sigma^2(\lambda-2)} & I_{\mu\sigma} = \frac{n}{\sigma^2} & I_{\mu\lambda} = \frac{n}{\sigma(\lambda-1)} \\ I_{\mu\sigma} & I_{\sigma\sigma} = \frac{n\lambda}{\sigma^2} & I_{\sigma\lambda} = \frac{n}{\sigma} \\ I_{\mu\lambda} & I_{\sigma\lambda} & I_{\lambda\lambda} = n\psi'(\lambda) \end{bmatrix}. \tag{13.20}$$

The minimum-variance-bounds are then obtained from the inverse of (13.20). Alternatively, the *local* information matrix (see Section 3.7) can be used, provided $\lambda > 1$. This matrix is identical to (13.20), except for the leading term, which is

$$I_{\mu\mu} = (\lambda - 1)\sum_{i=1}^{n}(x_i - \mu)^{-2}. \tag{13.21}$$

For the three-parameter model all location measures are displaced to the right by μ relative to their two-parameter equivalents, while central moments remain the same. Thus the expected value of the three-parameter variable X is

$$\mu_1'(X) = \mu + \sigma\lambda. \tag{13.22}$$

The mode value is

$$x_m = \mu + \sigma(\lambda - 1). \tag{13.23}$$

The shape factors γ_1 and γ_2 are unchanged, but the coefficient of variation is now

$$cv(X) = \frac{\sigma\sqrt{\lambda}}{\mu + \sigma\lambda}. \tag{13.24}$$

The reproductive property of the two-parameter model holds, provided the values of both μ and σ are common to the variables X_i.

SPECIAL CASES: CHI-SQUARED, ERLANG, AND EXPONENTIAL MODELS

13.6 Definition and Properties

When the shape parameter λ is replaced by $\nu/2$ and the scale parameter is restricted to the value 2, the Gamma pdf (13.1) becomes the Chi-squared (χ^2) pdf:

$$f_X(x; \nu) = \frac{1}{2\Gamma\left(\frac{\nu}{2}\right)}\left(\frac{x}{2}\right)^{\frac{\nu}{2}-1} \exp\left\{-\frac{x}{2}\right\}; \quad x \geq 0, \quad \nu > 0. \tag{13.25}$$

The χ^2 distribution is the exact sampling model for several frequently used statistics, chiefly for Normal and Exponential variables. In that context, the parameter ν is termed the number of degrees of freedom, designating the number of unconstrained ways in which data values enter the statistic in question.

The properties of the χ^2 variable follow directly from those of the two-parameter Gamma variable. For instance, the mean value and variance are

$$\mu_1'(X) = \nu \quad \text{and} \quad \mu_2(X) = 2\nu. \tag{13.26}$$

Figure 13.2 shows the χ^2 pdf for several values of ν. For selected values of q,

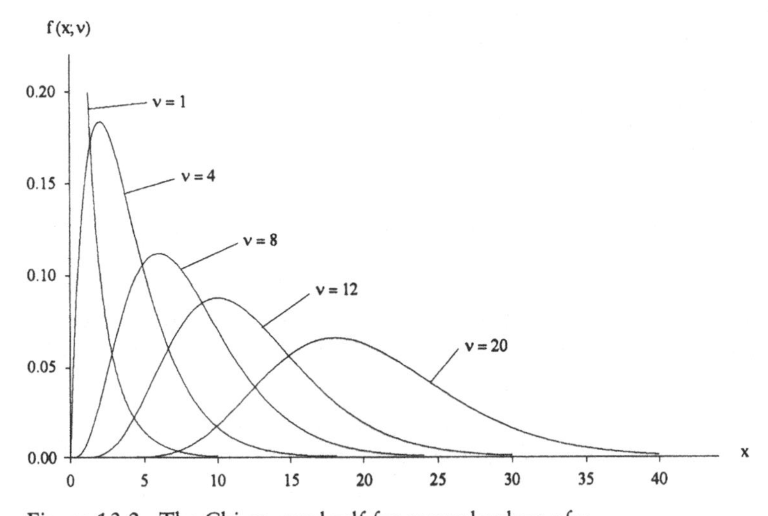

Figure 13.2. The Chi-squared pdf for several values of ν.

quantiles of the χ^2 variable are extensively tabulated in many statistics texts and statistical tables. However, with a modern computational aid it is straightforward to compute the χ^2 cdf or its inverse directly. Note that Mathcad 6+ has built-in functions that compute the χ^2 cdf and quantiles.

When the shape parameter λ is restricted to positive integers, the Gamma pdf (13.1) becomes the *Erlang model*. This distribution arises chiefly in *waiting-line* problems, where it models the waiting time to the occurrence of the λth Poisson event (see Section 13.16).

When the shape parameter takes the value $\lambda = 1$, the Gamma distribution specializes to the Exponential distribution (see Chapter 12).

PROBABILITY PLOT

13.7 Complete Samples

A probability plot of ordered data (see Section 3.13) should always be used to check the distributional assumption of the Gamma model. Since the Gamma cdf (13.4) cannot be expressed in closed form, it is not possible to linearize it algebraically. However, once the model has been estimated, it can be linearized numerically between any two chosen values a and b. The data plotting positions can then be adjusted by the same amounts as the estimated model. The result is a linearized graphical comparison of data and estimated model that provides visual information on how well the model fits the data. The plotting range (a, b) is of course chosen to include all given data x_i.

For the *estimated* Gamma cdf F, the straight-line model ordinate r_i for each ordered observation $x_{(i)}$ is given by

$$r_i = F(a) + \frac{F(b) - F(a)}{b - a}(x_{(i)} - a). \tag{13.27}$$

The adjustment from the cdf to the linear relation (13.27) is therefore

$$\Delta_i = F(x_{(i)}) - r_i. \tag{13.28}$$

The median data plotting position (see Section 3.13) is then adjusted to

$$p_i = \frac{i - 0.3}{n + 0.4} - \Delta_i. \tag{13.29}$$

The plot of p_i versus $x_{(i)}$ will roughly follow the linear model plot (13.27), if $X_{(i)}$ came from the estimated two-parameter Gamma process (see Section 3.14). See Example 13.1 for an illustration of this plotting procedure (following model estimation) for a complete sample.

When the appearance of the data plot suggests the presence of a threshold parameter μ (the lower end of the data plot curves downward), a *preliminary* estimate of its value may be obtained by subtracting trial values μ_k from the observations $x_{(i)}$ until the corresponding plot has straightened.

13.8 Censored Samples

The Gamma distribution is sometimes used as a model for the operating life of equipment. One difficulty with this application is that often the data are *multiply censored* (see also Sections 12.2 and 17.11). Consider, then, a multiply right-censored sample of size n, with r failure times x_i and $m = n - r$ survival times c_j. Label the combined and *ordered* set of failure and survival data as $\{t_{(k)}\}_n$. Thus, some of the $t_{(k)}$ are failure data x_i and some are survival data c_j.

For multiply censored samples, the plotting position (13.29) needs to be modified to include the information implicit in the survival times c_j. A suitable modification is based on the distribution-free *product limit* estimate of the reliability function (see Section 3.15). The modified median plotting position is then

$$p_k = 1 - \frac{n + 0.7}{n + 0.4} \prod_{q=1}^{k} \frac{n - q + 0.7}{n - q + 1.7} - \Delta_i, \tag{13.30}$$

where $q \leq k$ is the subscript of a *failure* time in the combined, ordered set $\{t_{(k)}\}_n$. When there is no censoring, (13.30) reduces to the complete-sample median plotting position (13.29). As in the preceding section, p_i is plotted versus the ordered failure data $x_{(i)}$. See Example 13.2 for an illustration of a Gamma probability plot for a censored sample.

POINT ESTIMATES: TWO-PARAMETER MODEL

13.9 Maximum Likelihood Estimates

The likelihood function of a complete sample of n independent observations on a Gamma variable X is

$$L(\sigma, \lambda) = \Gamma(\lambda)^{-n} \sigma^{-n\lambda} \prod_{i=1}^{n} x_i^{\lambda-1} \exp\left\{ -\frac{1}{\sigma} \sum_{i=1}^{n} x_i \right\}. \tag{13.31}$$

The maximum likelihood (ML) equations (see Section 3.6) are

$$\widehat{\sigma}\, \widehat{\lambda} = \bar{x} \tag{13.32}$$

and

$$\ln(\widehat{\sigma}) + \psi(\widehat{\lambda}) = \ln(g), \tag{13.33}$$

where $g = \prod_i x_i^{1/n}$ is the *geometric sample mean*. We see that the statistics $\bar{x}$ and g are *sufficient* in that they hold all information relevant to the estimation of the parameters. Combining the above two equations gives an equation in only $\widehat{\lambda}$:

$$\ln(\widehat{\lambda}) - \psi(\widehat{\lambda}) = \ln\left(\frac{\bar{x}}{g}\right). \tag{13.34}$$

The solution value $\widehat{\lambda}$ is readily computed with an equation solver. The ML estimate $\widehat{\sigma}$ is then computed from (13.32). The ML estimate $\widehat{\lambda}$ is biased upward, the bias being a function of both n and λ. Although various bias-correction formulas are available, the uncorrected estimate $\widehat{\lambda}$ is recommended, because the resulting (biased) model often provides a superior fit to the data, as indicated by the probability plot.

Asymptotic standard errors of these estimates are obtained from the minimum-variance-bounds (13.14):

$$\mathrm{se}(\widehat{\sigma}) = \sigma\sqrt{\frac{\psi'(\lambda)}{n[\lambda\psi'(\lambda) - 1]}} \tag{13.35}$$

and

$$\mathrm{se}(\widehat{\lambda}) = \sqrt{\frac{\lambda}{n[\lambda\psi'(\lambda) - 1]}}. \tag{13.36}$$

The above ML estimates are recommended for point estimates of the Gamma parameters, whenever a complete sample of Gamma observations is available. Approximate standard errors of estimated parameter functions $g(\sigma, \lambda)$ are obtained from the error propagation formula, using the elements of the covariance matrix (13.14):

$$\mathrm{Var}(g) \doteq \left(\frac{\partial g}{\partial \sigma}\right)^2 V_{\sigma\sigma} + \left(\frac{\partial g}{\partial \lambda}\right)^2 V_{\lambda\lambda} + 2\left(\frac{\partial g}{\partial \sigma}\right)\left(\frac{\partial g}{\partial \lambda}\right) V_{\sigma\lambda}; \tag{13.37}$$

refer to Section 3.3. See Example 13.1 where a percentile function is estimated.

For a statistical test-of-fit the *Anderson–Darling* statistic is recommended:

$$A = -n - \frac{1}{n}\sum_{i=1}^{n}(2i - 1)[\ln(w_i) + \ln(1 - w_{n-i+1})]. \tag{13.38}$$

Here w_i is the Gamma cdf (13.4), evaluated at the order statistic $x_{(i)}$ and the MLEs $\widehat{\sigma}$ and $\widehat{\lambda}$.

Critical values for A are given in Table 13.1 for a range of values of λ. These critical values apply to all sample sizes n. Example 13.1 demonstrates the estimation process.

13.10 Moment Estimates

Sometimes a sample of observations on a variable X is unavailable but instead the engineer has some prior information on its average $\bar{x}$ and its coefficient of variation cv. Design loads on a structure are an example. A model of the variable can be specified in accordance with this information by calculating *moment* estimates of the parameters. Using expression (13.9), a moment estimate of λ is

$$\tilde{\lambda} = \frac{1}{cv^2}. \tag{13.39}$$

Table 13.1[a]. *Critical values of the Anderson–Darling statistic A for the two-parameter Gamma model. Interpolate values if necessary.*

	Significance Level α			
λ	0.100	0.050	0.025	0.010
1	0.657	0.786	0.917	1.092
2	0.643	0.768	0.894	1.062
3	0.639	0.762	0.886	1.052
4	0.637	0.759	0.883	1.048
5	0.635	0.758	0.881	1.045
6	0.635	0.757	0.880	1.043
8	0.634	0.755	0.878	1.041
10	0.633	0.754	0.877	1.040
12	0.633	0.754	0.876	1.039
15	0.632	0.754	0.876	1.038

[a]Extracted from R. B. D' Agostino, M. A. Stephens, *Goodness-of-Fit Techniques*, Marcel Dekker Inc., New York, 1986, by courtesy of the publisher.

The mean value $\bar{x}$ is then used to calculate a moment estimate $\widetilde{\sigma}$ from (13.7). Because the efficiency of moment estimates is relatively low, and ML estimates are easily computed, moment estimates are not recommended when a sample of Gamma observations is on hand. See Example 13.2, where moment estimates provide starting values for the maximum likelihood computation.

13.11 Censored Samples

The likelihood function of a multiply censored sample, defined in Section 13.8, from a two-parameter Gamma process is

$$L(\sigma, \lambda) = \sigma^{-r\lambda}\Gamma^{-r}(\lambda)g^{r(\lambda-1)}\exp\left\{-\frac{r}{\sigma}\bar{x}\right\}\prod_{j=1}^{m}[1 - F(c_j; \sigma, \lambda)], \tag{13.40}$$

where $\bar{x} = \frac{1}{r}\sum_i x_i$ is the mean of the failure data and $g = \prod_i x_i^{1/r}$ is the *geometric* mean of these data. Working with the log of L, the maximum likelihood equations can be expressed in terms of reduced censoring data $z_j = c_j/\sigma$ as

$$\frac{r\bar{x}}{\widehat{\sigma}} - r\widehat{\lambda} + \sum_{j=1}^{m}\frac{z_j^{\widehat{\lambda}}\exp\{-z_j\}}{\Gamma(\widehat{\lambda}) - G(z_j, \widehat{\lambda})} = 0 \tag{13.41}$$

and

$$r\ln(g/\widehat{\sigma}) - n\psi(\widehat{\lambda}) + \sum_{j=1}^{m}\frac{\Gamma(\widehat{\lambda})\psi(\widehat{\lambda}) - J(z_j, \widehat{\lambda})}{\Gamma(\widehat{\lambda}) - G(z_j, \widehat{\lambda})} = 0, \tag{13.42}$$

where

$$G(x, a) = \int_0^x t^{a-1} \exp\{-t\}\, dt \tag{13.43}$$

is the incomplete Gamma function, and $J(x, a)$ is defined as

$$J(x, a) = \int_0^x t^{a-1} \ln(t) \exp\{-t\}\, dt. \tag{13.44}$$

The solution values $\widehat{\lambda}$ and $\widehat{\sigma}$ are readily computed with an equation solver; see Example 13.2.

In the general case of multiply censored samples it is difficult to calculate the *expected* information matrix for the ML estimates. Instead, the *local* information matrix can be computed from the observed values of the log-likelihood function (see Section 3.7). For the two-parameter Gamma model that matrix simplifies to

$$\begin{bmatrix} I_{\sigma\sigma} = -\frac{r}{\sigma^2}\left(\frac{\bar{x}}{\sigma}(\lambda - 1) - \lambda^2\right) - \frac{s_1}{\sigma^2} & I_{\sigma\lambda} = \frac{1}{\sigma}(r - s_2 + s_3) \\ I_{\sigma\lambda} = \frac{1}{\sigma}(r - s_2 + s_3) & I_{\lambda\lambda} = n\psi'(\lambda) - s_4 + s_5 \end{bmatrix}, \tag{13.45}$$

where

$$s_1 = \sum_{j=1}^{m} T_j(z_j - T_j), \qquad s_2 = \sum_{j=1}^{m} T_j \ln(z_j),$$

$$s_3 = \sum_{j=1}^{m} T_j \frac{\Gamma(\lambda)\psi(\lambda) - J(z_j, \lambda)}{\Gamma(\lambda) - G(z_j, \lambda)},$$

$$s_4 = \sum_{j=1}^{m} \frac{\Gamma(\lambda)[\psi^2(\lambda) + \psi'(\lambda)] - J_2(z_j)}{\Gamma(\lambda) - G(z_j, \lambda)},$$

$$s_5 = \sum_{j=1}^{m} \left(\frac{\Gamma(\lambda)\psi(\lambda) - J(z_j, \lambda)}{\Gamma(\lambda) - G(z_j, \lambda)}\right)^2$$

and

$$T_j = \frac{z_j^{\lambda} \exp\{-z_j\}}{\Gamma(\lambda) - G(z_j, \lambda)}, \qquad J_2(x) = \int_0^x t^{\lambda-1} \ln^2(t) \exp\{-t\}\, dt.$$

Although these formulas look somewhat intimidating, they are easily evaluated numerically at the ML estimates $\widehat{\sigma}, \widehat{\lambda}$, using a modern computational tool (e.g., Mathcad). The inverse of matrix I gives the covariance matrix of the parameter estimates $\widehat{\sigma}, \widehat{\lambda}$, from which approximate standard errors of the estimates are obtained. See Example 13.2 for an illustration of these computations.

Statistical tests-of-fit are not available for multiply censored Gamma samples. Probability plotting is recommended as a reliable check on the distributional postulate and the model fit.

INTERVAL ESTIMATES: TWO-PARAMETER MODEL

13.12 Normal Approximation

For large samples, the asymptotic sampling distributions of ML estimators are convenient to construct approximate confidence intervals on parameters and parameter functions. Recall from Section 3.7 that the sampling pdf of the ML estimate $\widehat{\theta}$ is asymptotically Normal, with mean value θ and variance MVB_θ:

$$f_{\mathrm{N}}(\widehat{\theta}; \theta, \sqrt{MVB_\theta}), \tag{13.46}$$

where θ stands for σ or λ. Thus, the $(1-\alpha)$-level confidence interval on θ is obtained as

$$(l_1, l_2) = \theta \pm z_{\frac{\alpha}{2}} \sqrt{MVB_\theta}, \tag{13.47}$$

such that

$$Pr(l_1 \le \theta \le l_2) = 1 - \alpha. \tag{13.48}$$

See Examples 13.3 and 13.4 for illustrations.

For parameter functions $g(\theta)$, the invariance property of ML estimators is used (see Section 3.7). That is, the sampling pdf of $\widehat{g} = g(\widehat{\theta})$ is asymptotically Normal, with mean value g and variance approximated by the error propagation formula (13.37).

13.13 Likelihood Ratio Approximation

For small to moderate-sized samples, likelihood ratio methods usually give more accurate results than the Normal method, particularly if the sample is censored. Recall from Section 3.9 that the statistic

$$LR(\theta) = 2\ln[L(\widehat{\theta})] - 2\ln[L(\theta)] \tag{13.49}$$

is approximately Chi-squared with $\nu = 1$ degree of freedom. Thus, a $(1-\alpha)$-level confidence interval on θ comprises those values θ for which $LR(\theta) \le \chi^2_{1,1-\alpha}$. For a two-parameter model, one parameter in (13.49) needs to be expressed in terms of the other parameter by its ML equation. Thus, a confidence interval on λ is obtained from

$$LR(\lambda) = 2\ln[L(\widehat{\sigma}, \widehat{\lambda})] - 2\ln[L(\sigma\{\lambda\}, \lambda)], \tag{13.50}$$

where $\sigma\{\lambda\}$ is defined as $\sigma\{\lambda\} = \bar{x}/\lambda$ for complete samples and by ML equation (13.41) for censored samples. Similarly, a confidence interval on σ is obtained from

$$LR(\sigma) = 2\ln[L(\widehat{\sigma}, \widehat{\lambda})] - 2\ln[L(\sigma, \lambda\{\sigma\})], \tag{13.51}$$

where $\lambda\{\sigma\}$ is defined by ML equation (13.33) for complete samples and by

(13.42) for censored samples. See Examples 13.3 and 13.4 for illustrations of these computations.

POINT ESTIMATES: THREE-PARAMETER MODEL

13.14 Complete Samples

For a complete sample of n independent Gamma observations, the likelihood function is

$$L(\mu,\sigma,\lambda)=\Gamma^{-n}(\lambda)\sigma^{-n\lambda}\prod_{i=1}^{n}(x_i-\mu)^{\lambda-1}\exp\left\{-\frac{1}{\sigma}\sum_{i=1}^{n}(x_i-\mu)\right\}. \tag{13.52}$$

The maximum likelihood equations (see Section 3.6) are then

$$\widehat{\sigma}=n\left[(\widehat{\lambda}-1)\sum_{i=1}^{n}(x_i-\widehat{\mu})^{-1}\right]^{-1}, \tag{13.53}$$

$$\widehat{\sigma}=\frac{1}{n\widehat{\lambda}}\sum_{i=1}^{n}(x_i-\widehat{\mu}), \tag{13.54}$$

and

$$\psi(\widehat{\lambda})+\ln(\widehat{\sigma})=\frac{1}{n}\sum_{i=1}^{n}\ln(x_i-\widehat{\mu}). \tag{13.55}$$

Equating the first two relations, solving for $\widehat{\lambda}$, and substituting into (13.55) gives a single equation in $\widehat{\mu}$ only:

$$\psi\left(\left\{1-\frac{1}{(\bar{x}-\widehat{\mu})s_1(\widehat{\mu})}\right\}^{-1}\right)+\ln\left(\bar{x}-\widehat{\mu}-\frac{1}{s_1(\widehat{\mu})}\right)-s_2(\widehat{\mu})=0, \tag{13.56}$$

where $s_1(\widehat{\mu})=\frac{1}{n}\sum_i(x_i-\widehat{\mu})^{-1}$ and $s_2(\widehat{\mu})=\frac{1}{n}\sum_i\ln(x_i-\widehat{\mu})$.

When a solution $0\le\widehat{\mu}\le x_{(1)}$ exists, it is easily computed with an equation solver, although it is prudent to first plot the function (13.56) to locate a good starting value for μ. The estimate $\widehat{\lambda}$ is then obtained from

$$\widehat{\lambda}=\left(1-\frac{1}{(\bar{x}-\widehat{\mu})s_1(\widehat{\mu})}\right)^{-1}, \tag{13.57}$$

and $\widehat{\sigma}$ is given by

$$\widehat{\sigma}=\frac{1}{\widehat{\lambda}}(\bar{x}-\widehat{\mu}). \tag{13.58}$$

Approximate standard errors of the estimates are then found from the inverse of information matrix (13.20) or (13.21). A statistical test-of-fit for the three-parameter Gamma model is not available. Probability plotting is recommended as a check of the model fit to the data. See Example 13.5.

For three-parameter Gamma models, convenient small-sample methods for *confidence intervals* are not available. The asymptotic formula (13.47) can be used for these models. However, caution is advised, since the sampling distributions of estimators deviate from Normality for small to moderate-sized samples.

When the threshold value μ is not of primary interest, one can construct inferences on σ and λ by conditioning on the estimated value $\hat{\mu}$: First transform the data to $y_i = x_i - \hat{\mu}$, and then apply two-parameter methods. Often such conditional inferences are insensitive to $\hat{\mu}$, since the likelihood function does not hold much information on μ. See Example 13.6 for an illustration.

APPLICATIONS

13.15 General

The Gamma distribution serves well as a general measurement model of engineering variables because of its shape flexibility and its positive sample space. That is, if nothing is known about the physics of the measurement phenomenon, the Gamma distribution is a good choice to simply *fit the data*. Since modern computational tools have removed the burdensome numerical difficulties that were associated with this model, the Gamma distribution is likely to rise in prominence.

A general motivation for *postulating* this model comes from its reproductive property (see Section 13.3). Suppose that a measured variable X can be considered as the *sum* of a *small* number of underlying causes of unknown distributions. Because the Gamma model is a highly flexible one, it may easily serve as an assumed model for the causes. The Gamma model is then the natural choice for the measured variable X (see Example 13.2). Note that, as the number of these causes increases, λ increases accordingly. The Gamma model then approaches the Normal distribution, and the above consideration shifts to a central limit theorem (see Section 2.4).

When the logarithm of the measurement X is treated as a Gamma variable, the resulting distribution of X is *log-Gamma*, sometimes termed the *log-Pearson type III distribution*. Both the two- and three-parameter versions of this distribution have found uses as a model for hydrological phenomena.

There are two particular application areas where the Gamma model has achieved prominence. One is for *waiting time* (see the following section), and the other is as a *life model* (see Sections 13.17 to 13.20).

13.16 Waiting Time

Consider a *counting* process N, modeled by a Poisson distribution with unit rate $\nu = 1/\sigma$ (see Chapter 8). That is, $N \sim p(n; t/\sigma)$. The time T to the occurrence of the $(n-1)$th Poisson event is termed a *waiting time*. Its distribution is shown in Section 8.4 to be Erlang with parameters σ and n. This result also follows from the intuitively obvious identity that Pr(the number of events in period t is $\leq n-1) = Pr$(the waiting time to the nth event is $\geq t$).

That is,

$$F_{\text{Poisson}}\left(n-1; \frac{t}{\sigma}\right) = 1 - F_{\text{Erlang}}(t; \sigma, n). \tag{13.59}$$

Hence, the Erlang waiting-time model is an equivalent description of a Poisson counting process. This application area was first studied by the Danish engineer Erlang in the context of telephone networks.

When $\lambda = 1$, T represents the waiting time to the *first* Poisson event, and its model is the Exponential distribution. Since this model is *memoryless* (see Section 12.4), T also models the waiting time between *any* two adjoining Poisson events. This waiting time is termed the *inter-arrival time* (IAT). Thus, the Exponential distribution is an alternative equivalent description of a Poisson process (see also Section 8.4). Which of these three equivalent descriptions is chosen to model the process is a matter of convenience and depends on the context of the problem.

An immediate application of (13.59) is to the time-to-failure X of a *standby redundant system* of k Exponentially distributed, critical components. In such a system, one component carries the system function at any time; when it fails, a standby component takes over the system function, and so on. Thus, the system life X is the sum of component lives. The variable X is therefore Erlang with shape parameter $\lambda = k$ and the same scale parameter σ as the components.

LIFE TESTING

13.17 Hazard Function

One of the major applications of statistical distributions in engineering is in the modeling of the time-to-failure (TTF) of operating equipment. Here the decision function of central interest is the reliability of the equipment at a specified operating time, or equivalently, the operating time at which the equipment is characterized by a specified reliability value. Chapter 12 details some of the basic concepts of reliability analysis.

In particular, Section 12.16 introduces the *hazard function*

$$h(x) = \frac{f(x)}{1 - F(x)}.$$

For the Exponential distribution this function is constant, which implies that the probability of failure is independent of equipment age. For other than electronic devices this implication is difficult to accept since most equipment is subject to

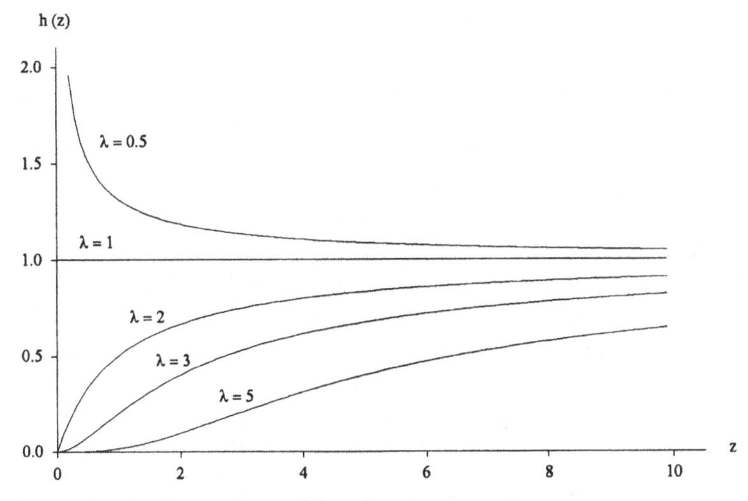

Figure 13.3. Gamma hazard functions for $\sigma = 1$.

wear during use, resulting in an age-dependent changing failure hazard. Thus there is a need to generalize the Exponential life model. The Gamma distribution is one such generalized life model, which includes the Exponential distribution as a special case; see also Sections 17.22 to 17.28 on the Weibull distribution. Figure 13.3 shows the Gamma hazard function for $\sigma = 1$ and several values of λ, indicating that the Gamma distribution can model a TTF variable for increasing-, decreasing-, and constant-hazard regimes. It can be shown that this function approaches the limit

$$\lim_{x \to \infty} h(x) = \frac{1}{\sigma}. \tag{13.60}$$

Hence, the initial rate of change in failure hazard eventually decreases with time, while the hazard function approaches the constant Exponential value.

13.18 Reliability

The reliability value is defined for specified operating conditions as

$$R \mid x = 1 - F(x; \sigma, \lambda). \tag{13.61}$$

This quantity can be estimated by maximum likelihood as

$$\widehat{R} \mid x = 1 - F(x; \widehat{\sigma}, \widehat{\lambda}), \tag{13.62}$$

with standard error $\sqrt{\mathrm{Var}(\widehat{R} \mid x)}$ computed from the error propagation formula (13.37). See Example 13.7 for an illustration.

It is often required in reliability engineering to construct a *lower* $(1-\alpha)$-level confidence limit $L_{1-\alpha} \mid x$ on the reliability value, rather than to obtain its point estimate and standard error. Asymptotically, that confidence limit is given by

$$L_{1-\alpha} \mid x = \widehat{R} + z_\alpha \sqrt{\mathrm{Var}(\widehat{R} \mid x)}, \tag{13.63}$$

where z_α is the standard Normal quantile of order α. This estimate must be used with caution unless the sample size is large.

An alternative, possibly superior, approach is to use the likelihood ratio method (see Section 13.13). For a given value x the reliability (13.61) can be numerically inverted to define the parameter λ in terms of R and σ as $\lambda\{R, \sigma\}$. Substituting $\lambda\{R, \sigma\}$ into the likelihood function (13.31), differentiating with respect to σ, and equating to zero yields a constraint relation in $\sigma\{R\}$. A lower confidence limit on R is then obtained from

$$LR(R) = 2\ln[L(\widehat{\sigma}, \widehat{\lambda})] - 2\ln[L(\sigma\{R\}, \lambda\{R, \sigma\{R\}\})]. \tag{13.64}$$

For censored samples the likelihood function (13.40) is used instead of (13.31). A computational solution is straightforward, although care must be exercised in choosing appropriate starting values for the several equation solvers. See Example 13.7 for an illustration.

13.19 Expected Number of Failures

It is sometimes required to estimate the *number* of failures for a group of operating equipments (see also Section 17.26). If n units are in operation, with current operating time t_i for the ith unit, then the *current* expected number of failures is

$$\mathrm{E}\{N\}_{\mathrm{cur}} = \sum_{i=1}^{n} F(t_i; \sigma, \lambda). \tag{13.65}$$

When it is required to estimate the number of failures (without replacement) up to a given *future* time T, that estimate is

$$\mathrm{E}\{N\}_{\mathrm{fut}} = \sum_{i=1}^{n} \frac{F(t_i + u_i; \sigma, \lambda) - F(t_i; \sigma, \lambda)}{1 - F(t_i; \sigma, \lambda)}, \tag{13.66}$$

where u_i is the anticipated usage time of the ith unit during T. The above expressions are easily evaluated with a computational aid such as Mathcad.

13.20 Age Replacement

It is often of economic advantage to replace an aged unit before it fails (see also Section 17.27). If the failure-replacement cost C_f is larger than the planned age-replacement cost C_r, the decision function to minimize is the average cost per unit operating time for the equipment, at the replacement time t:

$$Ca(t) = \frac{C_f F(t; \sigma, \lambda) + C_r[1 - F(t; \sigma, \lambda)]}{\int_0^t [1 - F(x; \sigma, \lambda)]\, dx}. \tag{13.67}$$

Minimizing (13.67) with respect to t gives the solution equation for the optimal replacement time t^*:

$$\frac{f(t^*;\sigma,\lambda)}{1-F(t^*;\sigma,\lambda)}\int_0^{t^*}[1-F(x;\sigma,\lambda)]\,dx+1-F(t^*;\sigma,\lambda)-\frac{C_f}{C_f-C_r}=0. \tag{13.68}$$

When units are replaced only upon failure (no planned replacements), the average cost per time unit for the equipment unit is

$$C=\frac{C_f}{\sigma\lambda}. \tag{13.69}$$

The difference between (13.69) and (13.68) gives an estimate of the average savings per time unit of planned replacement for an equipment unit. See Example 13.8 for an illustration of the computations.

EXAMPLE 13.1

The following annual maximum discharge values (in 1,000 cubic feet per second) were recorded on the S. Mary's river at Stillwater, Nova Scotia, from 1950 to 1974:

16.0, 11.6, 19.9, 18.6, 18.0, 13.1, 29.1, 10.3, 12.2, 15.6, 12.7, 13.1,
19.2, 19.5, 23.0, 6.7, 7.1, 14.3, 20.6, 25.6, 8.2, 34.4, 16.1, 10.2, 12.3.

Assuming a two-parameter Gamma model for this discharge, estimate the parameters and their standard errors, and verify the estimated model graphically. Test the model fit at the 5% level of significance. Estimate the 80-percentile of the 10-year maximum discharge.

$n := 25$ $\quad i := 1..n \quad$ digamma function: $\psi(\lambda) := \frac{d}{d\lambda}\ln(\Gamma(\lambda))$

$y_i :=$

16
11.6
19.9
18.6
18
13.1
29.1
10.3
12.2
15.6
12.7
13.1
19.2
19.5
23
6.7

1. Estimation

$x := \mathrm{sort}(y) \qquad xb := \frac{1}{n}\cdot\sum_i x_i \qquad xb = 16.296$

$g := \prod_i (x_i)^{\left(\frac{1}{n}\right)} \qquad g = 15.042 \qquad L := \ln\left(\frac{xb}{g}\right)$

$t := 5 \qquad \lambda := \mathrm{root}(\ln(t)-\psi(t)-L,\,t) \qquad \lambda = 6.40$

$\sigma := \frac{xb}{\lambda} \qquad \sigma = 2.54$

2. Standard errors

trigamma function: $\psi'(\lambda) := \frac{d}{d\lambda}\psi(\lambda)$

7.1
14.3
20.6
25.6
8.2
34.4
16.1
10.2
12.3

Let

$$C := n \cdot (\lambda \cdot \psi'(\lambda) - 1)$$

$$V := \frac{1}{C} \cdot \begin{pmatrix} \sigma^2 \cdot \psi'(\lambda) & -\sigma \\ -\sigma & \lambda \end{pmatrix}$$

$$V = \begin{pmatrix} 0.533 & -1.239 \\ -1.239 & 3.119 \end{pmatrix} \qquad \begin{aligned} SE\sigma &:= \sqrt{V_{1,1}} \\ SE\lambda &:= \sqrt{V_{2,2}} \end{aligned} \qquad \begin{aligned} SE\sigma &= 0.73 \\ SE\lambda &= 1.77 \end{aligned}$$

3. Linearized model/data plot

Gamma cdf: $F(x) := \int_0^{\frac{x}{\sigma}} \exp((\lambda - 1) \cdot \ln(t) - t - \ln(\Gamma(\lambda)))\, dt$

Choose the plotting range to cover the sample:

$\min(x) = 6.70 \qquad a := 5 \qquad L := F(a)$

$\max(x) = 34.40 \qquad b := 40 \qquad U := F(b)$

$r_i := L + \frac{(U - L)}{(b - a)} \cdot (x_i - a)$: linearized model ordinates,

$\Delta_i := F(x_i) - r_i$: linearizing adjustment for the data plot,

$p_i := \frac{i - 0.3}{n + 0.4} - \Delta_i$: ordinates for data plot

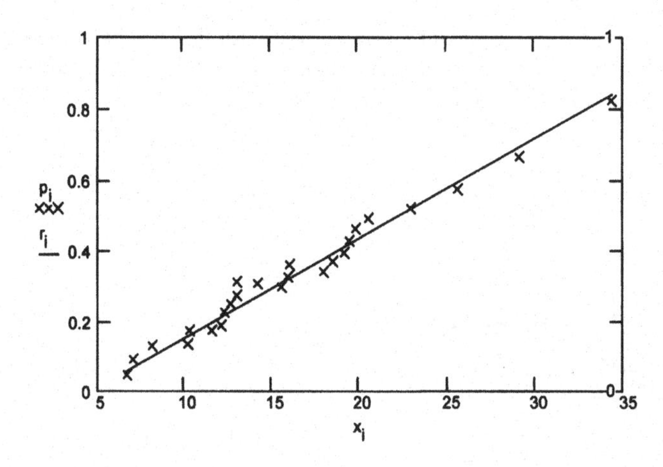

The model fit looks reasonable.

4. Test-of-fit

$$A := -n - \frac{1}{n} \cdot \sum_i (2 \cdot i - 1) \cdot \left[\ln(F(x_i)) + \ln\left[1 - F\left[x_{(n-i+1)}\right]\right]\right] \qquad A = 0.16$$

From Table 13.1 the 5% critical value is 0.756. There is therefore no reason to reject the estimated model.

5. pdf plot

$$j := 1..200 \quad t_j := 0.2 \cdot j \quad f(x) := \frac{1}{\sigma \cdot \Gamma(\lambda)} \cdot \left(\frac{x}{\sigma}\right)^{\lambda-1} \cdot \exp\left(-\frac{x}{\sigma}\right)$$

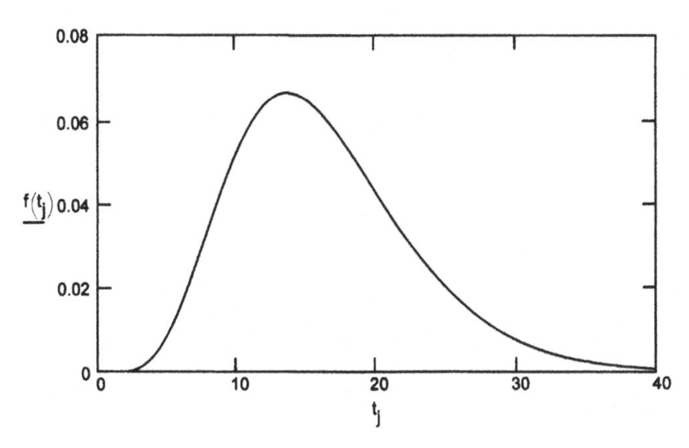

6. Percentile of 10-year maximum flow

$$F(x, \sigma, \lambda) := \int_0^{\frac{x}{\sigma}} \exp((\lambda - 1) \cdot \ln(t) - t - \ln(\Gamma(\lambda)))\, dt$$

$x := 32 \qquad T(\sigma, \lambda) := \text{root}(F(x, \sigma, \lambda)^{10} - 0.80, x) \qquad T(\sigma, \lambda) = 31.7$

standard error:

$j := 1..2$

$k := 1..2 \qquad d_1 := \frac{d}{d\sigma} T(\sigma, \lambda) \qquad d_2 := \frac{d}{d\lambda} T(\sigma, \lambda)$

$\text{Var}T := \sum_j \sum_k d_j \cdot d_k \cdot V_{j,k} \qquad \text{se}T := \sqrt{\text{var T}} \qquad \text{se}T = 3.5$

EXAMPLE 13.2

An accelerated life test on the corrosion resistance of a multilayer coating was performed. Since the corrosive medium penetrates one layer after another, the life of the coating can be considered the sum of coating lives. Hence a Gamma life model was asssumed. Three of the following data (*) were censored when the test apparatus malfunctioned:

161, 334, 378, 485*, 514, 646, 665, 718*, 867*, 947, 1081, 1284, 1391, 1789.

Estimate the parameters and verify the model graphically.

$n := 14 \qquad r := 11 \qquad m := n - r$

$k := 1..n \qquad i := 1..r \qquad j := 1..m$

$t_k :=$ $x_i :=$ $c_j :=$

t_k	x_i	c_j
161	161	485
334	334	718
378	378	867
485	514	
514	646	
646	665	
665	947	
718	1081	
867	1284	
947	1391	
1081	1789	
1284		
1391		
1789		

$xb := \frac{1}{r} \cdot \sum_i x_i \quad xb = 835.455$

$g := \prod_i (x_i)^{\frac{1}{r}} \quad g = 681.305$

approximate (moment) estimates:

$sd := \sqrt{\frac{1}{r} \cdot \sum_i (x_i - xb)^2} \qquad cv := \frac{sd}{xb}$

$am := 3 \qquad a := \mathrm{root}\left(\frac{1}{\sqrt{am}} - cv, am\right) \qquad s := \frac{xb}{a}$

Choose starting values: $s = 280 \quad a = 2.98$

1. Maximum likelihood estimation

$$G(x, a) := \int_0^x t^{a-1} \cdot \exp(-t)\, dt$$

$$J(x, a) := \int_0^x t^{a-1} \cdot \ln(t) \cdot \exp(-t)\, dt \qquad \psi(a) := \frac{d}{da} \ln(\Gamma(a))$$

GIVEN

$$\frac{r \cdot xb}{s} - r \cdot a + \sum_j \frac{\left(\frac{c_j}{s}\right)^a \cdot \exp\left(-\frac{c_j}{s}\right)}{\Gamma(a) - G\left(\frac{c_j}{s}, a\right)} = 0$$

$$r \cdot \ln\left(\frac{g}{s}\right) - n \cdot \psi(a) + \sum_j \frac{\psi(a) \cdot \Gamma(a) - J\left(\frac{c_j}{s}, a\right)}{\Gamma(a) - G\left(\frac{c_j}{s}, a\right)} = 0$$

$$\begin{pmatrix} \sigma \\ \lambda \end{pmatrix} := \mathrm{FIND}(s, a) \qquad \sigma = 332 \quad \lambda = 2.78$$

2. Standard errors

$$z_j := \frac{c_j}{\sigma} \qquad T_j := \frac{(z_j)^\lambda \cdot \exp(-z_j)}{\Gamma(\lambda) - G(z_j, \lambda)} \qquad \psi'(a) := \frac{d}{da}\psi(a)$$

$$J2(x) := \int_0^x t^{\lambda-1} \cdot \ln(t)^2 \cdot \exp(-t)\, dt \qquad s1 := \sum_j T_j \cdot (z_j - T_j)$$

$$s2 := \sum_j T_j \cdot \ln(z_j) \qquad s3 := \sum_j T_j \cdot \frac{\Gamma(\lambda) \cdot \psi(\lambda) - J(z_j, \lambda)}{\Gamma(\lambda) - G(z_j, \lambda)}$$

$$s4 := \sum_j \frac{\Gamma(\lambda) \cdot (\psi(\lambda)^2 + \psi'(\lambda)) - J2(z_j)}{\Gamma(\lambda) - G(z_j, \lambda)}$$

$$s5 := \sum_j \left(\frac{\Gamma(\lambda) \cdot \psi(\lambda) - J(z_j, \lambda)}{\Gamma(\lambda) - G(z_j, \lambda)}\right)^2$$

$$I := \begin{bmatrix} -\frac{r}{\sigma^2}\cdot\left[\frac{xb}{\sigma}\cdot(\lambda-1)-\lambda^2\right]-\frac{s1}{\sigma^2} & \frac{1}{\sigma}\cdot(r-s2+s3) \\ \frac{1}{\sigma}\cdot(r-s2+s3) & n\cdot\psi'(\lambda)-s4+s5 \end{bmatrix}$$

$$I = \begin{pmatrix} 2.944\cdot10^{-4} & 0.038 \\ 0.038 & 5.661 \end{pmatrix}$$

$V := I^{-1}$ $\qquad$ $SE\sigma := \sqrt{V_{0,0}}$ $\qquad$ $SE\sigma = 149$

$$V = \begin{pmatrix} 2.222\cdot10^{4} & -147.48 \\ -147.48 & 1.155 \end{pmatrix}$$ $\qquad$ $SE\lambda := \sqrt{V_{1,1}}$ $\qquad$ $SE\lambda = 1.07$

3. Linearized model/data plot

Choose $a := 100$ and $b := 1800$

cdf: $F(x) := \frac{1}{\Gamma(\lambda)}\cdot G\left(\frac{x}{\sigma},\lambda\right)$

$L := F(a) \qquad U := F(b)$

$r_i := L + \frac{U-L}{b-a}\cdot(x_i-a)$: linearized model ordinates,

$\Delta_i := F(x_i) - r_i$: linearizing adjustment for the data plot

$q_i :=$ (q_i is the order of failures in the total sample.)

1
2
3
5
6
7
10
11
12
13
14

ordinates for data plot:

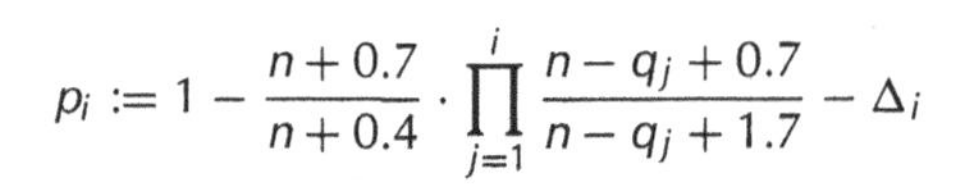

$$p_i := 1 - \frac{n+0.7}{n+0.4}\cdot\prod_{j=1}^{i}\frac{n-q_j+0.7}{n-q_j+1.7} - \Delta_i$$

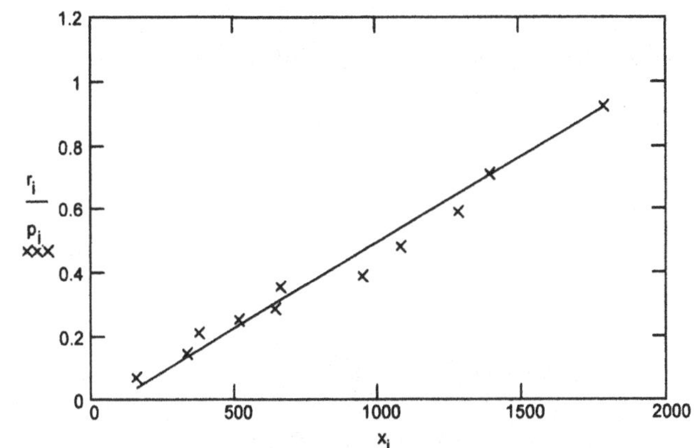

The plot indicates a reasonable model fit.

4. pdf plot

$j := 1..200 \qquad t_j := 15 \cdot j \qquad f(x) := \dfrac{1}{\sigma \cdot \Gamma(\lambda)} \cdot \left(\dfrac{x}{\sigma}\right)^{\lambda-1} \cdot \exp\left(-\dfrac{x}{\sigma}\right)$

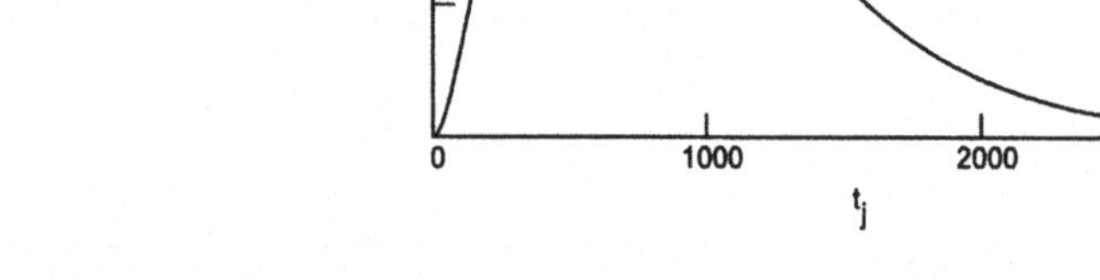

EXAMPLE 13.3

For the data of Example 13.1, calculate 90% confidence intervals on the parameters and on the 5-percentile.

From Example 13.1:

$$\begin{matrix} & \lambda := 6.40 & \sigma := 2.54 & xb := 16.296 \\ n := 25 & SE\lambda := 1.77 & SE\sigma := 0.73 & g := 15.042 \end{matrix} \qquad V := \begin{pmatrix} 0.533 & -1.239 \\ -1.239 & 3.119 \end{pmatrix}$$

1. Parameters

i) Normal approximation: Standard Normal 95-percentile is $z95 := 1.645$

$L\lambda := \lambda - z95 \cdot SE\lambda \qquad L\lambda = 3.49$

$U\lambda := \lambda + z95 \cdot SE\lambda \qquad U\lambda = 9.31$

$L\sigma := \sigma - z95 \cdot SE\sigma \qquad L\sigma = 1.34$

$U\sigma := \sigma + z95 \cdot SE\sigma \qquad U\sigma = 3.74$

ii) Likelihood ratio method: Chi-squared 90-percentile at $v = 1$ is $K := 2.71$

$$\psi(\lambda) := \frac{d}{d\lambda} \ln(\Gamma(\lambda))$$

log of likelihood function:

$$LL(\sigma, \lambda) := -n \cdot \ln(\Gamma(\lambda)) - n \cdot \lambda \cdot \ln(\sigma) + n \cdot (\lambda - 1) \cdot \ln(g) - n \cdot \frac{xb}{\sigma}$$

shape parameter:

$$LR(a) := 2 \cdot LL(\sigma, \lambda) - 2 \cdot LL\left(\frac{xb}{a}, a\right)$$

$a := \lambda - 2 \qquad \lambda l := \text{root}(LR(a) - K, a) \qquad \lambda l = 3.93$

$a := \lambda + 2 \qquad \lambda u := \text{root}(LR(b) - K, a) \qquad \lambda u = 9.78$

scale parameter:

$a := \lambda$ $\quad l(s) := \text{root}(\ln(s) + \psi(a) - \ln(g), a)$

$LR(s) := 2 \cdot LL(\sigma, \lambda) - 2 \cdot LL(s, l(s))$

$b := \sigma - 1$ $\quad \sigma l := \text{root}(LR(b) - K, b)$ $\quad \sigma l = 1.64$

$b := \sigma + 1$ $\quad \sigma u := \text{root}(LR(b) - K, b)$ $\quad \sigma u = 4.25$

2. Percentile

Normal approximation:

$q := 0.05$ $\quad F(x, \sigma, \lambda) := \dfrac{1}{\Gamma(\lambda)} \cdot \displaystyle\int_0^{\frac{x}{\sigma}} t^{\lambda-1} \cdot \exp(-t)\, dt$

$x := 10$ $\quad x05(\sigma, \lambda) := \text{root}(F(x, \sigma, \lambda) - q, x)$ $\quad x05(\sigma, \lambda) = 7.31$

standard error: $j := 1..2$ $\quad k := 1..2$

$d_1 := \dfrac{d}{d\sigma} x05(\sigma, \lambda)$ $\quad d_2 := \dfrac{d}{d\lambda} x05(\sigma, \lambda)$

$\text{var}x := \displaystyle\sum_j \sum_k d_j \cdot d_k \cdot V_{j,k}$ $\quad \text{SE}x := \sqrt{\text{var}x}$ $\quad \text{SE}x = 1.14$

$Lx := x05(\sigma, \lambda) - z95 \cdot \text{SE}x$ $\quad Lx = 5.43$

$Ux := x05(\sigma, \lambda) + z95 \cdot \text{SE}x$ $\quad Ux = 9.19$

EXAMPLE 13.4

For the data of Example 13.2, calculate 90% confidence intervals on the parameters.

From Example 13.2:

$\lambda := 2.78$ $\quad \sigma := 332$ $\quad xb := 835.46$

$\text{SE}\lambda := 1.07$ $\quad \text{SE}\sigma := 149$ $\quad g := 681.31$

i) Normal approximation: Standard Normal 96-percentile is $z95 := 1.645$

$L\lambda := \lambda - z95 \cdot \text{SE}\lambda$ $\quad L\lambda = 1.02$

$U\lambda := \lambda + z95 \cdot \text{SE}\lambda$ $\quad U\lambda = 4.54$

$L\sigma := \sigma - z95 \cdot \text{SE}\sigma$ $\quad L\sigma = 87$

$U\sigma := \sigma + z95 \cdot \text{SE}\sigma$ $\quad U\sigma = 577$

ii) Likelihood ratio method: Chi-squared 90-percentile at $v = 1$ is $K := 2.71$

$n := 14$ $\quad r := 11$ $\quad m := n - r$ $\quad j := 1..m$

$c_j :=$

485
718
867

$G(x, a, b) := \displaystyle\int_0^{\frac{x}{b}} t^{a-1} \cdot \exp(-t)\, dt$ $\quad F(x, a, b) := \dfrac{G(x, a, b)}{\Gamma(a)}$

log of likelihood function:

$$LL(\sigma, \lambda) := r \cdot \lambda \cdot \ln(\sigma) - r \cdot \ln(\Gamma(\lambda)) + r \cdot (\lambda - 1) \cdot \ln(g) - \frac{r \cdot xb}{\sigma} + \sum_j \ln(1 - F(c_j, \lambda, \sigma))$$

shape parameter:

$$b := \sigma \qquad s(a) := \text{root}\left[\frac{r \cdot xb}{b} - r \cdot a + \sum_j \frac{\left(\frac{c_j}{b}\right)^a \cdot \exp\left(\frac{c_j}{b}\right)}{\Gamma(a) - G(c_j, a, b)}, b\right]$$

$$LR(a) := 2 \cdot LL(\sigma, \lambda) - 2 \cdot LL(s(a), a)$$

$a := \lambda - 1 \qquad \lambda l := \text{root}(LR(a) - K, a) \qquad \lambda l = 1.38$

$a := \lambda + 2 \qquad \lambda u := \text{root}(LR(b) - K, a) \qquad \lambda u = 4.97$

scale parameter:

$$\psi(\lambda) := \frac{d}{d\lambda} \ln(\Gamma(\lambda)) \qquad J(x, a, b) := \int_0^{\frac{x}{b}} t^{a-1} \cdot \ln(t) \cdot \exp(-t)\, dt$$

$$a := \lambda \quad l(b) := \text{root}\left(r \cdot \ln\left(\frac{g}{b}\right) - n \cdot \psi(a) + \sum_j \frac{\Gamma(a) \cdot \psi(a) - J(c_j, a, b)}{\Gamma(a) - G(c_j, a, b)}, a\right)$$

$$LR(s) := 2 \cdot LL(\sigma, \lambda) - 2 \cdot LL(s, l(s))$$

$b := \sigma - 150 \qquad \sigma l := \text{root}(LR(b) - K, b) \qquad \sigma l = 173$

$b := \sigma + 150 \qquad \sigma u := \text{root}(LR(b) - K, b) \qquad \sigma u = 782$

EXAMPLE 13.5

Generate 25 observations from a Gamma distribution with $\mu = 40$, $\sigma = 10$, and $\lambda = 1.65$. Estimate the parameters and their standard errors.

$n := 25 \qquad i := 1..n \qquad u_i := rnd(1) \qquad m := 40 \qquad s := 10 \qquad l := 1.65$

$$F(x) := \frac{1}{\Gamma(l)} \cdot \int_0^{\frac{x-m}{s}} z^{l-1} \cdot \exp(-z)\, dz \qquad x := 50 \qquad x_i := \text{root}(F(x) - u_i, x)$$

$x_i :=$

56.8
55.2
51.4
56.7
50
47.9
72

1. Sample check

$xb := \frac{1}{n} \cdot \sum_i x_i \qquad xb = 58.8$ (should be close to $\mu + \sigma \cdot \lambda = 56.5$)

$S2 := \frac{1}{n} \cdot \sum_i (x_i - xb)^2 \qquad S2 = 133.7$ (should be close to $\sigma^2 \cdot \lambda = 165$)

78.1
68.4
52.2
67.4
63
45.1
50.3
47.6
47.8
52.9
66.6
58.5
84.9
63.1
64
81.9
44.9
43.3

2. Estimation

digamma function: $\psi(a) := \frac{d}{da}\ln(\Gamma(a))$

$$s1(m) := \frac{1}{n}\cdot\sum_i (x_i - m)^{-1} \qquad s2(m) := \frac{1}{n}\cdot\sum_i \ln(x_i - m)$$

$$(13.56):\quad g(m) := \psi\left[\left[1 - \frac{1}{(xb - m)\cdot s1(m)}\right]^{-1}\right] + \ln\left(xb - m - \frac{1}{s1(m)}\right) - s2(m)$$

Plot this function: $j := 1..20 \qquad t_j := 38 + 0.25\cdot j$

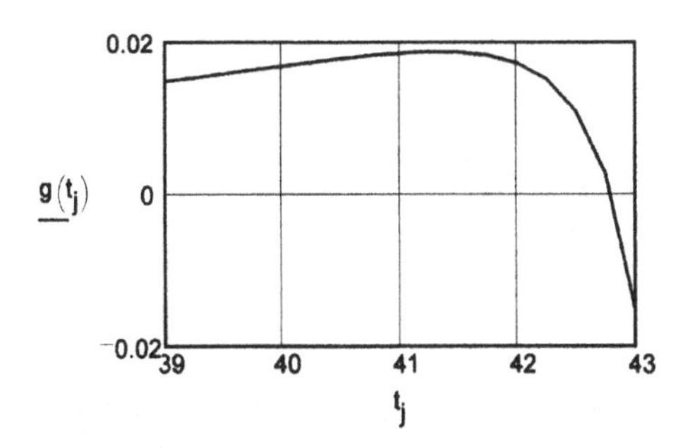

$$a := 42.5 \qquad \mu := \mathrm{root}(g(a), a) \qquad \mu = 42.8$$

$$\lambda := \left[1 - \frac{1}{(xb - \mu)\cdot s1(\mu)}\right]^{-1} \qquad \lambda = 1.47$$

$$\sigma := \frac{1}{\lambda}\cdot(xb - \mu) \qquad \sigma = 10.9$$

3. Standard errors

$$s := \frac{1}{n}\cdot\sum_i (x_i - \mu)^{-2} \qquad I := n\cdot\begin{bmatrix} s\cdot(\lambda - 1) & \frac{1}{\sigma^2} & \frac{1}{\sigma\cdot(\lambda-1)} \\ \frac{1}{\sigma^2} & \frac{\lambda}{\sigma^2} & \frac{1}{\sigma} \\ \frac{1}{\sigma\cdot(\lambda-1)} & \frac{1}{\sigma} & \frac{d}{d\lambda}\psi(\lambda) \end{bmatrix} \qquad V := I^{-1}$$

$$V = \begin{pmatrix} 1.537 & 4.327 & -0.728 \\ 4.327 & 23.189 & -3.109 \\ -0.728 & -3.109 & 0.489 \end{pmatrix}$$

$$SE\mu := \sqrt{V_{1,1}} \qquad SE\mu = 1.2$$

$$SE\sigma := \sqrt{V_{2,2}} \qquad SE\sigma = 4.8$$

$$SE\lambda := \sqrt{V_{3,3}} \qquad SE\lambda = 0.7$$

4. Linearized model/data plot

$$z := \mathrm{sort}(x - \mu) \qquad F(z) := \frac{1}{\Gamma(\lambda)}\cdot\int_0^{\frac{z}{\sigma}} t^{\lambda-1}\cdot\exp(-t)\,dt$$

Choose the plotting range to cover the sample:

$\min(z) = 0.5 \qquad a := .3 \qquad L := F(a)$

$\max(z) = 42.1 \qquad b := 45 \qquad U := F(b)$

$r_i := L + \dfrac{U - L}{b - a} \cdot (z_i - a)$: linearized model ordinates,

$\Delta_i := F(z_i) - r_i$: linearizing adjustment for the data,

$p_i := \dfrac{i - 0.3}{n + 0.4} - \Delta_i$: ordinates for the data plot

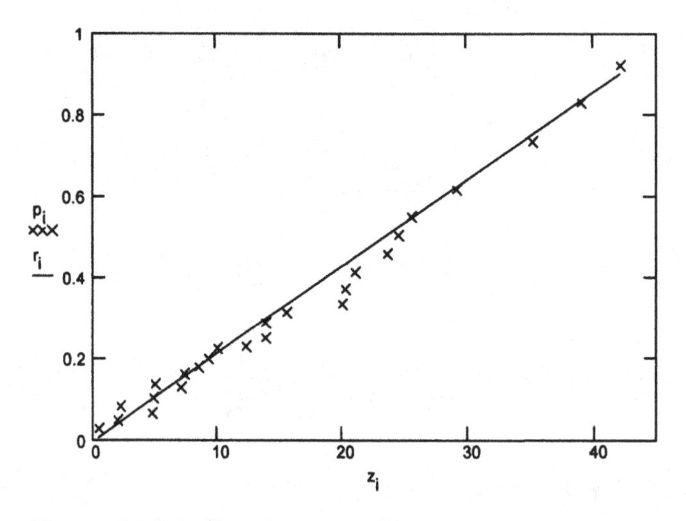

The model fit looks quite reasonable.

EXAMPLE 13.6

For the data generated in Example 13.5, estimate the 20-percentile and its 90% confidence interval.

From Example 13.5:

$$\mu := 42.8 \qquad \sigma := 10.9 \qquad \lambda := 1.47 \qquad V := \begin{pmatrix} 1.537 & 4.327 & -0.728 \\ 4.327 & 23.189 & -3.109 \\ -0.728 & -3.109 & 0.489 \end{pmatrix}$$

$$\text{cdf:} \quad F(x, \mu, \sigma, \lambda) := \frac{1}{\Gamma(\lambda)} \cdot \int_0^{\frac{x-\mu}{\sigma}} t^{\lambda-1} \cdot \exp(-t)\, dt$$

true 20-percentile:

$x := 50 \qquad x\text{true} := \text{root}(F(x, 40, 10, 1.65) - 0.2, x) \qquad x\text{true} = 46.0$

1. Percentile estimate

$x(\mu, \sigma, \lambda) := \text{root}(F(x, \mu, \sigma, \lambda) - 0.2, x) \qquad x(\mu, \sigma, \lambda) = 48.1$

2. Confidence interval

$i := 1..3 \qquad j := 1..3$

$d_1 := 1 \qquad d_2 := \frac{d}{d\sigma} x(\mu, \sigma, \lambda) \qquad d_3 := \frac{d}{d\lambda} x(\mu, \sigma, \lambda)$

$\mathrm{var}x := \sum_i \sum_j d_i \cdot d_j \cdot V_{i,j} \qquad SEx := \sqrt{\mathrm{var}x} \qquad SEx = 1.72$

a. Normal approximation:
Standard Normal 95-percentile: $z95 := 1.645$

$Lx := x(\mu, \sigma, \lambda) - z95 \cdot SEx \qquad Lx = 45.3$

$Ux := x(\mu, \sigma, \lambda) + z95 \cdot SEx \qquad Ux = 50.9$

b. Conditional inference, given $\mu = 42.8$:
Transforming the data of Example 13.5 to $xi - \mu$
and applying the methods of Example 13.1 gives $\lambda := 1.48 \qquad \sigma := 10.81$

Hence the percentile estimate is

$x := 45 \qquad xx := \mathrm{root}(F(x, \mu, \sigma, \lambda) - 0.2, x) \qquad xx = 48.1$

Apply the methods of Example 13.3 to the transformed data: $SEx := 1.43$

Hence the (Normal) interval estimate is

$Lx := xx - z95 \cdot SEx \qquad Lx = 45.7$

$Ux := xx + z95 \cdot SEx \qquad Ux = 50.5$

EXAMPLE 13.7

For the data of Example 13.2, estimate the coating reliability at 250 accelerated time units. Calculate a 90% lower confidence limit on the reliability.

From Example 13.2:

$r := 11 \qquad m := 3 \qquad xb := 835.455 \qquad V := \begin{pmatrix} 22220 & -147.48 \\ -147.48 & 1.155 \end{pmatrix}$

$\sigma := 332 \qquad \lambda := 2.78 \qquad g := 681.305$

$j := 1..3$

$c_j :=$

485
718
867

1. Reliability estimate

$xo := 250$

$R(x, \sigma, \lambda) := 1 - \frac{1}{\Gamma(\lambda)} \cdot \int_0^{\frac{x}{\sigma}} t^{\lambda - 1} \cdot \exp(-t)\, dt \qquad R(xo, \sigma, \lambda) = 0.942$

2. Standard error

$i := 1..2 \qquad k := 1..2 \qquad d_1 := \frac{d}{d\sigma} R(xo, \sigma, \lambda) \qquad d_2 := \frac{d}{d\lambda} R(xo, \sigma, \lambda)$

$\mathrm{var}R := \sum_i \sum_k d_i \cdot d_k \cdot V_{i,k} \qquad \mathrm{se}R := \sqrt{\mathrm{var}R} \qquad \mathrm{se}R := 0.047$

3. Lower confidence limit

i) Normal approximation: Standard Normal 90-percentile is $z90 := 1.278$

$Rl := R(xo, \sigma, \lambda) - z90 \cdot \mathrm{se}R \qquad Rl = 0.882$

ii) likelihood ratio method: Chi-squared 80-percentile at $v = 1$ is $K := 1.642$

log of likelihood function:

$$LL(\sigma, \lambda) := -r \cdot \lambda \cdot \ln(\sigma) - r \cdot \ln(\Gamma(\lambda)) + r \cdot (\lambda - 1) \cdot \ln(g) - \frac{r \cdot xb}{\sigma} + \sum_j \ln(R(c_j, \sigma, \lambda))$$

$a := 2.7 \qquad s := \sigma \qquad l(Ro, s) := \mathrm{root}(R(xo, s, a) - Ro, a)$

$dLL(Ro, b) := \frac{d}{db} LL(b, l(Ro, b)) \qquad ss(Ro) := \mathrm{root}(dLL(Ro, s), s)$

$LR(Ro) := 2 \cdot LL(\sigma, \lambda) - 2 \cdot LL(ss(Ro), l(Ro, ss(Ro)))$

$Ro := .86 \qquad Rl := \mathrm{root}(LR(Ro) - K, Ro) \qquad Rl = 0.856$

EXAMPLE 13.8

For the coating-life data of Example 13.2, estimate the cost-optimal (accelerated) replacement time, if the cost parameters are $Cf = \$20{,}000$ and $Cr = \$5{,}000$.

From Example 13.2:

$\sigma := 332 \qquad \lambda := 2.78 \qquad V := \begin{pmatrix} 22220 & -147.48 \\ -147.48 & 1.155 \end{pmatrix}$

$Cf := 20000 \qquad Cr := 5000$

pdf: $f(x, \sigma, \lambda) := \frac{1}{\sigma \cdot \Gamma(\lambda)} \cdot \left(\frac{x}{\sigma}\right)^{\lambda - 1} \cdot \exp\left(-\frac{x}{\sigma}\right)$

cdf: $F(x, \sigma, \lambda) := \frac{1}{\Gamma(\lambda)} \cdot \int_0^{\frac{x}{\sigma}} t^{t-1} \cdot \exp(-t)\, dt$

1. Optimum replacement time

$$G(x,\sigma,\lambda) := \frac{f(x,\sigma,\lambda)}{1-F(x,\sigma,\lambda)} \cdot \int_0^x (1-F(t,\sigma,\lambda))\,dt + 1 - F(x,\sigma,\lambda) - \frac{Cf}{Cf-Cr}$$

$x := 600 \qquad T(\sigma,\lambda) := \mathrm{root}(G(x,\sigma,\lambda), x) \qquad T(\sigma,\lambda) = 576$

2. Standard error

$i := 1..2 \qquad j := 1..2$

$d_1 := \frac{d}{d\sigma} T(\sigma,\lambda) \qquad d_2 := \frac{d}{d\lambda} T(\sigma,\lambda)$

$\mathrm{var}T := \sum_i \sum_j d_i \cdot d_j \cdot V_{i,j} \qquad \mathrm{se}T := \sqrt{\mathrm{var}T} \qquad \mathrm{se}T = 170$

3. Age-replacement cost/unit time

$$Ca(t) := \frac{Cf \cdot F(t,\sigma,\lambda) + Cr \cdot (1-F(t,\sigma,\lambda))}{\int_0^t (1-F(x,\sigma,\lambda))\,dx} \qquad Ca(T(\sigma,\lambda)) = 18$$

4. Failure-replacement cost/unit time

$$C := \frac{Cf}{\sigma \cdot \lambda} \qquad C = 22$$

5. Cost function

$k := 1..40 \qquad t_k := 25 \cdot k$

$$\mathrm{cost}_k := \frac{Cf \cdot F(t_k,\sigma,\lambda) + Cr \cdot (1-F(t_k,\sigma,\lambda))}{\int_0^{t_k} (1-F(x,\sigma,\lambda))\,dx}$$

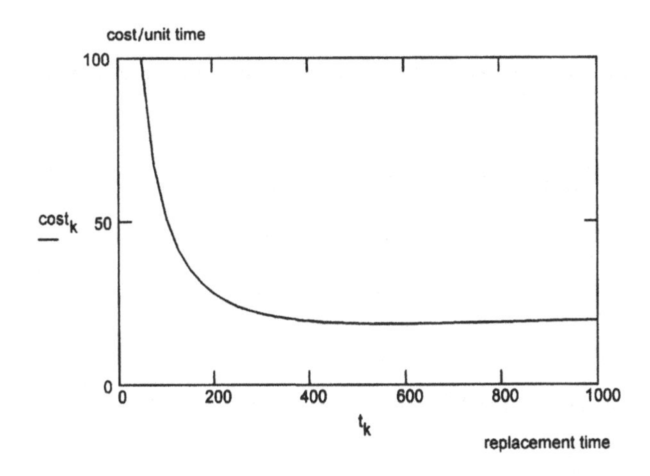

CHAPTER FOURTEEN

Beta Distributions

INTRODUCTION

14.1 Definition

A continuous random variable X has a *Beta distribution* if its pdf has the form

$$f(x; \lambda_1, \lambda_2) = \frac{\Gamma(\lambda_1 + \lambda_2)}{\Gamma(\lambda_1)\Gamma(\lambda_2)} x^{\lambda_1 - 1}(1-x)^{\lambda_2 - 1}; \quad 0 < x < 1; \quad \lambda_1, \lambda_2 > 0; \tag{14.1}$$

where Γ represents the *gamma function* (see Section 13.1). This model does not have location-scale structure (see Section 9.2). Hence, both λ_1 and λ_2 are *shape* parameters, which are symmetrically related by:

$$f(x; \lambda_1, \lambda_2) = f(1-x; \lambda_2, \lambda_1). \tag{14.2}$$

This distribution arose as the theoretical model of various statistics and statistical functions. It is now an important statistical model of random variables whose values are restricted to the unit range.

PROPERTIES: TWO-PARAMETER MODEL

14.2 Beta Variable

The cdf of a Beta variable, as defined by (14.1), cannot be expressed in closed form:

$$F(x; \lambda_1, \lambda_2) = \frac{\Gamma(\lambda_1 + \lambda_2)}{\Gamma(\lambda_1)\Gamma(\lambda_2)} \int_0^x z^{\lambda_1 - 1}(1-z)^{\lambda_2 - 1}\, dz. \tag{14.3}$$

The inverse of the ratio of gamma functions in the above expression is called the *beta function*:

$$B(\lambda_1, \lambda_2) = \frac{\Gamma(\lambda_1)\Gamma(\lambda_2)}{\Gamma(\lambda_1 + \lambda_2)}. \tag{14.4}$$

The integral in (14.3) is termed the *incomplete beta function*:

$$B_x(\lambda_1, \lambda_2) = \int_0^x z^{\lambda_1 - 1}(1-z)^{\lambda_2 - 1}\, dz. \tag{14.5}$$

Note that $B_x \to B$ as $x \to 1$. The cdf can thus be expressed as the *incomplete beta function ratio*

$$F(x; \lambda_1, \lambda_2) = \frac{B_x(\lambda_1, \lambda_2)}{B(\lambda_1, \lambda_2)}. \tag{14.6}$$

From (14.2) it follows that

$$F(x; \lambda_1, \lambda_2) = 1 - F(1 - x; \lambda_2, \lambda_1).$$

Tables[1] are available for beta functions. However, these functions are readily computed numerically. Note that Mathcad 6+ features a built-in function that returns the Beta cdf for positive shape parameters. When λ_1 and λ_2 are both integer valued, the Beta cdf can be evaluated as a Binomial sum by using the identity

$$F(x; \lambda_1, \lambda_2) = 1 - \sum_{i=0}^{\lambda_1 - 1} \binom{\lambda_1 + \lambda_2 - 1}{i} x^i (1 - x)^{\lambda_1 + \lambda_2 - 1 - i}. \tag{14.7}$$

14.3 Properties

The rth moment of X about the origin is given by

$$\mu_r'(X) = \frac{\Gamma(\lambda_1 + \lambda_2)\Gamma(\lambda_1 + r)}{\Gamma(\lambda_1)\Gamma(\lambda_1 + \lambda_2 + r)}. \tag{14.8}$$

Thus, the expected value of X is

$$\mu_1'(X) = \frac{\lambda_1}{\lambda_1 + \lambda_2}. \tag{14.9}$$

The variance of X is

$$\mu_2(X) = \frac{\lambda_1 \lambda_2}{(\lambda_1 + \lambda_2)^2(\lambda_1 + \lambda_2 + 1)}. \tag{14.10}$$

The coefficient of variation of X is thus

$$cv(X) = \sqrt{\frac{\lambda_2}{\lambda_1(\lambda_1 + \lambda_2 + 1)}}. \tag{14.11}$$

For $\lambda_1 + \lambda_2 > 2$ and $\lambda_1 \geq 1$, the mode value is

$$x_{\mathrm{m}} = \frac{\lambda_1 - 1}{\lambda_1 + \lambda_2 - 2}. \tag{14.12}$$

[1] Pearson, E. S., Johnson, N. L., *Tables of the Incomplete Beta Function*, Cambridge University Press, Cambridge, UK, 1968.

The quantile of order q is defined by

$$F(x_q; \lambda_1, \lambda_2) = q \tag{14.13}$$

and is easily computed with an equation solver. Note that Mathcad 6+ has a built-in function that gives the Beta quantile for positive shape parameters.

The first shape factor (see Section 1.10) is

$$\gamma_1 = \frac{2(\lambda_2 - \lambda_1)}{\lambda_1 + \lambda_2 + 2}\sqrt{\frac{\lambda_1 + \lambda_2 + 1}{\lambda_1 \lambda_2}}. \tag{14.14}$$

If $\lambda_1 = \lambda_2$, then $\gamma_1 = 0$, and the pdf is symmetrical. If $\lambda_2 > \lambda_1$ then $\gamma_1 > 0$, and the pdf is skewed to the right. Similarly, $\lambda_2 < \lambda_1$ gives $\gamma_1 < 0$ for left skew. The second shape factor is

$$\gamma_2 = \frac{3(\lambda_1 + \lambda_2 + 1)[2(\lambda_1 + \lambda_2)^2 + \lambda_1\lambda_2(\lambda_1 + \lambda_2 - 6)]}{\lambda_1\lambda_2(\lambda_1 + \lambda_2 + 2)(\lambda_1 + \lambda_2 + 3)}. \tag{14.15}$$

Because both shape factors are symmetrical functions of λ_1 and λ_2, interchanging the parameters in a pdf yields its mirror image. For the symmetrical case ($\lambda_1 = \lambda_2$), $\gamma_2 \to 3$ as λ becomes large, and the Beta pdf approaches the Normal model.

Since the Beta distribution features two shape parameters, there is high shape flexibility. Figures 14.1 to 14.5 indicate this flexibility for various combinations of parameter values. The resulting shapes include pdfs that are symmetrical: $\lambda_1 = \lambda_2$; skewed: $\lambda_1 \neq \lambda_2$; U-shaped: $\lambda_1, \lambda_2 < 1$; and J-shaped: $(\lambda_1 - 1) \cdot(\lambda_2 - 1) < 0$.

The matrix of minimum-variance-bounds (see Section 3.2) for estimators of λ_1 and λ_2 is

$$\begin{bmatrix} V_{\lambda_1\lambda_1} & V_{\lambda_1\lambda_2} \\ V_{\lambda_1\lambda_2} & V_{\lambda_2\lambda_2} \end{bmatrix} = \frac{1}{n\{ab - c(a+b)\}} \begin{bmatrix} b - c & c \\ c & a - c \end{bmatrix}, \tag{14.16}$$

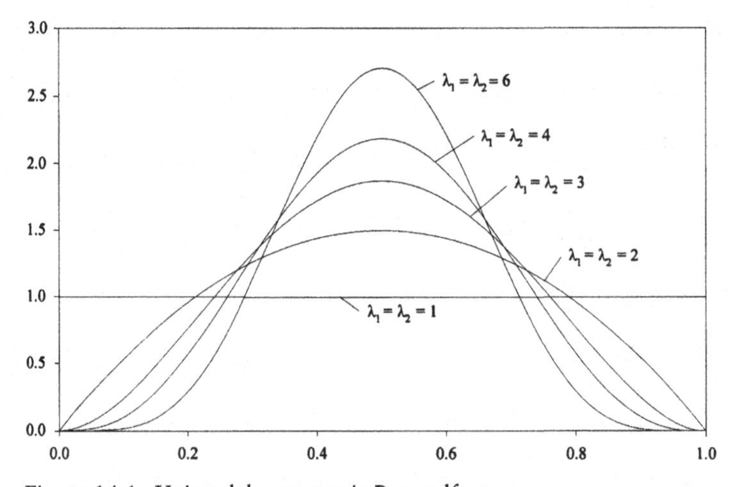

Figure 14.1. Unimodal, symmetric Beta pdfs.

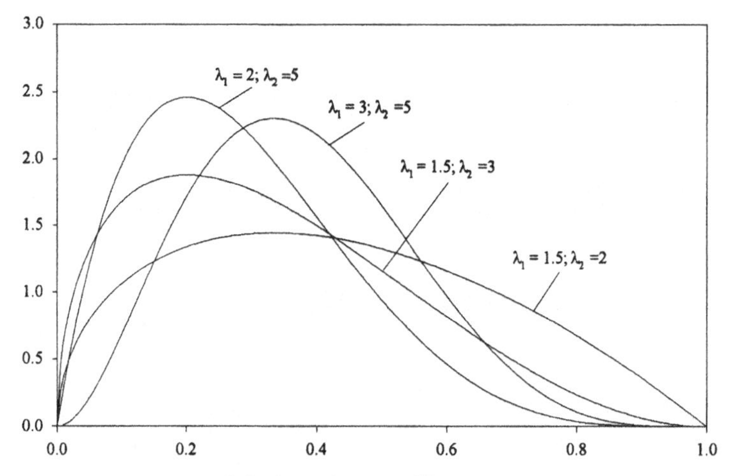

Figure 14.2. Unimodal, skewed Beta pdfs.

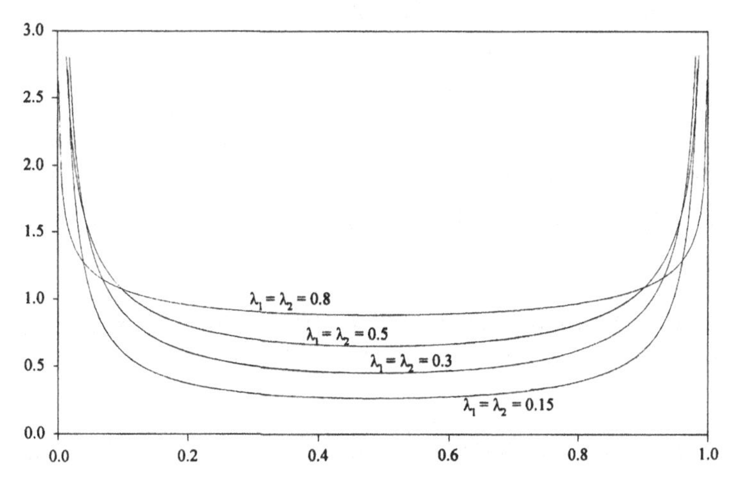

Figure 14.3. U-shaped, symmetrical Beta pdfs.

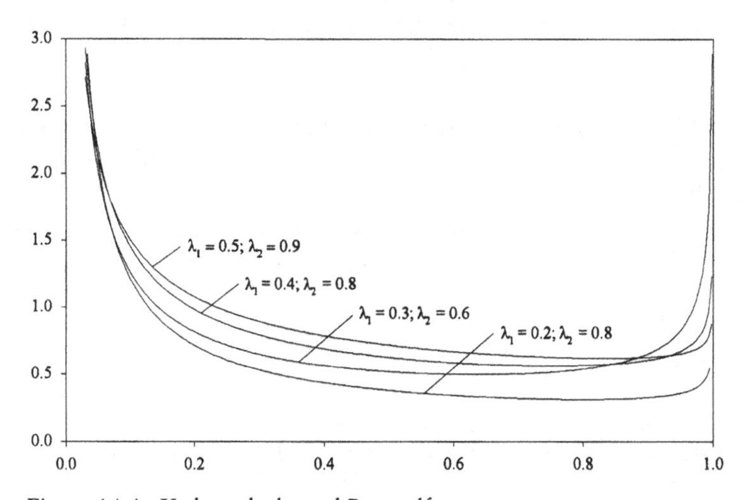

Figure 14.4. U-shaped, skewed Beta pdfs.

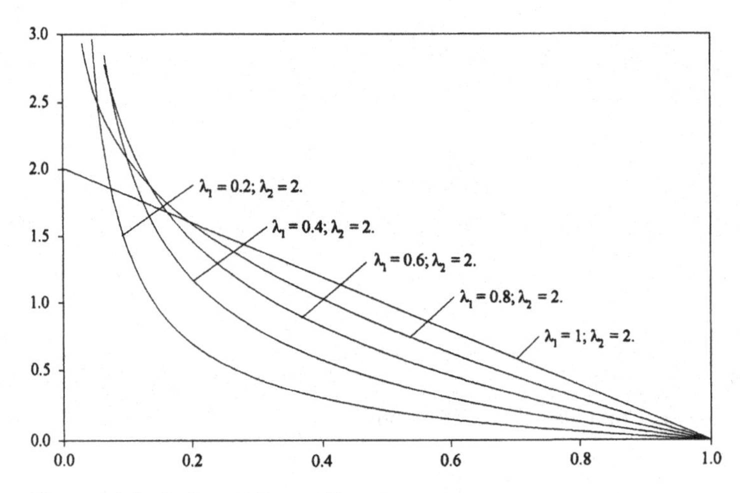

Figure 14.5. J-shaped Beta pdfs.

where $a = \psi'(\lambda_1), b = \psi'(\lambda_2)$, and $c = \psi'(\lambda_1 + \lambda_2)$. See Section 13.2 for the definition of the *trigamma* function $\psi'(\lambda)$.

14.4 Simulation

Random observations x_i from a Beta process with known parameters λ_1, λ_2 can be simulated by a simple *rejection technique*[2] for unimodal pdfs with $\lambda_1, \lambda_2 > 1$:

1. Calculate the mode x_m from (14.12) and its density value $M = f(x_m; \lambda_1, \lambda_2)$.
2. Generate two random numbers u_1, u_2 on the interval (0, 1).
3. If $u_2 \leq f(u_1; \lambda_1, \lambda_2)/M$, then accept u_1 as a random observation from f; otherwise reject u_1 and repeat the process.

This scheme is based on the recognition that the probability of u_2 being less than or equal to $f(x)/M$ is equal to $f(x)/M$. Hence, the pdf of accepted observations x will be $f(x)$.

For $\lambda_1 \leq 1$ and $\lambda_2 \leq 1$, the following scheme can be used.[3] Generate two Uniform random numbers u and v on the interval $(0, 1)$ until the condition

$$u^{1/\lambda_1} + v^{1/\lambda_2} \leq 1$$

[2] More efficient but more complicated rejection schemes are given in "Beta Variate Generation via Exponential Majorizing Functions," Schmeiser, B. and Babu, A. J. G., *Operations Research*, Vol. 28, pp. 917–926, 1980.

[3] Jöhnk, M. D., "Erzeugung von Betaverteilten und Gammaverteilten Zufallszahlen." *Metrika*, Vol. 8, pp. 5–15, 1964.

is satisfied. Then

$$x = \frac{u^{1/\lambda_1} + v^{1/\lambda_2}}{u^{1/\lambda_1}}$$

is a random observation from the Beta distribution $F(x; \lambda_1, \lambda_2)$. This scheme can also be used when $\lambda_1 > 1$ and/or $\lambda_2 > 1$, but it is inefficient for $(\lambda_1 + \lambda_2)$ large.

For small simulated samples it is perhaps more convenient to invert the given cdf directly: A simulated observation x_i from the Beta cdf (14.3) is simply the u_i-quantile defined by $F(x_i; \lambda_1, \lambda_2) = u_i$. In any case, it is advisable to check the adequacy of a simulated sample by comparing at least its first two moments with those of the given model. See Example 14.1 for an illustration. Note that Mathcad 6+ has a built-in function that generates random Beta observations for all values $\lambda_1 > 0$ and $\lambda_2 > 0$.

PROPERTIES: FOUR-PARAMETER MODEL

14.5 Definition and Properties

The two-parameter beta pdf (14.1) can model engineering variables on the unit range (0, 1); a *proportion* is an important example. More generally, the finite sample space is arbitrary.

For example, the cost of engineering projects of a given type, scope, and complexity is a random variable that clearly exhibits a minimum bound different from zero. As well, an upper bound necessarily exists, usually dictated by the available budget, and is different from 1.

The Beta pdf (14.1) is generalized to accommodate different finite sample spaces by introducing two location parameters μ_1 and μ_2, with $\mu_1 < \mu_2$, to give

$$f(x; \mu_1, \mu_2, \lambda_1, \lambda_2) = \frac{\Gamma(\lambda_1 + \lambda_2)}{\Gamma(\lambda_1)\Gamma(\lambda_2)} \left(\frac{x - \mu_1}{\mu_2 - \mu_1}\right)^{\lambda_1 - 1} \left(1 - \frac{x - \mu_1}{\mu_2 - \mu_1}\right)^{\lambda_2 - 1} \frac{1}{\mu_2 - \mu_1}, \quad (14.17)$$

for $\mu_1 \leq x \leq \mu_2$. We see that $\sigma = \mu_2 - \mu_1$ is a scale parameter. Thus, two-parameter location measures are relocated by μ_1 and rescaled by σ. The expected value is

$$\mu_1'(X) = \mu_1 + (\mu_2 - \mu_1)\frac{\lambda_1}{\lambda_1 + \lambda_2}, \quad (14.18)$$

and for $\lambda_1 + \lambda_2 > 2$, $\lambda_1 \geq 1$ the mode is

$$x_m = \mu_1 + (\mu_2 - \mu_1)\frac{\lambda_1 - 1}{\lambda_1 + \lambda_2 - 2}. \quad (14.19)$$

Two-parameter dispersion measures are rescaled by σ, so that the variance is

$$\mu_2(X) = \frac{(\mu_2 - \mu_1)^2 \lambda_1 \lambda_2}{(\lambda_1 + \lambda_2)^2 (\lambda_1 + \lambda_2 + 1)} \tag{14.20}$$

and the coefficient of variation becomes

$$cv(X) = \frac{(\mu_2 - \mu_1)\sqrt{\lambda_1 \lambda_2}}{\sqrt{\lambda_1 + \lambda_2 + 1}(\mu_1 \lambda_2 + \mu_2 \lambda_1)}. \tag{14.21}$$

Provided that $\lambda_1, \lambda_2 > 2$, the *expected* information matrix (see Section 3.1) is

$$\begin{bmatrix} I_{\mu_1\mu_1} = \frac{n\lambda_2(\lambda_1+\lambda_2-1)}{(\lambda_1-2)\sigma^2} & I_{\mu_1\mu_2} = \frac{n(\lambda_1+\lambda_2-1)}{\sigma^2} & I_{\mu_1\lambda_1} = \frac{n\lambda_2}{(\lambda_1-1)\sigma} & I_{\mu_1\lambda_2} = \frac{-n}{\sigma} \\ I_{\mu_1\mu_2} & I_{\mu_2\mu_2} = \frac{n\lambda_1(\lambda_1+\lambda_2-1)}{(\lambda_2-2)\sigma^2} & I_{\mu_2\lambda_1} = \frac{n}{\sigma} & I_{\mu_2\lambda_2} = \frac{-n\lambda_1}{(\lambda_2-1)\sigma} \\ I_{\mu_1\lambda_1} & I_{\mu_2\lambda_1} & I_{\lambda_1\lambda_1} = n[\psi'(\lambda_1) - \psi'(\lambda_1+\lambda_2)] & I_{\lambda_1\lambda_2} = -n\psi'(\lambda_1+\lambda_2) \\ I_{\mu_1\lambda_2} & I_{\mu_2\lambda_2} & I_{\lambda_1\lambda_2} & I_{\lambda_2\lambda_2} = n[\psi'(\lambda_2) - \psi'(\lambda_1+\lambda_2)] \end{bmatrix}, \tag{14.22}$$

where $\psi'(\lambda)$ is the *trigamma function* (see Section 13.2) and $\sigma = \mu_2 - \mu_1$. The minimum-variance-bounds are obtained from the inverse of (14.22). Alternatively, the *local* information matrix can be used (see Section 3.7), provided $\lambda_1, \lambda_2 > 1$. This matrix is identical to (14.22), except for two diagonal elements:

$$\begin{aligned} I_{\mu_1\mu_1} &= n(\lambda_1 - 1)s_1 - \frac{n(\lambda_1 + \lambda_2 - 1)}{\sigma^2} \quad \text{and} \\ I_{\mu_2\mu_2} &= n(\lambda_2 - 1)s_2 - \frac{n(\lambda_1 + \lambda_2 - 1)}{\sigma^2}, \end{aligned} \tag{14.23}$$

where

$$s_1 = \frac{1}{n}\sum_i (x_i - \mu_1)^{-2} \quad \text{and} \quad s_2 = \frac{1}{n}\sum_i (\mu_2 - x_i)^{-2}.$$

SPECIAL CASE: UNIFORM DISTRIBUTIONS

14.6 Definition and Properties

When the shape parameters of the Beta pdf (14.1) take the value $\lambda_1 = \lambda_2 = 1$, the *Uniform* or *Rectangular* distribution on (0, 1) results, meaning that all possible values x are equally likely:

$$f(x) = 1; \quad 0 \le x \le 1, \tag{14.24}$$

with cdf

$$F(x) = x. \tag{14.25}$$

The expected value of X is

$$\mu_1'(X) = \frac{1}{2}, \tag{14.26}$$

and the variance is

$$\mu_2(X) = \frac{1}{12}. \tag{14.27}$$

The quantile of order q is

$$x_q = q. \tag{14.28}$$

This is the distribution of "random numbers." The above properties can be used as a first check on the adequacy of a simulated random number set or of a random number generator itself.

The pdf (14.24) generalizes to cover an arbitrary interval (μ_1, μ_2):

$$f(x; \mu_1, \mu_2) = \frac{1}{\mu_2 - \mu_1}; \quad \mu_1 \leq x \leq \mu_2, \quad 0 \leq \mu_1 < \mu_2. \tag{14.29}$$

The cdf is

$$F(x; \mu_1, \mu_2) = \frac{x - \mu_1}{\mu_2 - \mu_1}. \tag{14.30}$$

The expected value of X is

$$\mu_1'(X) = \frac{\mu_1 + \mu_2}{2}, \tag{14.31}$$

and the variance is

$$\mu_2(X) = \frac{(\mu_2 - \mu_1)^2}{12}. \tag{14.32}$$

The quantile of order q is

$$x_q = \mu_1 + (\mu_2 - \mu_1)q. \tag{14.33}$$

Neither distribution is useful as a *measurement* model of engineering random variables, since such variables practically never exhibit equiprobable values over their sample spaces. The importance of the Uniform distribution rests on the fact that *any* cdf, considered as a random function, is itself distributed according to (14.24) (see Section 2.8). This pdf is therefore at the core of all *Monte Carlo* simulation work.

PROBABILITY PLOT

14.7 Plotting Procedure

A probability plot of ordered data (see Section 3.13) should always be used to check the distributional assumption of the Beta model. Since the Beta cdf (14.3) cannot be expressed in closed form, it is not possible to linearize it algebraically. However, once the model has been estimated, it can be linearized numerically over the sample space (0, 1), and the data plotting positions can be adjusted by the same amounts as the estimated model. The result is a linearized graphical comparison of data and estimated model that provides visual information on how well the model fits the data.

Linearizing the *estimated* Beta cdf over (0, 1) gives the straight-line model ordinate as $x_{(i)}$ from expression (13.27) for the general linearization scheme, since for $a = 0$ and $b = 1$ we have $F(a) = 0$ and $F(b) = 1$. The adjustment from the estimated cdf to its linearized version is therefore

$$\Delta_i = F\left(x_{(i)}\right) - x_{(i)}, \tag{14.34}$$

so that the median data plotting position is adjusted to

$$p_i = \frac{i - 0.3}{n + 0.4} - \Delta_i. \tag{14.35}$$

The plot of p_i versus $x_{(i)}$ will roughly follow the linear model plot, if $X_{(i)}$ came from the estimated two-parameter Beta process (see Section 3.14). See Example 14.2 for an illustration (following model estimation).

For the four-parameter model, the observations x_i are reduced to the two-parameter case by the estimated location parameters $\widehat{\mu}_1$ and $\widehat{\mu}_2$:

$$z_i = \frac{x_i - \widehat{\mu}_1}{\widehat{\mu}_2 - \widehat{\mu}_1}, \tag{14.36}$$

and z takes the place of x in the above plotting procedure.

POINT ESTIMATES: TWO-PARAMETER MODEL

14.8 Maximum Likelihood Estimates

The likelihood function of a sample of n independent observations on a Beta variable X is

$$L(\lambda_1, \lambda_2) = B^{-n}(\lambda_1, \lambda_2) \prod_{i=1}^{n} x_i^{\lambda_1 - 1} \prod_{i=1}^{n} (1 - x_i)^{\lambda_2 - 1}, \tag{14.37}$$

where $B(\lambda_1, \lambda_2)$ is the beta function (14.4). The maximum likelihood equations (see Section 3.6) are

$$\psi(\widehat{\lambda}_1) - \psi(\widehat{\lambda}_1 + \widehat{\lambda}_2) = s_3 \tag{14.38}$$

and

$$\psi(\widehat{\lambda}_2) - \psi(\widehat{\lambda}_1 + \widehat{\lambda}_2) = s_4, \tag{14.39}$$

where $s_3 = \frac{1}{n}\sum_i \ln(x_i)$, $s_4 = \frac{1}{n}\sum_i \ln(1 - x_i)$, and $\psi(\lambda)$ is the *digamma function*, defined in Section 13.2.

The solution is easily obtained with an equation solver. To locate starting values for the solution process, one can use *moment* estimates (next section) as ballpark values. A display of the log-likelihood function (or its contour plot) in the neighborhood of these values may then show the location of the maximum, which serves as the starting point for the numerical solution. The desired plotting function is

$$\begin{aligned} LLF(\lambda_1, \lambda_2) &= n\ln[\Gamma(\lambda_1 + \lambda_2)] - n\ln[\Gamma(\lambda_1)] - n\ln[\Gamma(\lambda_2)] \\ &\quad + n(\lambda_1 - 1)s_3 + n(\lambda_2 - 1)s_4 + C, \end{aligned} \tag{14.40}$$

where C is an arbitrary constant, chosen to give small positive values of the function near the maximum. See Example 14.2 for an illustration of the computations.

Approximate standard errors of these estimates are obtained from the matrix of minimum-variance-bounds (14.16). These error values are used as well to determine approximate standard errors of parameter functions $g(\lambda_1, \lambda_2)$ using the error propagation formula (see Section 3.3):

$$\mathrm{Var}(g) \doteq \left(\frac{\partial g}{\partial \lambda_1}\right)^2 V_{\lambda_1\lambda_1} + \left(\frac{\partial g}{\partial \lambda_2}\right)^2 V_{\lambda_2\lambda_2} + 2\left(\frac{\partial g}{\partial \lambda_1}\right)\left(\frac{\partial g}{\partial \lambda_2}\right) V_{\lambda_1\lambda_2}. \tag{14.41}$$

Statistical tests-of-fit are not available for Beta models. A linearized model/data plot (see Section 14.7) is recommended as a reliable graphical check on the Beta postulate.

14.9 Moment Estimates

In order to provide starting values for the ML solution process (preceding section), moment estimates may be obtained by equating the first two distribution moments, from (14.8), to corresponding data moments about the origin, and solving for the parameters

$$\widetilde{\lambda}_1 = \frac{m_1^2 - m_1 m_2}{m_2 - m_1^2} \tag{14.42}$$

and

$$\widetilde{\lambda}_2 = \frac{m_1 - m_2}{m_2 - m_1^2} - \widetilde{\lambda}_1, \tag{14.43}$$

where $m_r = \frac{1}{n}\sum_i x_i^r$. See Example 14.2 for an illustration.

INTERVAL ESTIMATES: TWO-PARAMETER MODEL

14.10 Normal Approximation

For large samples, approximate confidence intervals on the parameters λ_1 and λ_2 can be constructed from the asymptotic sampling distributions of the ML estimates (see Section 3.7). That is, the sampling pdf of $\widehat{\theta}$ is asymptotically Normal with mean θ and variance MVB_θ:

$$f_{\mathrm{N}}(\widehat{\theta}; \theta, \sqrt{MVB_\theta}), \tag{14.44}$$

where θ stands for λ_1 or λ_2. Thus, the $(1-\alpha)$-level confidence interval on θ is obtained as

$$(l_1, l_2) = \theta \pm z_{\frac{\alpha}{2}} \sqrt{MVB_\theta}, \tag{14.45}$$

such that

$$Pr(l_1 \leq \theta \leq l_2) = 1 - \alpha.$$

See Example 14.3 for an illustration.

Approximate confidence intervals for parameter functions $g(\lambda_1, \lambda_2)$ are similarly obtained, with the error propagation formula (14.41) providing the function's variance estimate. Again see Example 14.3.

14.11 Likelihood Ratio Approximation

For small to moderate-sized samples, likelihood ratio methods tend to give more accurate results than the above Normal method. Recall from Section 3.9 that the statistic

$$LR(\theta) = 2\ln[L(\widehat{\theta})] - 2\ln[L(\theta)] \tag{14.46}$$

is approximately Chi-squared distributed with $\nu = 1$ degree of freedom. Thus, a $(1-\alpha)$-level confidence interval on θ comprises those values θ for which $LR(\theta) \leq \chi^2_{1,1-\alpha}$. For a two-parameter model, one parameter in (14.46) must be expressed in the other parameter by its ML equation. Thus, a confidence interval on λ_1 is obtained from

$$LR(\lambda_1) = 2\ln[L(\widehat{\lambda}_1, \widehat{\lambda}_2)] - 2\ln[L(\lambda_1, \lambda_2\{\lambda_1\})], \tag{14.47}$$

where $\lambda_2\{\lambda_1\}$ is defined by ML equation (14.39). Similarly, a confidence interval on λ_2 is obtained from

$$LR(\lambda_2) = 2\ln[L(\widehat{\lambda}_1, \widehat{\lambda}_2)] - 2\ln[L(\lambda_1\{\lambda_2\}, \lambda_2)], \tag{14.48}$$

where $\lambda_1\{\lambda_2\}$ is defined by the ML equation (14.38). See Example 14.3 for an illustration.

POINT ESTIMATES: FOUR-PARAMETER MODEL

14.12 Maximum Likelihood Estimates

The likelihood function of a sample of n independent observations from a four-parameter Beta process is

$$L(\mu_1, \mu_2, \lambda_1, \lambda_2) = B^{-n}(\lambda_1, \lambda_2) \prod_{i=1}^{n} (x_i - \mu_1)^{\lambda_1 - 1} \prod_{i=1}^{n} (\mu_2 - x_i)^{\lambda_2 - 1} (\mu_2 - \mu_1)^{n(1-\lambda_1-\lambda_2)}, \tag{14.49}$$

where $B(\lambda_1, \lambda_2)$ is the beta function (14.4). The ML equations (see Section 3.6), are

$$\psi(\widehat{\lambda}_1) - \psi(\widehat{\lambda}_1 + \widehat{\lambda}_2) = \frac{1}{n}\sum_{i=1}^{n} \ln\left(\frac{x_i - \widehat{\mu}_1}{\widehat{\mu}_2 - \widehat{\mu}_1}\right), \tag{14.50}$$

$$\psi(\widehat{\lambda}_2) - \psi(\widehat{\lambda}_1 + \widehat{\lambda}_2) = \frac{1}{n}\sum_{i=1}^{n} \ln\left(\frac{\widehat{\mu}_2 - x_i}{\widehat{\mu}_2 - \widehat{\mu}_1}\right), \tag{14.51}$$

$$\frac{1 - \widehat{\lambda}_1 - \widehat{\lambda}_2}{\widehat{\mu}_2 - \widehat{\mu}_1} + \frac{\widehat{\lambda}_1 - 1}{n}\sum_{i=1}^{n} (x_i - \widehat{\mu}_1)^{-1} = 0, \tag{14.52}$$

$$\frac{1 - \widehat{\lambda}_1 - \widehat{\lambda}_2}{\widehat{\mu}_2 - \widehat{\mu}_1} + \frac{\widehat{\lambda}_2 - 1}{n}\sum_{i=1}^{n} (\widehat{\mu}_2 - x_i)^{-1} = 0, \tag{14.53}$$

where $\psi(\lambda)$ is the *digamma function* (see Section 13.2). Solving (14.52) and (14.53) for $\widehat{\lambda}_1$ and $\widehat{\lambda}_2$ gives

$$\widehat{\lambda}_1 = \frac{s_5(\sigma s_6 - 1)}{s_6(\sigma s_5 - 1) - s_5} \tag{14.54}$$

and

$$\widehat{\lambda}_2 = \frac{s_6(\sigma s_5 - 1)}{s_6(\sigma s_5 - 1) - s_5}, \tag{14.55}$$

where $s_5 = \frac{1}{n}\sum_i (x_i - \widehat{\mu}_1)^{-1}$, $s_6 = \frac{1}{n}\sum_i (\widehat{\mu}_2 - x_i)^{-1}$, and $\sigma = \widehat{\mu}_2 - \widehat{\mu}_1$. Substituting (14.54) and (14.55) into (14.50) and (14.51) gives two expressions in $\widehat{\mu}_1$ and $\widehat{\mu}_2$ only:

$$\psi\left(\frac{s_5(\sigma s_6 - 1)}{s_6(\sigma s_5 - 1) - s_5}\right) - \psi\left(1 + \frac{\sigma s_5 s_6}{s_6(\sigma s_5 - 1) - s_5}\right) = s_7 - \ln(\sigma) \tag{14.56}$$

and

$$\psi\left(\frac{s_6(\sigma s_5 - 1)}{s_6(\sigma s_5 - 1) - s_5}\right) - \psi\left(1 + \frac{\sigma s_5 s_6}{s_6(\sigma s_5 - 1) - s_5}\right) = s_8 - \ln(\sigma), \tag{14.57}$$

where $s_7 = \frac{1}{n}\sum_i \ln(x_i - \widehat{\mu}_1)$ and $s_8 = \frac{1}{n}\sum_i \ln(\widehat{\mu}_2 - x_i)$.

In principle, the solution $(\widehat{\mu}_1, \widehat{\mu}_2)$ is obtained with an equation solver, followed by $(\widehat{\lambda}_1, \widehat{\lambda}_2)$ from (14.54) and (14.55). However, a solution may not exist, as typically happens when the sample size is small. Even when a solution does exist, it may not be simple to locate it: Good starting values for μ_1 and μ_2 are essential. Moment estimates (next section) may provide ballpark values, although they are often some distance from the ML solution.

It is convenient to display the log-likelihood function, expressed in μ_1 and μ_2, so that one can visually search for the maximum and obtain close starting values for the ML solution process:

$$\begin{aligned} LLF(\mu_1, \mu_2) = &-n \ln[B(\lambda_1, \lambda_2)] + n(1 - \lambda_1 - \lambda_2) \ln(\sigma) \\ &+ n(\lambda_1 - 1)s_7 + n(\lambda_2 - 1)s_8 + C. \end{aligned} \tag{14.58}$$

The terms λ_1 and λ_2 are given by (14.54) and (14.55), $\sigma = \mu_2 - \mu_1$, and C is an arbitrary constant, chosen to produce small positive values of the function near its maximum. The difficulty with finding a solution is that the likelihood function does not appear to hold much information on the location parameters. This is indicated by the typically shallow surface of that function.

Approximate standard errors of the estimates can be obtained from the inverse of the *expected* information matrix (14.22), provided $\lambda_1 > 2$ and $\lambda_2 > 2$, or the *local* version (14.23), provided $\lambda_1 > 1$ and $\lambda_2 > 1$. Although these formulas look intimidating, they are readily evaluated numerically at the ML estimates, using a modern computational tool (e.g., Mathcad). See Example 14.4. for an illustration of these computations. Approximate standard errors of parameter functions $g(\mu_1, \mu_2, \lambda_1, \lambda_2)$ are obtained from the error propagation formula (see Section 3.3).

Approximate confidence intervals on parameters can be constructed from the asymptotic formula (14.45) to give a rough idea of the uncertainties associated with estimated quantities. Small-sample methods for confidence intervals are not available.

Statistical tests-of-fit are not available for the four-parameter Beta model. A linearized model/data plot is recommended as a reliable check on the distributional postulate and the model fit; see Example 14.4.

14.13 Moment Estimates

When ML estimation fails, moment estimates may be used. The resulting model fit is usually acceptable, although standard errors are difficult to obtain for these estimates.

It is convenient to use data moments M_r about the mean, which are defined in terms of moments m_r about the origin (see Section 1.8) as

$$\begin{aligned} M_2 &= m_2 - m_1^2, \\ M_3 &= m_3 - 3m_1m_2 + 2m_1^3, \\ M_4 &= m_4 - 4m_1m_3 + 6m_1^2m_2 - 3m_1^4, \end{aligned} \tag{14.59}$$

where $m_r = \frac{1}{n}\sum_i x_i^r$. To estimate the shape parameters the shape factors (14.14) and (14.15) provide two equations in $\tilde{\lambda}_1$ and $\tilde{\lambda}_2$:

$$\frac{M_3}{M_2^{3/2}} = \frac{2(\tilde{\lambda}_2 - \tilde{\lambda}_1)}{(\tilde{\lambda}_1 + \tilde{\lambda}_2 + 2)}\sqrt{\frac{\tilde{\lambda}_1 + \tilde{\lambda}_2 + 1}{\tilde{\lambda}_1\tilde{\lambda}_2}} \tag{14.60}$$

and

$$\frac{M_4}{M_2^2} = \frac{3(\tilde{\lambda}_1 + \tilde{\lambda}_2 + 1)[2(\tilde{\lambda}_1 + \tilde{\lambda}_2)^2 + \tilde{\lambda}_1\tilde{\lambda}_2(\tilde{\lambda}_1 + \tilde{\lambda}_2 - 6)]}{\tilde{\lambda}_1\tilde{\lambda}_2(\tilde{\lambda}_1 + \tilde{\lambda}_2 + 2)(\tilde{\lambda}_1 + \tilde{\lambda}_2 + 3)}. \tag{14.61}$$

The estimates $\tilde{\lambda}_1$ and $\tilde{\lambda}_2$ are readily computed with an equation solver. Location-parameter estimates are then obtained from the mean value (14.18), $m_1 = \mu_1'(X)$, and the variance (14.20), $M_2 = \mu_2(X)$, as

$$\tilde{\mu}_1 = m_1 - \sqrt{M_2}\sqrt{\frac{\tilde{\lambda}_1(\tilde{\lambda}_1 + \tilde{\lambda}_2 + 1)}{\tilde{\lambda}_2}} \tag{14.62}$$

and

$$\tilde{\mu}_2 = m_1 + \sqrt{M_2}\sqrt{\frac{\tilde{\lambda}_2(\tilde{\lambda}_1 + \tilde{\lambda}_2 + 1)}{\tilde{\lambda}_1}}. \tag{14.63}$$

See Example 14.5 where a small sample did not lead to a ML solution, but the moment estimates produced a reasonable model fit.

14.14 Conditional Inferences

Sometimes the values of the location parameters are known: μ_1^* and μ_2^*. It is then simple to transform the data x_i to their reduced equivalents by

$$z_i = \frac{x_i - \mu_1^*}{\mu_2^* - \mu_1^*} \tag{14.64}$$

and to use two-parameter methods on the reduced data z_i. See Example 14.6.

APPLICATIONS

14.15 Engineering

Because of its limited sample space, $\mu_1 \leq x \leq \mu_2$, the general Beta distribution (14.17) serves as a useful measurement model for engineering variables for which the assumptions of an unlimited upper tail and the lower tail terminating at the origin are inappropriate. Applications include cost variables, task completion times, and load variables subject to inherent or imposed (e.g., legal) limits. The two-parameter distribution (14.1), however, is a natural candidate for modeling engineering *ratios*, for example efficiency measures, which vary over the unit range. Furthermore, the exceptional shape flexibility of the Beta distribution makes it attractive as a general measurement model and thus it is increasingly used in engineering work for variables subject to range limitations.

An interesting application of the Beta distribution occurs in the coordination of complex engineering projects that involve tasks of uncertain durations X_i. Such projects are often controlled by PERT (Project Evaluation and Control Technique). PERT considers task durations X_i as Beta random variables and requires input estimates of the expected value and variance for each project task i. Practically, these estimates are difficult to obtain directly. Hence, intuitively more accessible estimates are obtained instead, namely a *most likely* task duration m_i, an *optimistic* time a_i, and a *pessimistic* time b_i. These quantities are equivalent to the mode value x_m and to the location parameters μ_1 and μ_2, respectively.

The expected value and variance of task duration are then calculated using the following assumptions:

$$\mathrm{E}\{X\} = \frac{1}{3}\left[2m + \frac{1}{2}(a+b)\right] \tag{14.65}$$

and

$$\mathrm{Var}(X) = \frac{(b-a)^2}{36}. \tag{14.66}$$

By expressing the expected value (14.18) in terms of the mode value (14.19) and comparing with (14.65), we see that the shape parameters are constrained by $\lambda_1 + \lambda_2 = 6$. Similarly, approximating the variance (14.20) by (14.66) implies that $\lambda_1 \lambda_2 = 7$. Hence, PERT implicitly specifies the Beta shape parameters as (1.59, 4.41) for positive skew or (4.41, 1.59) for negative skew. Figure 14.6 shows these two Beta pdfs, reduced to the unit range. Despite this restriction, practice with

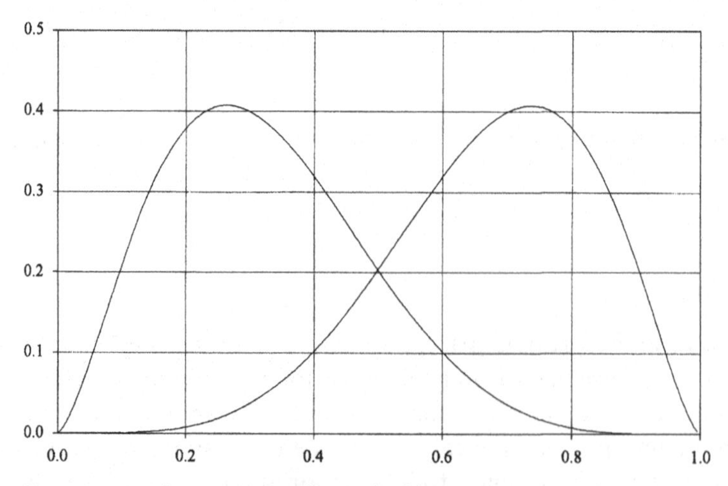

Figure 14.6. The two Beta pdfs implied by PERT.

PERT has shown that (14.65) and (14.66) give reasonable results, particularly in view of the inaccuracies that are inherent in the elicited estimates *m*, *a*, and *b*.

The expected value (14.65) and variance (14.66) of *task* durations are then used to construct a Normal probability distribution of *project* duration. This is done by invoking a central limit theorem (see Section 2.4) on the string of critical tasks that control the project duration. For large projects there are many such critical tasks, and thus the Normal distribution proves to be a valid assumption.

EXAMPLE 14.1

Generate 20 observations from a Beta distribution with $\lambda_1 = 3$ and $\lambda_2 = 5$.

$\lambda 1 := 3 \qquad \lambda 2 := 5 \qquad$ Beta function: $B := \dfrac{\Gamma(\lambda 1) \cdot \Gamma(\lambda 2)}{\Gamma(\lambda 1 + \lambda 2)}$

$n := 20 \qquad i := 1 \,..\, n \qquad u_i := rnd(1)$

$$F(x) := \frac{1}{B} \cdot \int_0^x z^{\lambda 1 - 1} \cdot (1 - z)^{\lambda 2 - 1}\, dz \qquad x := 0.5 \qquad x_i := \text{root}(F(x) - u_i, x)$$

Sample check:

mean value: $xb := \dfrac{1}{n} \cdot \sum_i x_i \qquad xb = 0.401$

This should be close to

$$\frac{\lambda 1}{\lambda 1 + \lambda 2} = 0.375$$

variance: $S2 := \dfrac{1}{n} \cdot \sum_i (x_i - xb)^2 \qquad S2 = 0.021$

This should be close to

$$\frac{\lambda 1 \cdot \lambda 2}{(\lambda 1 + \lambda 2)^2 \cdot (\lambda 1 + \lambda 2 + 1)} = 0.026$$

x_i
0.461
0.432
0.237
0.113
0.526
0.278
0.275
0.309
0.67
0.428
0.556
0.402
0.472
0.226
0.632
0.533
0.309
0.417
0.495
0.241

EXAMPLE 14.2

For the simulated data of Example 14.1, estimate the Beta parameters and their standard errors. Also, estimate the 5-percentile and its standard error.

From Example 14.1:

$n := 20 \qquad i := 1 \,..\, n$

$x_i :=$ $\quad y := \mathrm{sort}(x)$

$$x = \begin{bmatrix} .461 \\ .432 \\ .237 \\ .113 \\ .526 \\ .278 \\ .275 \\ .309 \\ .67 \\ .428 \\ .556 \\ .402 \\ .472 \\ .226 \\ .632 \\ .533 \\ .309 \\ .417 \\ .495 \\ .241 \end{bmatrix}$$

1. Moment estimates

$$m1 := \frac{1}{n} \cdot \sum_i x_i \qquad m2 := \frac{1}{n} \cdot \sum_i (x_i)^2$$

$$L1 := \frac{m1^2 - m1 \cdot m2}{m2 - m1^2} \qquad L1 = 4.222$$

$$L2 := \frac{m1 - m2}{m2 - m1^2} - L1 \qquad L2 = 6.317$$

2. Display the log-likelihood function

$$s3 := \frac{1}{n} \cdot \sum_i \ln(x_i) \qquad s4 := \frac{1}{n} \cdot \sum_i \ln(1 - x_i)$$

$$k := 1 .. 10 \qquad p := 1 .. 10$$

$$r_k := 3.6 + 0.1 \cdot k \qquad l_p := 6 + 0.1 \cdot p$$

$$B_{k,p} := \frac{\Gamma(r_k) \cdot \Gamma(l_p)}{\Gamma(r_k + l_p)} \qquad s3 = -0.9935$$

$$s4 = -0.5428$$

$$L_{k,p} := -n \cdot \ln(B_{k,p}) + n \cdot (r_k - 1) \cdot s3 + n \cdot (l_p - 1) \cdot s4 - 10$$

$$l^T = (6.100 \quad 6.200 \quad 6.300 \quad 6.400 \quad 6.500 \quad 6.600 \quad 6.700 \quad 6.800 \quad 6.900 \quad 7.000)$$

$$L = \begin{bmatrix} 0.32 & 0.24 & 0.146 & 0.04 & -0.08 & -0.213 & -0.357 & -0.514 & -0.683 & -0.863 \\ 0.439 & 0.381 & 0.308 & 0.222 & 0.123 & 0.011 & -0.113 & -0.25 & -0.399 & -0.559 \\ 0.52 & 0.483 & 0.431 & 0.366 & 0.287 & 0.195 & 0.09 & -0.027 & -0.156 & -0.297 \\ 0.563 & 0.547 & 0.516 & 0.471 & 0.412 & 0.34 & 0.255 & 0.158 & 0.048 & -0.073 \\ 0.571 & 0.575 & 0.564 & 0.539 & 0.501 & 0.449 & 0.383 & 0.305 & 0.215 & 0.112 \\ 0.543 & 0.568 & 0.577 & 0.573 & 0.554 & 0.521 & 0.475 & 0.417 & 0.345 & 0.261 \\ 0.483 & 0.528 & 0.557 & 0.572 & 0.573 & 0.56 & 0.533 & 0.493 & 0.441 & 0.376 \\ 0.39 & 0.455 & 0.504 & 0.539 & 0.559 & 0.565 & 0.557 & 0.536 & 0.503 & 0.456 \\ 0.266 & 0.351 & 0.42 & 0.474 & 0.513 & 0.538 & 0.549 & 0.547 & 0.532 & 0.504 \\ 0.113 & 0.217 & 0.305 & 0.378 & 0.437 & 0.481 & 0.511 & 0.527 & 0.53 & 0.52 \end{bmatrix} \qquad r = \begin{bmatrix} 3.7 \\ 3.8 \\ 3.9 \\ 4 \\ 4.1 \\ 4.2 \\ 4.3 \\ 4.4 \\ 4.5 \\ 4.6 \end{bmatrix}$$

$$M_{p,11-k} := L_{k,p}$$

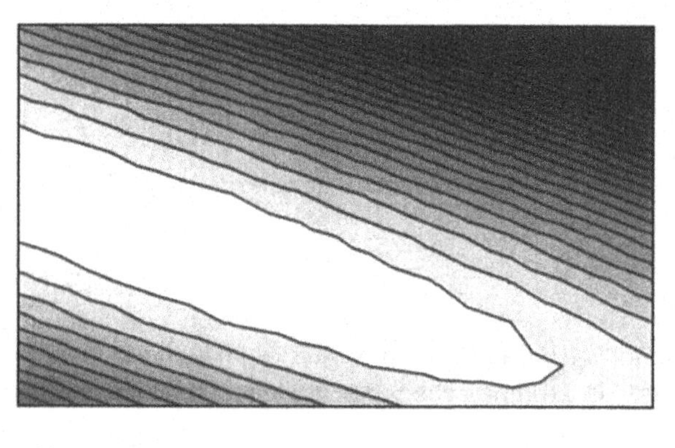

M

starting values: $\quad L1 := 4.2 \qquad L2 := 6.3$

3. Maximum likelihood estimates

digamma function: $\psi(L) := \frac{d}{dL} \ln(\Gamma(L))$

GIVEN $\psi(L1) - \psi(L1 + L2) = s3$

$\psi(L2) - \psi(L1 + L2) = s4$ $\quad \begin{pmatrix} \lambda 1 \\ \lambda 2 \end{pmatrix} := \text{FIND}\,(L1, L2)$

ML parameter estimates $\lambda 1 = 4.192$

$\lambda 2 = 6.305$

4. Standard errors

trigamma function: $\psi'(L) := \frac{d}{dL}\psi(L)$

$a := \psi'(\lambda 1) \quad b := \psi'(\lambda 2) \quad c := \psi'(\lambda 1 + \lambda 2) \quad d := n \cdot (a \cdot b - c \cdot (a + b))$

$SE\lambda 1 := \sqrt{\frac{b - c}{d}} \quad SE\lambda 2 := \sqrt{\frac{a - c}{d}} \quad SE\lambda 12 := \sqrt{\frac{c}{d}}$ $\quad SE\lambda 1 = 1.283$

$SE\lambda 2 = 1.969$

$SE\lambda 12 = 1.513$

5. Linearized model/data plot

$$F(x, \lambda 1, \lambda 2) := \frac{\Gamma(\lambda 1 + \lambda 2)}{\Gamma(\lambda 1) \cdot \Gamma(\lambda 2)} \cdot \int_0^x z^{\lambda 1 - 1} \cdot (1 - z)^{\lambda 2 - 1}\, dz$$

model ordinates: y_i

data ordinates: $p_i := \frac{i - 0.3}{n + 0.4} - F(y_i, \lambda 1, \lambda 2) + y_i$

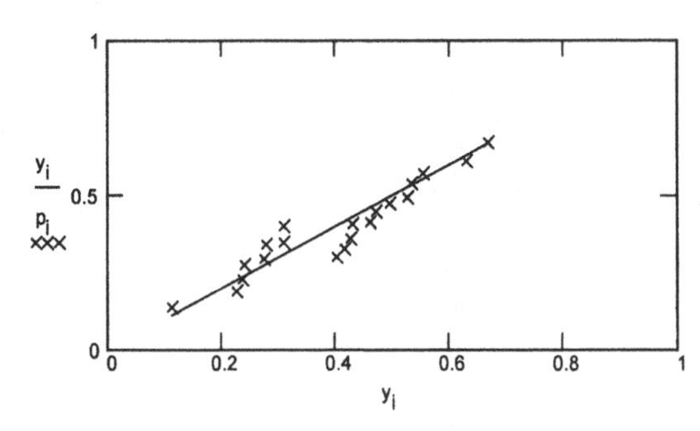

The estimated model fits the data reasonably well.

6. Density plot

$j := 1..199 \quad t_j := \frac{j}{200} \quad f_j := \frac{\Gamma(\lambda 1 + \lambda 2)}{\Gamma(\lambda 1) \cdot \Gamma(\lambda 2)} \cdot (t_j)^{\lambda 1 - 1} \cdot (1 - t_j)^{\lambda 2 - 1}$

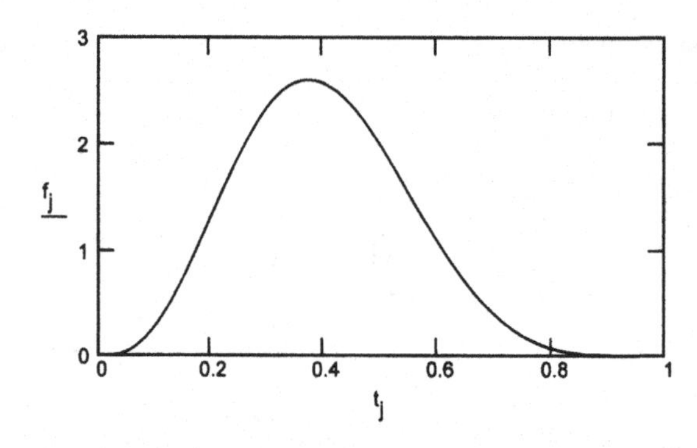

7. Percentile

$x := 0.2 \qquad x05 := \mathrm{root}(F(x, \lambda1, \lambda2) - 0.05, x) \qquad\qquad x05 = 0.173$

standard error: $d1 := \dfrac{d}{d\lambda1} F(x05, \lambda1, \lambda2) \qquad d2 := \dfrac{d}{d\lambda2} F(x05, \lambda1, \lambda2)$

$\mathrm{SE}x := \sqrt{(d1 \cdot \mathrm{SE}\lambda1)^2 + (d2 \cdot \mathrm{SE}\lambda2)^2 + 2 \cdot d1 \cdot d2 \cdot \mathrm{SE}\lambda12^2} \qquad\qquad \mathrm{SE}x = 0.036$

EXAMPLE 14.3

For the data of Example 14.1, calculate 90% confidence limits on the shape parameters and on the 5-percentile.

From Example 14.2:

$\lambda1 := 4.192 \qquad \lambda2 := 6.305 \qquad\qquad s3 := -0.9935$

$n := 20 \quad \mathrm{SE}\lambda1 := 1.283 \quad \mathrm{SE}\lambda2 := 1.969 \quad \mathrm{SE}\lambda12 := 1.513 \quad s4 := -0.5428$

1. Parameters

i) Normal approximation: Standard Normal 95-percentile is $z95 := 1.645$

$L\lambda1 := \lambda1 - z95 \cdot \mathrm{SE}\lambda1 \qquad\qquad L\lambda1 = 2.08$

$U\lambda1 := \lambda1 + z95 \cdot \mathrm{SE}\lambda1 \qquad\qquad U\lambda1 = 6.30$

$L\lambda2 := \lambda2 - z95 \cdot \mathrm{SE}\lambda2 \qquad\qquad L\lambda2 = 3.07$

$U\lambda2 := \lambda2 + z95 \cdot \mathrm{SE}\lambda2 \qquad\qquad U\lambda2 = 9.54$

ii) Likelihood ratio method: Chi-squared 90-percentile at $\nu = 1$ is $K := 2.71$

$B(a, b) := \dfrac{\Gamma(a) \cdot \Gamma(b)}{\Gamma(a + b)} \qquad \psi(a) := \dfrac{d}{da} \ln(\Gamma(a))$

log of likelihood function:

$LL(a, b) := -n \cdot \ln(B(a, b)) + n \cdot (a - 1) \cdot s3 + n \cdot (b - 1) \cdot s4$

shape parameter λ_1: $ML(a, b) := \psi(b) - \psi(a+b) - s4$

$b := \lambda 2$ $\quad l(a) := \text{root}(ML(a, b), b)$

$LR(a) := 2 \cdot LL(\lambda 1, \lambda 2) - 2 \cdot LL(a, l(a))$

$a := 2$ $\quad \lambda 1L := \text{root}(LR(a) - K, a)$ $\quad \lambda 1L = 2.43$

$a := 6$ $\quad \lambda 1U := \text{root}(LR(a) - K, a)$ $\quad \lambda 1U = 6.68$

shape parameter λ_2: $ML(a, b) := \psi(a) - \psi(a+b) - s3$

$a := \lambda 1$ $\quad l(b) := \text{root}(ML(a, b), a)$

$LR(b) := 2 \cdot LL(\lambda 1, \lambda 2) - 2 \cdot LL(l(b), b)$

$b := 3$ $\quad \lambda 2L := \text{root}(LR(b) - K, b)$ $\quad \lambda 2L = 3.60$

$b := 10$ $\quad \lambda 2U := \text{root}(LR(b) - K, b)$ $\quad \lambda 2U = 10.13$

2. Percentile

estimate:

$q := 0.05$ $\quad F(x, \lambda 1, \lambda 2) := \dfrac{1}{B(\lambda 1, \lambda 2)} \cdot \displaystyle\int_0^x t^{\lambda 1 - 1} \cdot (1-t)^{\lambda 2 - 1}\, dt$

$x := 0.1$ $\quad x05(\lambda 1, \lambda 2) := \text{root}(F(x, \lambda 1, \lambda 2) - q, x)$ $\quad x05(\lambda 1, \lambda 2) = 0.17$

standard error:

$d1 := \dfrac{d}{d\lambda 1} x05(\lambda 1, \lambda 2)$ $\quad d2 := \dfrac{d}{d\lambda 2} x05(\lambda 1, \lambda 2)$

$\text{var}x := (d1 \cdot \text{SE}\lambda 1)^2 + (d2 \cdot \text{SE}\lambda 2)^2 + 2 \cdot d1 \cdot d2 \cdot \text{SE}\lambda 12^2$ $\quad \text{SE}x := \sqrt{\text{var}x}$

$\text{SE}x = 0.04$

Normal interval approximation:

$Lx := x05(\lambda 1, \lambda 2) - z95 \cdot \text{SE}x$ $\quad Lx = 0.11$

$Ux := x05(\lambda 1, \lambda 2) + z95 \cdot \text{SE}x$ $\quad Ux = 0.23$

EXAMPLE 14.4

A sample of 182 observations is available from a test program on the compressive strength (kg/cm^2) of a concrete. The sample is too large to present in print; it is read from a computer file. Assuming a four-parameter Beta distribution for strength, estimate the parameters and their standard errors.

$n := 182$ $\quad i := 1..n$

Read the data file called "*df*": $\quad y_i := \text{READ}(df)$

$x := \text{sort}(y)$ $\quad x_1 = 209.000$ $\quad mx := 209$ $\quad x_n = 335.000$ $\quad Mx := 335$

1. Display of log-likelihood function

$j := 1..10 \qquad a_j := 206.6 + 0.2 \cdot j \qquad k := 1..10 \qquad b_k := 338 + k$

$$s5_j := \frac{1}{n} \cdot \sum_i (x_i - a_j)^{-1} \qquad s6_k := \frac{1}{n} \cdot \sum_i (b_k - x_i)^{-1}$$

$$s7_j := \frac{1}{n} \cdot \sum_i \ln(x_i - a_j) \qquad s8_k := \frac{1}{n} \cdot \sum_i \ln(b_k - x_i)$$

$$(14.54): \quad L1_{j,k} := \frac{s5_j \cdot [(b_k - a_j) \cdot s6_k - 1]}{s6_k \cdot [(b_k - a_j) \cdot s5_j - 1] - s5_j}$$

$$(14.55): \quad L2_{j,k} := \frac{s6_k \cdot [(b_k - a_j) \cdot s5_j - 1]}{s6_k \cdot [(b_k - a_j) \cdot s5_j - 1] - s5_j} \qquad B_{j,k} := \frac{\Gamma(L1_{j,k}) \cdot \Gamma(L2_{j,k})}{\Gamma(L1_{j,k} + L2_{j,k})}$$

$$LLF_{j,k} := -n \cdot \ln(B_{j,k}) + n \cdot (1 - L1_{j,k} - L2_{j,k}) \cdot \ln(b_k - a_j) + n \cdot (L1_{j,k} - 1) \cdot s7_j + n \cdot (L2_{j,k} - 1) \cdot s8_k + 866$$

$$b^T = (339.0 \;\; 340.0 \;\; 341.0 \;\; 342.0 \;\; 343.0 \;\; 344.0 \;\; 345.0 \;\; 346.0 \;\; 347.0 \;\; 348.0)$$

$$LLF = \begin{bmatrix} -0.044 & 0.255 & 0.445 & 0.568 & 0.645 & 0.690 & 0.711 & 0.715 & 0.704 & 0.683 \\ 0.043 & 0.327 & 0.505 & 0.617 & 0.684 & 0.720 & 0.734 & 0.730 & 0.712 & 0.684 \\ 0.127 & 0.394 & 0.559 & 0.659 & 0.716 & 0.742 & 0.747 & 0.734 & 0.710 & 0.675 \\ 0.206 & 0.455 & 0.604 & 0.691 & 0.737 & 0.753 & 0.748 & 0.727 & 0.693 & 0.651 \\ 0.277 & 0.507 & 0.639 & 0.711 & 0.744 & 0.748 & 0.732 & 0.701 & 0.659 & 0.607 \\ 0.338 & 0.544 & 0.657 & 0.713 & 0.731 & 0.722 & 0.694 & 0.652 & 0.599 & 0.537 \\ 0.380 & 0.560 & 0.651 & 0.687 & 0.688 & 0.663 & 0.621 & 0.565 & 0.500 & 0.427 \\ 0.392 & 0.539 & 0.602 & 0.615 & 0.594 & 0.551 & 0.491 & 0.419 & 0.339 & 0.252 \\ 0.342 & 0.446 & 0.473 & 0.455 & 0.407 & 0.338 & 0.256 & 0.163 & 0.063 & -0.043 \\ 0.140 & 0.181 & 0.155 & 0.091 & 0.002 & -0.103 & -0.219 & -0.343 & -0.472 & -0.605 \end{bmatrix} \qquad a = \begin{bmatrix} 206.8 \\ 207.0 \\ 207.2 \\ 207.4 \\ 207.6 \\ 207.8 \\ 208.0 \\ 208.2 \\ 208.4 \\ 208.6 \end{bmatrix}$$

$M_{j,11-k} := LLF_{k,j}$

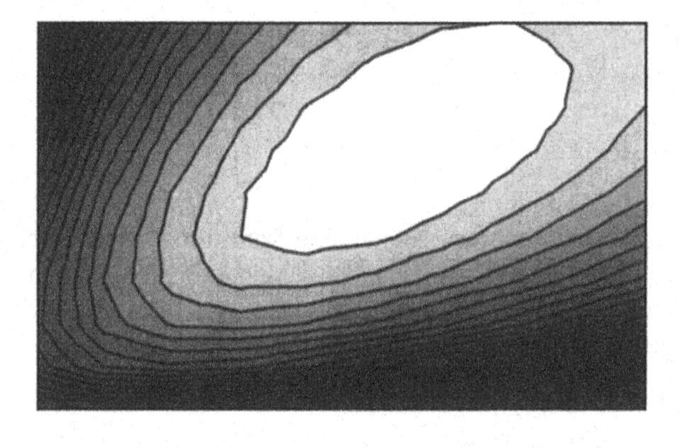

M

starting values: $\quad a := 207.4 \qquad b := 344$

2. Maximum likelihood estimates

digamma function: $\psi(t) := \frac{d}{dt}\ln(\Gamma(t))$

$$s5(a) := \frac{1}{n}\cdot\sum_i (x_i - a)^{-1} \qquad s6(b) := \frac{1}{n}\cdot\sum_i (b - x_i)^{-1}$$

$$s7(a) := \frac{1}{n}\cdot\sum_i \ln(x_i - a) \qquad s8(b) := \frac{1}{n}\cdot\sum_i \ln(b - x_i)$$

GIVEN

$$\psi\left[\frac{s5(a)\cdot((b-a)\cdot s6(b)-1)}{s6(b)\cdot((b-a)\cdot s5(a)-1)-s5(a)}\right] - \psi\left[1+\frac{(b-a)\cdot s5(a)\cdot s6(b)}{s6(b)\cdot((b-a)\cdot s5(a)-1)-s5(a)}\right] = s7(a) - \ln(b-a)$$

$$\psi\left[\frac{s6(b)\cdot((b-a)\cdot s5(a)-1)}{s6(b)\cdot((b-a)\cdot s5(a)-1)-s5(a)}\right] - \psi\left[1+\frac{(b-a)\cdot s5(a)\cdot s6(b)}{s6(b)\cdot((b-a)\cdot s5(a)-1)-s5(a)}\right] = s8(b) - \ln(b-a)$$

constraints on the location parameters: $a < mx \qquad b > Mx$

$$\begin{pmatrix}\mu 1\\ \mu 2\end{pmatrix} := \text{FIND}(a, b)$$

location parameters: $\mu 1 = 207.4$

$\mu 2 = 344.1$

$$\lambda 1 := \frac{s5(\mu 1)\cdot((\mu 2 - \mu 1)\cdot s6(\mu 2) - 1)}{s6(\mu 2)\cdot((\mu 2 - \mu 1)\cdot s5(\mu 1) - 1) - s5(\mu 1)}$$

shape parameters: $\lambda 1 = 1.625$

$$\lambda 2 := \frac{s6(\mu 2)\cdot((\mu 2 - \mu 1)\cdot s5(\mu 1) - 1)}{s6(\mu 2)\cdot((\mu 2 - \mu 1)\cdot s5(\mu 1) - 1) - s5(\mu 1)}$$

$\lambda 2 = 2.400$

3. Standard errors

$$s1 := \frac{1}{n}\cdot\sum_i (x_i - \mu 1)^{-2} \qquad s2 := \frac{1}{n}\cdot\sum_i (\mu 2 - x_i)^{-2}$$

trigamma function: $\psi'(t) := \frac{d}{dt}\psi(t) \qquad \sigma := \mu 2 - \mu 1$

local information matrix:

$$I := n\cdot\begin{bmatrix} (\lambda 1 - 1)\cdot s1 - \frac{\lambda 1 + \lambda 2 - 1}{\sigma^2} & \frac{\lambda 1 + \lambda 2 - 1}{\sigma^2} & \frac{\lambda 2}{(\lambda 1 - 1)\cdot\sigma} & -\frac{1}{\sigma} \\ \frac{\lambda 1 + \lambda 2 - 1}{\sigma^2} & (\lambda 2 - 1)\cdot s2 - \frac{\lambda 1 + \lambda 2 - 1}{\sigma^2} & \frac{1}{\sigma} & -\frac{\lambda 1}{(\lambda 2 - 1)\cdot\sigma} \\ \frac{\lambda 2}{(\lambda 1 - 1)\cdot\sigma} & \frac{1}{\sigma} & \psi'(\lambda 1) - \psi'(\lambda 1 + \lambda 2) & -\psi'(\lambda 1 + \lambda 2) \\ -\frac{1}{\sigma} & -\frac{\lambda 1}{(\lambda 2 - 1)\cdot\sigma} & -\psi'(\lambda 1 + \lambda 2) & \psi'(\lambda 2) - \psi'(\lambda 1 + \lambda 2) \end{bmatrix}$$

$$V := 1^{-1} \qquad V = \begin{pmatrix} 5.612 & -10.849 & -0.639 & -0.991 \\ -10.849 & 89.772 & 2.160 & 5.536 \\ -0.639 & 2.160 & 0.110 & 0.192 \\ -0.991 & 5.536 & 0.192 & 0.426 \end{pmatrix}$$

$$SE\mu1 := \sqrt{V_{1,1}} \qquad SE\mu1 = 2.369$$
$$SE\mu2 := \sqrt{V_{2,2}} \qquad SE\mu2 = 9.475$$
$$SE\lambda1 := \sqrt{V_{3,3}} \qquad SE\lambda1 = 0.332$$
$$SE\lambda2 := \sqrt{V_{4,4}} \qquad SE\lambda2 = 0.652$$

4. Linearized model/data plot

$$B := \frac{\Gamma(\lambda1) \cdot \Gamma(\lambda2)}{\Gamma(\lambda1 + \lambda2)} \qquad \text{model ordinates: } z_i := \frac{x_i - \mu1}{\mu2 - \mu1}$$

$$F(t) := \frac{1}{B} \cdot \int_0^t z^{\lambda1-1} \cdot (1-z)^{\lambda2-1} dz \qquad \text{data ordinates: } p_i := \frac{i - 0.3}{n + 0.4} - F(z_i) + z_i$$

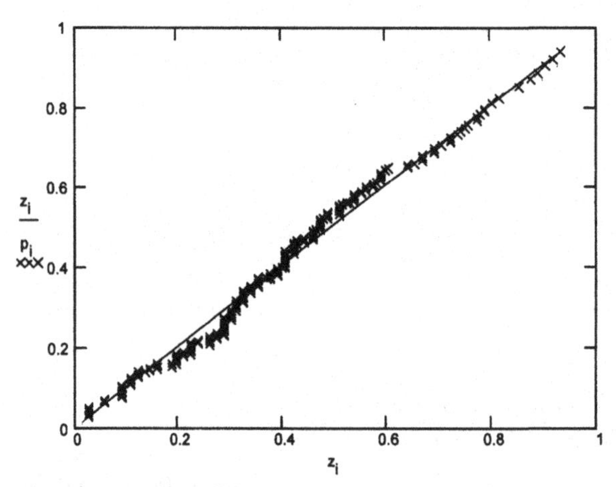

The estimated model fits the data quite well.

5. pdf plot

$$j := 1..999 \qquad t_j := \mu1 + \frac{\sigma \cdot j}{1000} \qquad f_j := \frac{1}{B \cdot \sigma} \cdot \left(\frac{j}{1000}\right)^{\lambda1-1} \cdot \left(1 - \frac{j}{1000}\right)^{\lambda2-1}$$

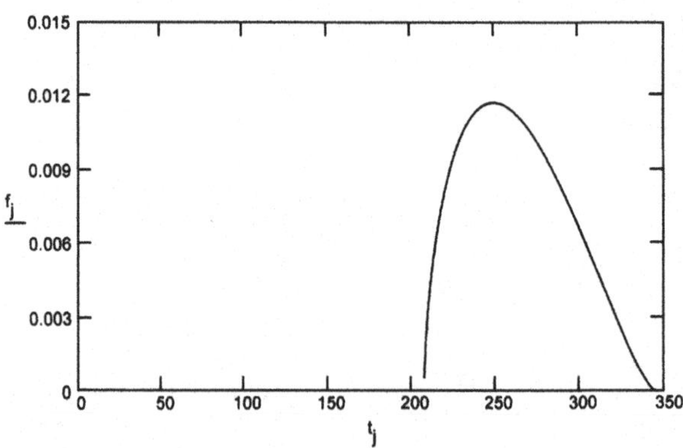

EXAMPLE 14.5

The following data are available on the modulus of elasticity (1,000,000 psi) of a certain size, grade, and species of lumber:

1.73, 1.50, 1.56, 1.89, 1.54, 1.68, 1.39, 1.64, 1.49, 1.43, 1.68, 1.61, 1.62.

Assuming a four-parameter Beta distribution for this quantity, estimate the parameters.

1. Maximum likelihood approach

$n := 13 \qquad i := 1..n$

$$y_i = \begin{bmatrix} 1.73 \\ 1.5 \\ 1.56 \\ 1.89 \\ 1.54 \\ 1.68 \\ 1.39 \\ 1.64 \\ 1.49 \\ 1.43 \\ 1.68 \\ 1.61 \\ 1.62 \end{bmatrix}$$

$x := \mathrm{sort}(y) \qquad x_1 = 1.390 \qquad x_n = 1.890$

Display of likelihood function:

$j := 1..10 \qquad a_j := 1.369 + 0.002 \cdot j$

$k := 1..10 \qquad b_k := 1.88 + 0.012 \cdot k$

$$s5_j := \frac{1}{n} \cdot \sum_i (x_i - a_j)^{-1} \qquad s6_k := \frac{1}{n} \cdot \sum_i (b_k - x_i)^{-1}$$

$$s7_j := \sum_i \ln(x_i - a_j) \qquad s8_k := \sum_i \ln(b_k - x_i)$$

$$(14.54): \quad L1_{j,k} := \frac{s5_j \cdot [(b_k - a_j) \cdot s6_k - 1]}{s6_k \cdot [(b_k - a_j) \cdot s5_j - 1] - s5_j}$$

$$(14.55): \quad L2_{j,k} := \frac{s6_k \cdot [(b_k - a_j) \cdot s5_j - 1]}{s6_k \cdot [(b_k - a_j) \cdot s5_j - 1] - s5_j}$$

$$B_{j,k} := \frac{\Gamma(L1_{j,k}) \cdot \Gamma(L2_{j,k})}{\Gamma(L1_{j,k} + L2_{j,k})}$$

$$LLF_{j,k} := -n \cdot \ln(B_{j,k}) + n \cdot (1 - L1_{j,k} - L2_{j,k}) \cdot \ln(b_k - a_j) + [(L1_{j,k} - 1) \cdot s7_j + (L2_{j,k} - 1) \cdot s8_k] - 9.1$$

$b^T = 1.892 \quad 1.904 \quad 1.916 \quad 1.928 \quad 1.940 \quad 1.952 \quad 1.964 \quad 1.976 \quad 1.988 \quad 2.000)$

$$LLF = \begin{bmatrix}
-1.193 & -0.735 & -0.499 & -0.377 & -0.309 & -0.271 & -0.250 & -0.240 & -0.235 & -0.236 \\
-1.128 & -0.684 & -0.458 & -0.342 & -0.280 & -0.247 & -0.229 & -0.221 & -0.220 & -0.222 \\
-1.058 & -0.629 & -0.414 & -0.306 & -0.250 & -0.221 & -0.207 & -0.203 & -0.204 & -0.209 \\
-0.981 & -0.569 & -0.367 & -0.267 & -0.218 & -0.194 & -0.184 & -0.184 & -0.188 & -0.196 \\
-0.897 & -0.505 & -0.315 & -0.226 & -0.184 & -0.166 & -0.161 & -0.164 & -0.172 & -0.183 \\
-0.803 & -0.433 & -0.259 & -0.181 & -0.147 & -0.136 & -0.137 & -0.144 & -0.156 & -0.171 \\
-0.697 & -0.353 & -0.197 & -0.132 & -0.107 & -0.104 & -0.111 & -0.124 & -0.140 & -0.159 \\
-0.573 & -0.261 & -0.126 & -0.075 & -0.063 & -0.068 & -0.082 & -0.101 & -0.123 & -0.146 \\
-0.425 & -0.150 & -0.040 & -0.008 & -0.008 & -0.024 & -0.047 & -0.073 & -0.101 & -0.130 \\
-0.232 & -0.001 & -0.079 & -0.090 & -0.074 & -0.046 & -0.013 & -0.022 & -0.057 & -0.092
\end{bmatrix} \qquad a = \begin{bmatrix} 1.371 \\ 1.373 \\ 1.375 \\ 1.377 \\ 1.379 \\ 1.381 \\ 1.383 \\ 1.385 \\ 1.387 \\ 1.389 \end{bmatrix}$$

$M_{j,11-k} := LLF_{k,j}$ Conclusion: It appears that the ML estimate of μ_1 converges to $x_1 = 1.39$.

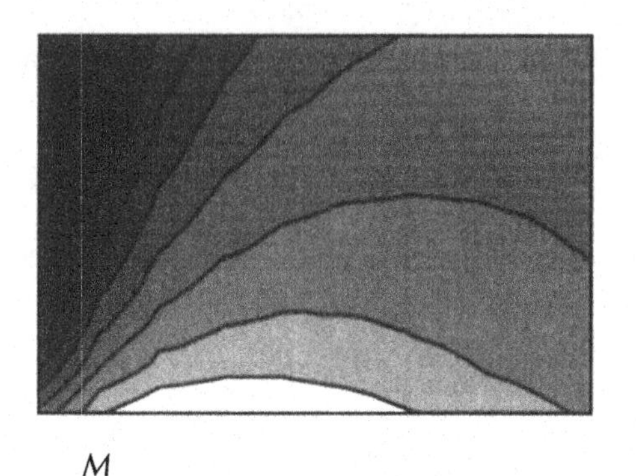

M

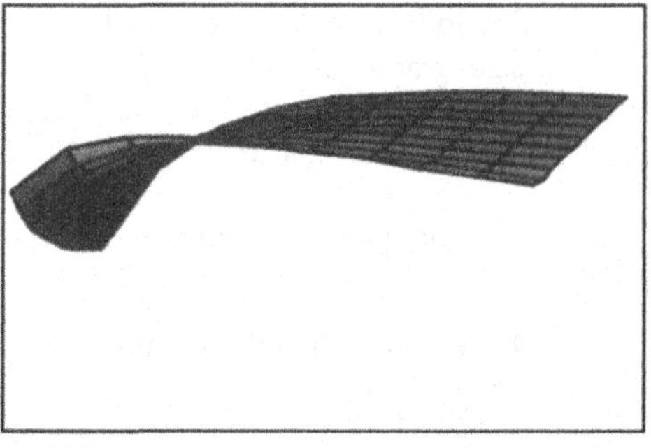

LLF

2. Moment estimates

$r := 1..4 \qquad m_r := \frac{1}{n} \cdot \sum_i (x_i)^r \qquad M2 := m_2 - (m_1)^2$

$M3 := m_3 - 3 \cdot m_1 \cdot m_2 + 2 \cdot (m_1)^3$

$M4 := m_4 - 4 \cdot m_1 \cdot m_3 + 6 \cdot (m_1)^2 \cdot m_2 - 3 \cdot (m_1)^4$

Shape parameters:

$$a := 1 \quad b := 1 \quad \text{GIVEN} \quad \frac{M3}{M2^{1.5}} = 2 \cdot \frac{b-a}{a+b+2} \cdot \sqrt{\frac{a+b+1}{a \cdot b}}$$

$$\frac{M4}{M2^2} = \frac{3 \cdot (a+b+1) \cdot [2 \cdot (a+b)^2 + a \cdot b \cdot (a+b-6)]}{a \cdot b \cdot (a+b+2) \cdot (a+b+3)}$$

$$\begin{pmatrix} \lambda 1 \\ \lambda 2 \end{pmatrix} := \text{FIND}(a, b) \qquad \lambda 1 = 4.088$$

$$\lambda 2 = 10.417$$

Location parameters:

$$\mu 1 := m_1 - \sqrt{M2} \cdot \sqrt{\frac{\lambda 1 \cdot (\lambda 1 + \lambda 2 + 1)}{\lambda 2}} \qquad \mu 1 = 1.279$$

$$\mu 2 := m_1 + \sqrt{M2} \cdot \sqrt{\frac{\lambda 2 \cdot (\lambda 1 + \lambda 2 + 1)}{\lambda 1}} \qquad \mu 2 = 2.407$$

3. Linearized model/data plot

$$B := \frac{\Gamma(\lambda 1) \cdot \Gamma(\lambda 2)}{\Gamma(\lambda 1 + \lambda 2)} \qquad \text{model ordinates:} \quad z_i := \frac{x_i - \mu 1}{\mu 2 - \mu 1}$$

$$F(t) := \frac{1}{B} \cdot \int_0^t z^{\lambda 1 - 1} \cdot (1-z)^{\lambda 2 - 1} dz \qquad \text{data ordinates:} \quad p_i := \frac{i - 0.3}{n + 0.4} - F(z_i) + z_i$$

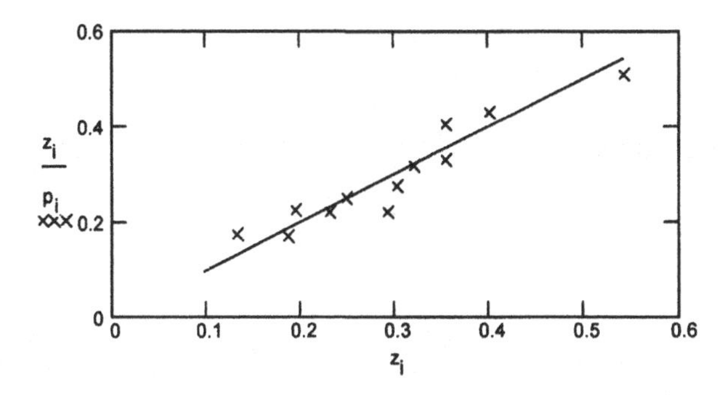

The estimated model fits the data reasonably well.

4. pdf plot

$$j := 1..199 \qquad t_j := \mu 1 + \frac{(\mu 2 - \mu 1) \cdot j}{200}$$

$$f_j := \frac{1}{B \cdot (\mu 2 - \mu 1)} \cdot \left(\frac{j}{200}\right)^{\lambda 1 - 1} \cdot \left(1 - \frac{j}{200}\right)^{\lambda 2 - 1}$$

f_j

2

0

0 0.5 1 1.5 2 2.5

t_j

EXAMPLE 14.6

A sample of 16 observations was obtained on the time it takes to assemble a certain product:

27.0, 28.7, 29.2, 28.6, 30.8, 27.5, 30.1, 31.2,
29.8, 28.3, 27.3, 29.1, 27.9, 26.5, 30.0, 31.4.

Assembly time is thought to be limited to the range (25,32). Estimate the Beta parameters and their standard errors. Estimate the coefficient of variation and its standard error.

$n := 16 \qquad i := 1..n$

$y_i :=$ $\quad x := \text{sort}(y) \qquad z_i := \dfrac{x_i - 25}{32 - 25}$

y_i
27.0
28.7
29.2
28.6
30.8
27.5
30.1
31.2
29.8
28.3
27.3
29.1
27.9
26.5
30.0
31.4

1. Moment estimates

$$m1 := \frac{1}{n} \cdot \sum_i z_i \qquad m2 := \frac{1}{n} \cdot \sum_i (z_i)^2$$

$$L1 := \frac{m1^2 - m1 \cdot m2}{m2 - m1^2} \qquad L1 = 2.635$$

$$L2 := \frac{m1 - m2}{m2 - m1^2} - L1 \qquad L2 = 2.020$$

2. Display the log-likelihood function

$$s3 := \sum_i \ln(z_i) \qquad s4 := \sum_i \ln(1 - z_i)$$

$$k := 1\,..\,10 \qquad p := 1\,..\,10$$

$$r_k := 2.6 + 0.05 \cdot k \qquad l_p := 1.8 + 0.05 \cdot p \qquad B_{k,p} := \frac{\Gamma(r_k) \cdot \Gamma(l_p)}{\Gamma(r_k + l_p)}$$

$$L_{k,p} := -n \cdot \ln(B_{k,p}) + (r_k - 1) \cdot s3 + (l_p - 1) \cdot s4 - 3$$

$$l^T = (1.850 \quad 1.900 \quad 1.950 \quad 2.000 \quad 2.050 \quad 2.100 \quad 2.150 \quad 2.200 \quad 2.250 \quad 2.300)$$

$$L = \begin{bmatrix} 0.183 & 0.230 & 0.259 & 0.271 & 0.267 & 0.247 & 0.213 & 0.164 & 0.101 & 0.025 \\ 0.153 & 0.210 & 0.249 & 0.270 & 0.276 & 0.265 & 0.240 & 0.200 & 0.147 & 0.080 \\ 0.115 & 0.182 & 0.230 & 0.261 & 0.276 & 0.275 & 0.259 & 0.228 & 0.184 & 0.126 \\ 0.070 & 0.146 & 0.204 & 0.244 & 0.268 & 0.276 & 0.269 & 0.248 & 0.212 & 0.162 \\ 0.017 & 0.102 & 0.169 & 0.219 & 0.252 & 0.270 & 0.271 & 0.259 & 0.232 & 0.191 \\ -0.044 & 0.051 & 0.128 & 0.187 & 0.229 & 0.255 & 0.266 & 0.262 & 0.243 & 0.211 \\ -0.111 & -0.007 & 0.079 & 0.147 & 0.198 & 0.233 & 0.252 & 0.257 & 0.247 & 0.223 \\ -0.185 & -0.072 & 0.022 & 0.099 & 0.159 & 0.203 & 0.231 & 0.244 & 0.243 & 0.228 \\ -0.266 & -0.144 & -0.041 & 0.045 & 0.114 & 0.166 & 0.203 & 0.225 & 0.232 & 0.225 \\ -0.354 & -0.223 & -0.110 & -0.016 & 0.062 & 0.123 & 0.168 & 0.198 & 0.213 & 0.215 \end{bmatrix} \qquad r = \begin{bmatrix} 2.650 \\ 2.700 \\ 2.750 \\ 2.800 \\ 2.850 \\ 2.900 \\ 2.950 \\ 3.000 \\ 3.050 \\ 3.100 \end{bmatrix}$$

$$M_{p,11-k} := L_{k,p}$$

M

starting values: $r_3 = 2.750 \qquad l_5 = 2.050$

3. Maximum likelihood estimates

$L1 := 2.75 \qquad L2 := 2.05$

digamma function: $\psi(L) := \frac{d}{dL}\ln(\Gamma(L))$

GIVEN

$\psi(L1) - \psi(L1 + L2) = \frac{s3}{n}$

$\psi(L2) - \psi(L1 + L2) = \frac{s4}{n}$

$\begin{pmatrix} \lambda 1 \\ \lambda 2 \end{pmatrix} := \text{FIND}(L1, L2)$ ML parameter estimates: $\lambda 1 = 2.754$
$\lambda 2 = 2.074$

4. Standard errors

trigamma function: $\psi'(L) := \frac{d}{dL}\psi(L)$

$a := \psi'(\lambda 1) \quad b := \psi'(\lambda 2) \quad c := \psi'(\lambda 1 + \lambda 2) \quad d := n \cdot (a \cdot b - c \cdot (a + b))$

$\text{SE}\lambda 1 := \sqrt{\frac{b-c}{d}} \quad \text{SE}\lambda 2 := \sqrt{\frac{a-c}{d}} \quad \text{SE}\lambda 12 := \sqrt{\frac{c}{d}}$ $\text{SE}\lambda 1 = 0.946$
$\text{SE}\lambda 2 = 0.692$
$\text{SE}\lambda 12 = 0.730$

5. Linearized model/data plot

$F(x, \lambda 1, \lambda 2) := \frac{\Gamma(\lambda 1 + \lambda 2)}{\Gamma(\lambda 1) \cdot \Gamma(\lambda 2)} \cdot \int_0^x z^{\lambda 1 - 1} \cdot (1 - z)^{\lambda 2 - 1} dz$

model ordinates: z_i

data ordinates: $p_i := \frac{i - 0.3}{n + 0.4} - F(z_i, \lambda 1, \lambda 2) + z_i$

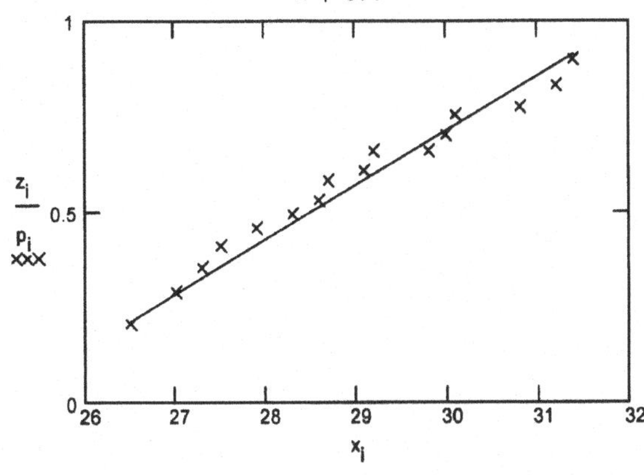

The estimated model fits the data quite well.

6. Density plot

$$j := 1..199 \quad t_j := 25 + \frac{7 \cdot j}{200}$$

$$f_j := \frac{\Gamma(\lambda 1 + \lambda 2)}{\Gamma(\lambda 1) \cdot \Gamma(\lambda 2) \cdot 7} \cdot \left(\frac{j}{200}\right)^{\lambda 1 - 1} \cdot \left(1 - \frac{j}{200}\right)^{\lambda 2 - 1}$$

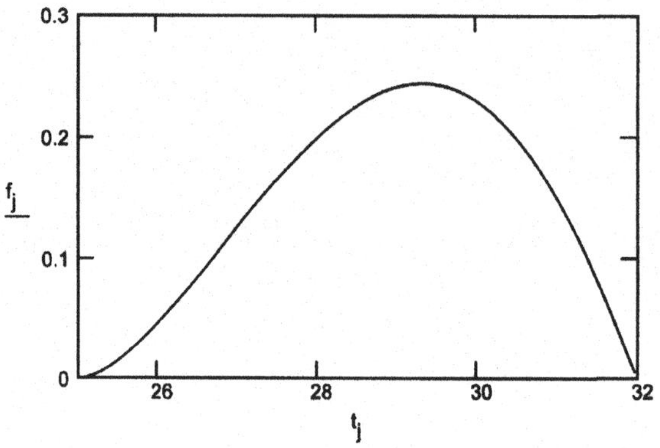

7. Coefficient of variation

(14.21): $$\mathrm{cv}(\lambda 1, \lambda 2) := \frac{7 \cdot \sqrt{\lambda 1 \cdot \lambda 2}}{\sqrt{\lambda 1 + \lambda 2 + 1} \cdot (25 \cdot \lambda 2 + 32 \cdot \lambda 1)} \qquad \mathrm{cv}(\lambda 1, \lambda 2) = 0.050$$

standard error: $$d1 := \frac{d}{d\lambda 1}\mathrm{cv}(\lambda 1, \lambda 2) \qquad d2 := \frac{d}{d\lambda 2}\mathrm{cv}(\lambda 1, \lambda 2)$$

$$\mathrm{SE}x := \sqrt{(d1 \cdot \mathrm{SE}\lambda 1)^2 + (d2 \cdot \mathrm{SE}\lambda 2)^2 + 2 \cdot d1 \cdot d2 \cdot \mathrm{SE}\lambda 12^2} \qquad \mathrm{SE}x = 0.007$$

CHAPTER FIFTEEN

Gumbel Distributions

INTRODUCTION

15.1 Definition

A continuous random variable X has a *Gumbel distribution* if its pdf has the form

$$f(x;\mu,\sigma)=\frac{1}{\sigma}\exp\left\{-\frac{x-\mu}{\sigma}-\exp\left\{-\frac{x-\mu}{\sigma}\right\}\right\};$$
$$\sigma>0,\quad -\infty<x,\mu<\infty. \qquad (15.1)$$

This distribution arises as the limiting form of the distribution (2.6) (see Section 2.5) of the largest order statistic in a sample of size n from some initial distribution with an unlimited upper tail.

For example, suppose the load L on a structure has been modeled as a Log-Normal variable and that during the structure's service life n such loads are expected to occur. Naturally, the largest load $L_{(n)}$ is of concern to the designer. Its distribution is given exactly by (2.6). However, if n is large, this distribution is closely approximated by (15.1). With increasing n, (2.6) approaches (15.1). The Gumbel model is therefore an asymptotic distribution of maxima.

15.2 Extreme Value Distributions

The largest, or smallest, observation in a finite sample is termed an *extreme value*. Its distribution converges with increasing sample size to one of three types. These three limiting distributions are of considerable interest in engineering because engineering decisions are often based on statistical inferences that focus on the extremes of data sets. To provide a context for such distributions, the following remarks briefly indicate under what conditions each distribution type arises.

Consider observing a sequence of samples, each of size n, from some *initial* distribution $f(y;\theta)$. The distribution of the extreme sample value $y_{(1)}$ or $y_{(n)}$ clearly depends on the nature of the corresponding tail of $f(y;\theta)$ from which the samples are drawn. If the tail of $f(y;\theta)$ is unbounded and decreases at least as rapidly as an exponential function, the asymptotic extreme value distribution is termed *type I*. For *maximum* extremes the pdf of that distribution is (15.1). Thus, the largest values in samples from Normal, Log-Normal, Exponential, Gamma, and Weibull processes give rise asymptotically to the Gumbel distribution.

Similar results are obtained for *minimum* extremes of data sets from the Normal and the Gumbel distributions. These give rise to the *type I* distribution of *minima*:

$$f(x; \mu, \sigma) = \frac{1}{\sigma}\exp\left\{+\frac{x-\mu}{\sigma} - \exp\left\{+\frac{x-\mu}{\sigma}\right\}\right\}, \qquad \sigma > 0. \tag{15.2}$$

This extreme value distribution is of some interest in reliability engineering where it models life data with an exponentially increasing hazard function:

$$h(x; \mu, \sigma) = \frac{1}{\sigma}\exp\left\{\frac{x-\mu}{\sigma}\right\}. \tag{15.3}$$

Refer to Chapters 12 and 17 for an introduction to life-data analysis. The properties of (15.2) are easily deduced from those of the Gumbel model, since one model is the mirror image of the other about the parameter μ.

If the initial distribution $f(y; \theta)$ has an unbounded upper tail, but not all of its moments are finite, then the asymptotic extreme value distribution is termed *type II* of *maxima*. This model is discussed in Chapter 16 on Frechet distributions.

If the initial distribution $f(y; \theta)$ features a bounded tail, then the asymptotic distribution of the corresponding extreme is termed *type III*. Thus, the smallest values in samples from the Log-Normal, Gamma, Beta, Frechet, and Weibull distributions give rise to the *type III* extreme value distribution of *minima*. This model is discussed in Chapter 17 on Weibull distributions.

Interest in the distribution of maxima arose chiefly from a need to explain larger than usual observations in astronomy. The theory of extreme values was developed in the 1920s by contributions from many researchers, chiefly among them von Mises, Frechet, Gnedenko, and Fisher and Tippet. Later these distributions were applied to a varity of problems in many fields of practical interest. Gumbel played a major role in further theoretical developments and in introducing extreme value statistics to important applications such as the prediction of floods. In engineering circles, therefore, the type I distribution of maxima (15.1) is usually called the *Gumbel model.*

The Gumbel variable X is related to a Weibull variable Y by $X = \ln(1/Y)$, with the Gumbel parameters μ and σ expressed by the Weibull parameters σ_w and λ_w as $\mu = \ln(1/\sigma_w)$ and $\sigma = 1/\lambda_w$.

PROPERTIES

15.3 Distribution Function

A Gumbel variable X, as defined by (15.1), has the cdf

$$F(x; \mu, \sigma) = \exp\left\{-\exp\left\{-\frac{x-\mu}{\sigma}\right\}\right\}; \qquad \sigma > 0, \quad -\infty < x, \mu < \infty. \tag{15.4}$$

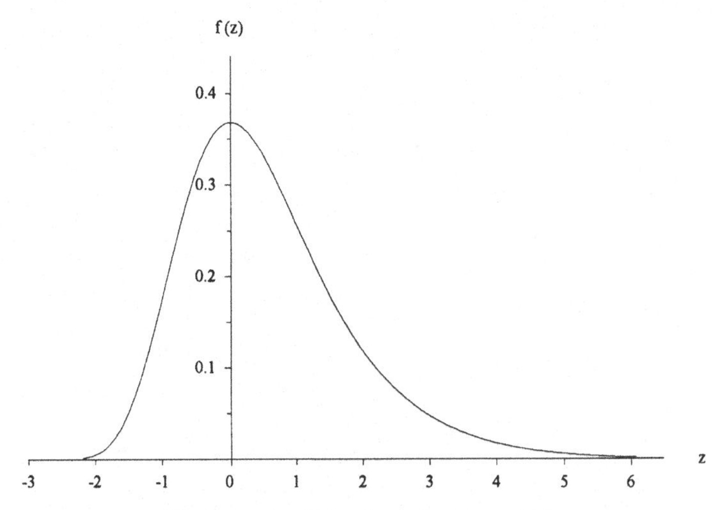

Figure 15.1. The Gumbel pdf for the reduced variable z.

This model has location-scale structure (see Section 9.2) with μ a location parameter and σ a scale parameter. The pdf of the reduced variable $Z = (x - \mu)/\sigma$ is therefore parameter free:

$$f(z) = \exp\{-z - \exp\{-z\}\}. \quad (15.5)$$

Figure 15.1 shows the *fixed* shape of the reduced Gumbel pdf.

Engineering applications of the Gumbel model $f(x; \mu, \sigma)$ usually involve maximum extremes. As a result the ratio μ/σ tends to be relatively large in practice. Hence, the pdf is located well away from the origin, so that the probability $F(0)$ is negligible. For practical purposes the sample space is then $x \geq 0$.

15.4 Reduced Variable

The moment-generating function (see Section 1.8) of the reduced variable Z is

$$M(t) = \Gamma(1 - t), \quad (15.6)$$

where Γ is the *gamma function* (see Section 13.1). The rth moment of Z about the origin is therefore

$$\mu'_r(Z) = \frac{d^r}{dt^r}\Gamma(1 - t)|_{t=0}. \quad (15.7)$$

Accordingly, the first few moments about the origin are

$$\mu'_1(Z) = \text{Euler's constant } \gamma = 0.57722\ldots, \quad (15.8)$$

$$\mu'_2(Z) = \gamma^2 + \frac{\pi^2}{6} = 1.97811\ldots, \quad (15.9)$$

$$\mu'_3(Z) = 5.44487\ldots, \quad (15.10)$$

$$\mu'_4(Z) = 23.56147\ldots. \quad (15.11)$$

The variance of Z is therefore

$$\mu_2(Z) = 1.64493\ldots, \tag{15.12}$$

and the next two moments about the mean are

$$\mu_3(Z) = 2.40411\ldots, \tag{15.13}$$

$$\mu_4(Z) = 14.61136\ldots. \tag{15.14}$$

The coefficient of variation is

$$cv(Z) = \frac{\sqrt{\mu_2}}{\mu'_1} = 2.22196\ldots. \tag{15.15}$$

The first shape factor (see Section 1.10) is

$$\gamma_1 = \frac{\mu_3}{(\mu_2)^{3/2}} = 1.13955\ldots, \tag{15.16}$$

indicating the fixed positive skew of the Gumbel model. The second shape factor is

$$\gamma_2 = \frac{\mu_4}{(\mu_2)^2} = 5.40000\ldots. \tag{15.17}$$

15.5 Gumbel Variable

The moments of the Gumbel variable X are obtained from the above expressions and the relation $X = \mu + \sigma Z$. The expected value and variance are therefore

$$\mu'_1(X) = \mu + 0.57722\,\sigma \tag{15.18}$$

and

$$\mu_2(X) = 1.64493\,\sigma^2. \tag{15.19}$$

The coefficient of variation of X is

$$cv(X) = \frac{1.28255}{0.57722 + \mu/\sigma}. \tag{15.20}$$

The mode value is

$$x_{\mathrm{m}} = \mu. \tag{15.21}$$

The quantile of order q is given by

$$x_q = \mu - \left[\ln\ln\left(\frac{1}{q}\right)\right]\sigma, \tag{15.22}$$

so that the median is $x_{0.5} = \mu + 0.36651\,\sigma$.

The matrix of minimum-variance-bounds (see Section 3.2) for estimators of μ and σ is

$$\begin{bmatrix} V_{\mu\mu} & V_{\mu\sigma} \\ V_{\mu\sigma} & V_{\sigma\sigma} \end{bmatrix} = \frac{\sigma^2}{n} \begin{bmatrix} 1.10867 & 0.25702 \\ 0.25702 & 0.60793 \end{bmatrix}, \tag{15.23}$$

so that

$$MVB_{\mu} = 1.10867 \frac{\sigma^2}{n}, \tag{15.24}$$

$$MVB_{\sigma} = 0.60793 \frac{\sigma^2}{n}, \tag{15.25}$$

$$MVB_{\mu\sigma} = 0.25702 \frac{\sigma^2}{n}. \tag{15.26}$$

The sampling variances of estimators can be compared with the above minimum values in order to indicate the efficiency of the estimators in extracting information from a sample (see Section 3.1).

15.6 Reproductive Property

The Gumbel distribution features a reproductive property for its own maximum sample extreme. That is, the distribution of $X_{(n)}$ is again Gumbel, with the same scale parameter but with the location parameter increased by the amount $\sigma \ln(n)$. This feature is verified by substituting (15.4) into (2.7). Thus, the pdf of $X_{(n)}$ is the same as that of X but shifted to the right by the above amount.

15.7 Simulation

Random observations from a Gumbel process with known parameter values μ and σ are easily simulated by inversion (see Section 2.8) as

$$x_i = \mu - \sigma \ln\left\{ \ln\left(\frac{1}{u_i}\right)\right\}, \tag{15.27}$$

where u_i is a Uniform random number on (0, 1) (see Section 14.6). It is advisable to check the adequacy of a simulated sample by comparing at least its first two moments with those of the given model.

PROBABILITY PLOT

15.8 Plotting Procedure

A probability plot of ordered data (see Section 3.13) is routinely used as a check on the distributional assumption of the Gumbel model. Taking natural logs twice of the cdf (15.4) gives a linear relation in $x_{(i)}$:

$$-\ln\{-\ln[F(x_{(i)})]\} = -\frac{\mu}{\sigma} + \frac{1}{\sigma} x_{(i)}.$$

Replacing $F(x_{(i)})$ by a plotting position such as the approximate median value of $F(x_{(i)})$,

$$p_i = \frac{i - 0.3}{n + 0.4},$$

gives

$$-\ln\{-\ln[p_i]\} = a + bx_{(i)}. \tag{15.28}$$

Thus, plotting $-\ln\{-\ln[p_i]\}$ versus the ordered data $x_{(i)}$ results in approximately a straight line, if the data came from a Gumbel process. A noticeable departure of the plot from a straight line would put the Gumbel postulate in question (see Section 3.14). The parameters μ and σ could be estimated from the intercept a and the slope b of the plot. Since the statistical properties of the resulting estimates are not known, this procedure is not recommended.

Probability paper for Gumbel data is commercially available, and allows direct entry of p_i and $x_{(i)}$ on the paper. However, the graphics capability of modern mathematics application programs makes these papers obsolete. See Example 15.1 for a Mathcad plot of some Gumbel data.

POINT ESTIMATES

15.9 Maximum Likelihood Estimates

The likelihood function of a sample of n independent observations on a Gumbel variable X is

$$L(\mu, \sigma) = \sigma^{-n} \exp\left\{ -\sum_{i=1}^{n} \frac{x_i - \mu}{\sigma} - \sum_{i=1}^{n} \exp\left\{ -\frac{x_i - \mu}{\sigma} \right\} \right\}. \tag{15.29}$$

The maximum likelihood equations (see Section 3.6) are

$$\sum_{i=1}^{n} \exp\left\{ -\frac{x_i - \widehat{\mu}}{\widehat{\sigma}} \right\} = n \tag{15.30}$$

and

$$\widehat{\sigma} + \widehat{\mu} + \frac{1}{n} \sum_{i=1}^{n} x_i \exp\left\{ -\frac{x_i - \widehat{\mu}}{\widehat{\sigma}} \right\} - \frac{\widehat{\mu}}{n} \sum_{i=1}^{n} \exp\left\{ -\frac{x_i - \widehat{\mu}}{\widehat{\sigma}} \right\} = \bar{x}, \tag{15.31}$$

where $\bar{x}$ is the sample average. Substituting the first equation into the second gives an equation in only $\widehat{\sigma}$:

$$\widehat{\sigma} - \bar{x} + \frac{\sum_{i=1}^{n} x_i \exp\{-\frac{x_i}{\widehat{\sigma}}\}}{\sum_{i=1}^{n} \exp\{-\frac{x_i}{\widehat{\sigma}}\}} = 0. \tag{15.32}$$

The solution value $\widehat{\sigma}$ is readily obtained with an equation solver. The ML estimate $\widehat{\mu}$ is then computed from (15.30) as

$$\widehat{\mu} = -\widehat{\sigma} \ln\left(\frac{1}{n}\sum_{i=1}^{n} \exp\left\{-\frac{x_i}{\widehat{\sigma}}\right\}\right). \tag{15.33}$$

The etimator $\widehat{\sigma}$ is biased downward. Using a convenient bias-correction formula of

$$b_n = 1 + \frac{2.2}{n^{1.13}}, \tag{15.34}$$

the unbiased estimate of the scale parameter becomes

$$\widehat{\sigma}_{\mathrm{u}} = b_n \widehat{\sigma}. \tag{15.35}$$

The values $\widehat{\mu}$ and $\widehat{\sigma}_{\mathrm{u}}$ are used for inferences on the Gumbel variable X. Hence, the *MVB*s (15.24) to (15.26) are corrected for b_n where appropriate, to give the following standard errors:

$$\mathrm{se}(\widehat{\mu}) = 1.05293\frac{\sigma}{\sqrt{n}}, \tag{15.36}$$

$$\mathrm{se}(\widehat{\sigma}_{\mathrm{u}}) = 0.77970\frac{\sigma b_n}{\sqrt{n}}, \tag{15.37}$$

$$\mathrm{se}(\widehat{\mu}, \widehat{\sigma}_{\mathrm{u}}) = 0.50697\sigma\sqrt{\frac{b_n}{n}}. \tag{15.38}$$

These standard errors are evaluated at the (biased) estimate $\widehat{\sigma}$.

The standard errors of parameter functions $g(\mu, \sigma)$ are approximated from the error propagation formula (see Section 3.3) using the above standard errors:

$$\mathrm{Var}(g) \doteq \left(\frac{\partial g}{\partial \mu}\mathrm{se}_{\mu}\right)^2 + \left(\frac{\partial g}{\partial \sigma}\mathrm{se}_{\sigma}\right)^2 + 2\left(\frac{\partial g}{\partial \mu}\right)\left(\frac{\partial g}{\partial \sigma}\right)\mathrm{se}_{\mu\sigma}^2. \tag{15.39}$$

For a statistical test-of-fit the *Anderson–Darling* statistic is recommended:

$$A = -n - \frac{1}{n}\sum_{i=1}^{n}(2i-1)[\ln(w_i) + \ln(1 - w_{n-i+1})], \tag{15.40}$$

where w_i is the Gumbel cdf (15.4), evaluated at the order statistic $x_{(i)}$ and the MLEs $\widehat{\mu}$ and $\widehat{\sigma}$ (biased). The value A is then modified to account for sample size:

$$A_{\mathrm{m}} = A\left(1 + \frac{0.2}{\sqrt{n}}\right). \tag{15.41}$$

Critical values[1] for A_m are:

α	0.100	0.050	0.025	0.010
A_c	0.637	0.757	0.877	1.038

Example 15.2 demonstrates the estimation process.

15.10 Moment Estimates

There are occasions when the engineer does not have a sample of observations on hand, but has prior information on the average value $\bar{x}$ of a random variable X and its coefficient of variation cv. A load process for a structural design is an example. A model of X can be deduced to accord with the given information by calculating *moment* estimators of μ and σ. From expression (15.18) and (15.19) these estimators are obtained in terms of $\bar{x}$ and the sample standard deviation $s = \bar{x} \cdot cv$ as:

$$\tilde{\mu} = \bar{X} - 0.45006\, S, \tag{15.42}$$

and

$$\tilde{\sigma} = 0.77970\, S \tag{15.43}$$

See Example 15.3 for an illustration.

It can be shown that these estimators have the following approximate sampling variances:

$$\text{Var}(\tilde{\mu}) \doteq 1.16781 \frac{\sigma^2}{n} \tag{15.44}$$

and

$$\text{Var}(\tilde{\sigma}) \doteq 1.10001 \frac{\sigma^2}{n}. \tag{15.45}$$

Relative to the minimum-variance-bounds (15.24) and (15.25), the efficiencies of these estimators are

$$e(\tilde{\mu}) = 0.95 \quad \text{and} \quad e(\tilde{\sigma}) = 0.55. \tag{15.46}$$

The efficiency of $\tilde{\sigma}$ is rather low. These estimators should therefore not be used when a sample of Gumbel observations is available.

[1] Extracted from R. B. D'Agostino, M. A. Stephens, *Goodness-of-Fit Techniques*, Marcel Dekker Inc., New York, 1986, by courtesy of the publisher.

INTERVAL ESTIMATES AND TESTS

15.11 Normal Approximation

For large samples, the asymptotic sampling distributions of ML estimators are convenient for constructing approximate confidence intervals on parameters and parameter functions. Recall from Section 3.7 that the sampling pdf of the ML estimate $\widehat{\theta}$ is asymptotically Normal, with mean value θ and variance MVB_θ:

$$f_N(\widehat{\theta}; \theta, \sqrt{MVB_\theta}), \tag{15.47}$$

where θ stands for μ or σ. Thus, the $(1-\alpha)$-level confidence interval on θ is obtained as

$$(l_1, l_2) = \theta \pm z_{\frac{\alpha}{2}} \sqrt{MVB_\theta}, \tag{15.48}$$

such that

$$Pr(l_1 \le \theta \le l_2) = 1 - \alpha. \tag{15.49}$$

See Example 15.4 for an illustration.

For parameter functions $g(\theta)$, the invariance property of ML estimators is used (see Section 3.7). That is, the sampling pdf of $\widehat{g} = g(\widehat{\theta})$ is asymptotically Normal, with mean value g and variance approximated by the error propagation formula (15.39). See Example 15.5, where two parameter functions are estimated.

15.12 Likelihood Ratio Approximation

The errors in the preceding Normal approximations increase as the sample size decreases, particularly for the lower confidence limit. The likelihood-ratio method (see Section 3.9) tends to give more accurate results.

Recall that the statistic

$$LR(\theta) = 2\ln[L(\widehat{\theta})] - 2\ln[L(\theta)] \tag{15.50}$$

is approximately Chi-squared distributed with $\nu = 1$ degree of freedom. Thus, a $(1-\alpha)$-level confidence interval on θ comprises those values θ for which $LR(\theta) \le \chi^2_{1,1-\alpha}$. For a two-parameter model, one parameter in (15.50) must be expressed in terms of the other parameter by its ML equation. Thus, a confidence interval on σ is obtained from

$$LR(\sigma) = 2\ln[L(\widehat{\mu}, \widehat{\sigma})] - 2\ln[L(\mu\{\sigma\}, \sigma)], \tag{15.51}$$

where $\mu\{\sigma\}$ is defined by the ML equation (15.33).

Similarly, a confidence interval on μ is obtained from

$$LR(\mu) = 2\ln[L(\widehat{\mu}, \widehat{\sigma})] - 2\ln[L(\mu, \sigma\{\mu\})], \tag{15.52}$$

where $\sigma\{\mu\}$ is defined by the ML equation (15.31). See Example 15.4 for an illustration.

A confidence interval on a parameter function $g(\theta)$ is more difficult to obtain since the relation among parameters, implied by $g(\theta)$, constrains the solution process. As an illustration, consider the quantile $x_q = \mu - \sigma[\ln\ln(1/q)]$, so that

$$\mu\{x_q, \sigma\} = x_q + \sigma[\ln\ln(1/q)]. \tag{15.53}$$

Substituting (15.53) into the likelihood function (15.29) and differentiating with respect to σ gives the constraint relation:

$$ML(\sigma) = \frac{n(\bar{x} - x_q)}{\sigma^2} + \frac{\ln(q)}{\sigma}\sum_{i=1}^{n}\frac{x_i - x_q}{\sigma}\exp\left\{-\frac{x_i - x_q}{\sigma}\right\} - \frac{n}{\sigma}. \tag{15.54}$$

A confidence interval on x_q is then obtained from

$$LR(x_q) = 2\ln[L(\widehat{\mu}, \widehat{\sigma})] - 2\ln[L(\mu\{x_q, \sigma\{x_q\}\}, \sigma\{x_q\})], \tag{15.55}$$

where $\mu\{x_q, \sigma\}$ is given by (15.53) and $\sigma\{x_q\}$ is the solution of $ML(\sigma) = 0$ according to (15.54). Again see Example 15.4 for an illustration of the computations. Note that the *biased* estimate $\widehat{\sigma}$ is used in formulas (15.51) to (15.55).

A test on the null hypothesis $H_o : \sigma = \sigma_o$ is based of the test statistic $LR(\sigma_o)$, given by (15.51). Similarly, a test on $H_o : \mu = \mu_o$ is based on the test statistic $LR(\mu_o)$, given by (15.52). Critical values are obtained as the α-quantiles of the Chi-squared distribution with $\nu = 1$.

15.13 Simulation Results

For small samples, several exact and approximate methods for obtaining confidence intervals on Gumbel parameters are available in the statistical literature. One of the simplest approaches is based on the Monte Carlo simulation[2] of samples from the standard Exponential model ($\mu_E = 0, \sigma_E = 1$). This approach is based on the following results.

If X is Gumbel-distributed with parameters μ and σ, then the transformed variable

$$Y = \exp\{-(X - \mu)/\sigma\}$$

is distributed as a standard Exponential variable. By transforming the ML equations (15.32) and (15.33) from X to Y, the Weibull ML equations (17.42) and (17.40) are obtained with $T_1 = \sigma/\widehat{\sigma}$ replacing $\widehat{\lambda}_w$ and $T_2 = \exp\{-(\widehat{\mu} - \mu)/\widehat{\sigma}\}$ replacing $\widehat{\sigma}_w^{\widehat{\lambda}_w}$. Thus, the sampling distributions of T_1 and T_2 are independent

[2] Thoman, D. R., Bain, L. J., Antle, C. E., "Inferences on the Parameters of the Weibull Distribution," *Technometrics*, Vol. 11, No. 3, pp. 445–460, 1969.

of μ and σ, and they can be generated from the Weibull estimates $\widehat{\sigma}_w$ and $\widehat{\lambda}_w$ of standard Exponential samples. Tables 17.1 and 17.2 present the simulated results. These are used for Gumbel parameters as follows.

For selected confidence levels $(1-\alpha)$ and sample sizes n, Table 17.1 gives the values l_1 and l_2 such that the confidence interval on σ is obtained from

$$Pr(\widehat{\sigma}l_1 \leq \sigma \leq \widehat{\sigma}l_2) = 1 - \alpha. \quad (15.56)$$

Table 17.2 gives the values l_1 and l_2 such that the confidence interval on μ is obtained from

$$Pr(\widehat{\mu} + \widehat{\sigma}l_1 \leq \mu \leq \widehat{\mu} + \widehat{\sigma}l_2) = 1 - \alpha. \quad (15.57)$$

The probability level γ of the tables relates to the above α values as $\gamma = \alpha/2$ for the lower limit and $\gamma = 1 - \alpha/2$ for the upper limit. See Example 15.4 for an illustration. Note that the tables are based on the (biased) MLE $\widehat{\sigma}$, which should therefore be used in (15.56) and (15.57).

APPLICATIONS

15.14 General

The Gumbel model is an appropriate *postulate* for measurement variables that arise as the maximum extreme of some underlying random process whose characteristics have not been quantified. Many problems of practical interest in engineering fall into that class. Most of these problems involve the modeling of maxima of load processes (e.g., gust loads, thermal loads) or phenomena causing such loads (e.g., earthquake magnitude, flood volume). These maxima are naturally of prime concern to the designer of engineering devices or structures. Most other applications relate to size phenomena affecting material strength (e.g., size of material flaws and surface imperfections) and to event magnitudes that impact engineering operations (e.g., queue length, order lead time). In all cases the knowledge that the observed variable X is a maximum of some underlying random process provides a strong basis for postulating the Gumbel model. See Example 15.5 for an illustration of working with this model as a predictive tool.

Because the Gumbel distribution is of fixed shape, its application as a general model for fitting a given data frequency pattern is not recommended. Other models with shape flexibility should be considered for that purpose.

15.15 Life Testing

The Gumbel distribution sometimes arises in reliability engineering as a lifetime model from a consideration of the physical mechanism that leads to a component's failure. Suppose that the component can be considered as a *system* of identical parallel *elements* (whose life characteristics cannot be readily

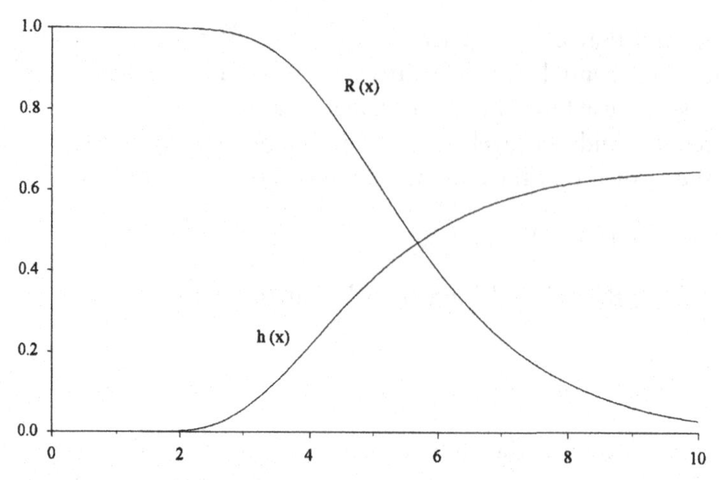

Figure 15.2. The Gumbel reliability and hazard functions for $\mu = 5$ and $\sigma = 1.5$.

measured). If one can argue that the life of the component equals the life of its longest-lived element, then it is reasonable to postulate the Gumbel distribution for component life.

For example, in many industrial situations a component's primary hazard is its corrosive environment. When redundant component elements are all exposed to this environment, and the component can be expected to continue to function until its last element fails, the Gumbel distribution is a good candidate to model component life. A multistrand electrical ground wire, subject to corrosion, is an example of the above situation.

The hazard function (Section 12.16) is a characterizing feature of a lifetime variable. For a Gumbel variable this function is

$$h(x;\,\mu,\sigma) = \frac{\exp\{-\frac{x-\mu}{\sigma}\}}{\sigma\left(\exp\left\{+\exp\left\{-\frac{x-\mu}{\sigma}\right\}\right\} - 1\right)}, \tag{15.58}$$

with limiting value $h = 1/\sigma$ as $x \to \infty$. Figure 15.2 shows the Gumbel reliability function $R(x) = 1 - F(x)$ and the hazard function $h(x)$ for $\mu = 5$ and $\sigma = 1.5$ time units. We see that a component with these particular life characteristics is highly reliable to about 2 time units and then deteriorates rapidly. Refer to Chapters 12 and 17 for a discussion of life-data analysis.

15.16 Flood Control

One of the major areas of application of the Gumbel distribution is the design of flood control structures, where the magnitude of an extreme load event is the controlling input to the design process. That event is the maximum daily flow rate X in a year, called the *annual flood*. Since X is a random variable, a given structure, designed to contain a flow rate x_D, may fall short of containing the maximum flow rate X occurring in any one year. Such an event is termed an *exceedance*. The designed structure is therefore associated with a *risk* that the annual flood X exceeds the design capacity x_D. Prudent design limits that risk to an economically and socially acceptable value α_D for the anticipated service life, N years, of the structure. Thus

$$\begin{aligned} 1 - \alpha_D &= Pr(X \leq x_D) \text{ for } N \text{ consecutive years} \\ &= [F(x_D)]^N \end{aligned}$$

(see Section 2.5). Here F is the Gumbel cdf (15.4), so that

$$1 - \alpha_D = \left[\exp\left\{ -\exp\left\{ -\frac{x_D - \mu}{\sigma} \right\} \right\} \right]^N.$$

The Gumbel parameters μ and σ are estimated from data on annual floods for the watershed in question. Expressing the preceding equation in x_D gives

$$\begin{aligned} x_D &= \mu - \sigma \ln\left[-\frac{1}{N} \ln(1 - \alpha_D) \right] \\ &= \mu + \sigma[\ln(N) - \ln\{-\ln(1 - \alpha_D)\}]. \end{aligned} \tag{15.59}$$

This is the flow rate to which a structure would be designed, given a risk level α_D and a service life of N years. We see that the design capacity x_D is proportional to the log of the service life N.

In Section 7.11 the *return period* of an exceedance is defined as the *average* number of years to the occurrence of an exceedance event:

$$t = \frac{1}{p_D},$$

where p_D is the annual probability of the event's occurrence. For the Gumbel postulate the return period becomes

$$t = \frac{1}{1 - F(x_D; \mu, \sigma)}.$$

Substituting x_D from (15.59) gives

$$t = \frac{1}{1 - (1 - \alpha_D)^{\frac{1}{N}}} \tag{15.60}$$

as the return period of an exceedance of the design capacity x_D for a Gumbel variable X. Expression (15.60) holds for other models F as well.

Recall from Section 7.11 that t is the *average* of the number of periods (of fixed length) between adjoining exceedances and that this variable (*Geometric*) is highly dispersed. Hence, the probability of observing an exceedance sooner than is indicated by t can be surprisingly high. As t increases, that probability approaches the constant 0.632. Example 15.6 provides an illustration.

PROBABILISTIC DESIGN

15.17 Design Reliability

The preceding remarks on exceedances can be expanded to encompass basic procedures of probabilistic engineering design of structures.

Failure of a structure occurs when the load L exceeds the strength S of the structure. Since the load L usually varies during the structure's service life or *mission*, and the strength S varies among structure specimens, failure is a probabilistic event. Its probability

$$P_{\mathrm{f}} = Pr(L \geq S)$$

determines the *risk* associated with the structure. The complement of P_{f} is the structure's *reliability*

$$R = Pr(L < S),$$

which is a design parameter, often fixed as a specification.

With $f_{\mathrm{L}}(l)$ and $f_{\mathrm{S}}(s)$ representing the load and strength pdfs, the structure's reliability becomes

$$R = \int_{l=0}^{s} \int_{s=0}^{\infty} f_{\mathrm{L}}(l)\, f_{\mathrm{S}}(s)\, dl\, ds. \tag{15.61}$$

Figure 15.3 shows the *interference* of the load and strength densities. This interference generates the failure probability P_{f}.

When the load cdf is of closed form, Equation (15.61) simplifies to

$$R = \int_{s=0}^{\infty} F_{\mathrm{L}}(s)\, f_{\mathrm{S}}(s)\, ds. \tag{15.62}$$

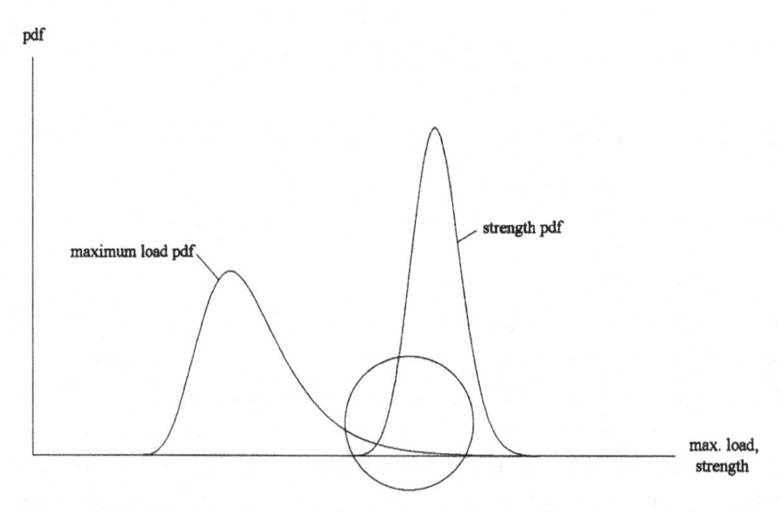

Figure 15.3. The interference of a maximum load pdf and a strength pdf.

Alternatively, if the strength cdf is of closed form, Equation (15.61) simplifies to

$$R = 1 - \int_{l=0}^{\infty} F_S(l)\, f_L(l)\, dl. \tag{15.63}$$

The above expressions imply that L is the *maximum* load value *during the mission* of the structure and that observations are available to model that maximum load directly as $f_L(l)$.

For example, suppose a critical piece of an aircraft structure is strain gauged on n aircraft and recordings are obtained on the maximum strain that occurred during the total flight time between mandatory overhauls. The stress engineer could then calculate and model the maximum load during the aircraft's mission (between overhauls) directly from the n data.

If the Gumbel distribution is an appropriate model for the maximum load, the reliability (15.62) becomes

$$R = \int_{s=0}^{\infty} \exp\left\{-\exp\left\{-\frac{s-\mu}{\sigma}\right\}\right\} f_S(s)\, ds. \tag{15.64}$$

The strength pdf $f_S(s)$ is determined from the material property M, which relates to the anticipated failure mode. Information on M is modeled by a pdf $f_M(m)$. The strength S of the device is then the *product* of a critical dimensional quantity A and the material property M. Hence, f_M transforms to the strength pdf f_S by the *scale* transformation $S = MA$. The goal of probabilistic design is to determine the dimension A such that R has a specified value. See Example 15.7 for an illustration.

15.18 Initial Loads

It is more usual that direct observations on a maximum load are not available. Instead, the designer may have information on the distribution of a general load variable X that his design will be subjected to. This load may be termed the *initial* load, with pdf $f(x)$. What is of concern to the designer is the *maximum* load L that might occur *during the mission* of the design. If n initial loads X can be expected to occur during the mission, the quantity of interest is therefore the order statistic $L = X_{(n)}$. Its distribution (see Section 2.5),

$$F_n(l) = [F(l)]^n, \tag{15.65}$$

is the appropriate load input to the probabilistic design process as described in the preceding section.

For example, suppose the distribution $F(x)$ of impact loads on an aircraft landing gear is known from strain-gauge recordings during a sequence of landing operations. If this aircraft is expected to go through n landings between inspections, the *maximum* load cdf for that mission is then given by (15.65).

When n is large, as it often is for realistic engineering designs, $F_n(l)$ is closely represented by a Gumbel model. Although the Gumbel parameters μ and σ can be calculated approximately from $F(x)$, modern computational tools allow the exact representation (15.65) to be used directly in the design process.

The occurrence of an initial load X is often a random event, and thus the *number* of occurrences N during the design mission is a random variable. Suppose that the occurrence process can be modeled by a Poisson pmf (see Chapter 8) as

$$p(n; \alpha t) = \frac{(\alpha t)^n \exp\{-\alpha t\}}{n!}.$$

Here the parameter α is the average occurrence rate, t is the period of the device mission, and αt is the expected number of events during t. The maximum load cdf (15.65) is now *conditional* on a given value n. The *joint* probability of $(L \leq l)$ and $(N = n)$ is then

$$J(l, n) = [F(l)]^n p(n; \alpha t),$$

and therefore the *marginal* probability of the event $(L \leq l)$ becomes

$$\begin{aligned} F_L(l) &= \sum_{i=0}^{\infty} [F(l)]^i p(i; \alpha t) \\ &= \exp\{-\alpha t\} \sum_{i=0}^{\infty} \frac{[\alpha t F(l)]^i}{i!}. \end{aligned}$$

This reduces to

$$F_L(l) = \exp\{-\alpha t[1 - F(l)]\}, \tag{15.66}$$

with pdf

$$f_L(l) = \alpha t \cdot f(l) \exp\{-\alpha t[1 - F(l)]\}. \tag{15.67}$$

This result holds for any initial general load distribution $F(l)$, and it provides the appropriate load input to the reliability expression (15.62), if the load occurrence process is Poisson. See Example 15.8 for an illustration.

A simpler approach is to set n in (15.65) to its Poisson mean value αt, so that

$$F_L(l) = [F(l)]^{\alpha t} \tag{15.68}$$

becomes the (simpler) load input to the design process. The corresponding pdf is

$$f_L(l) = \alpha t \cdot f(l)[F(l)]^{\alpha t - 1}. \tag{15.69}$$

By expanding (15.66) and (15.68) as power series in $F(l)$, it is easily shown that both expressions are approximately the same for αt large.

EXAMPLE 15.1

The following annual maximum temperatures (in degrees centigrade) were recorded at a weather station:

32.7, 30.4, 31.8, 33.2, 33.8, 35.3, 34.6, 33.0,
32.0, 35.7, 35.5, 36.8, 40.8, 38.7, 36.7.

A Gumbel model has been postulated for this variable. Check graphically if this postulate is reasonable.

$n := 15 \qquad i := 1..n$

$t_i :=$ $\quad x := \text{sort}(t) \qquad p_i := \dfrac{i-0.3}{n+0.4}$ $\qquad$ data ordinates: $r_i := -\ln(-\ln(p_i))$

t_i
32.7
30.4
31.8
33.2
33.8
35.3
34.6
33.0
32.0
35.7
35.5
36.8
40.8
38.7
36.7

The data plot looks fairly linear. The Gumbel postulate is therefore not unreasonable.

EXAMPLE 15.2

Given the annual maximum temperature of Example 15.1, assume a Gumbel model, estimate the ML parameters and their standard errors, test the model for fit at the 5% level of significance, and produce a probability plot of the data and the ML model. Estimate the 95-percentile and its standard error.

$n := 15 \qquad i := 1..n$

$t_i := \qquad x := \text{sort}(t)$

t_i
32.7
30.4
31.8
33.2

1. Estimation

$$S1(s) := \sum_i x_i \cdot \exp\left(-\frac{x_i}{s}\right) \qquad S2(s) := \sum_i \exp\left(-\frac{x_i}{s}\right) \qquad xb := \frac{1}{n} \cdot \sum_i x_i$$

33.8
35.3
34.6
33.0
32.0
35.7
35.5
36.8
40.8
38.7
36.7

$s := 1 \qquad \sigma := \text{root}\left(\frac{S1(s)}{S2(s)} + s - xb, s\right)$ $\qquad \sigma = 2.241$

$\mu := -\sigma \cdot \ln\left(\frac{S2(\sigma)}{n}\right)$ $\qquad \mu = 33.5$

$bn := 1 + \frac{2.2}{n^{1.13}} \qquad \sigma_u := \sigma \cdot bn$ $\qquad \sigma_u = 2.47$

2. Standard errors

$SE\mu := 1.05293 \cdot \frac{\sigma}{\sqrt{n}} \qquad SE\sigma := 0.7797 \cdot \frac{bn \cdot \sigma}{\sqrt{n}}$

$SE\mu\sigma := 0.50697 \cdot \sqrt{\frac{bn}{n}} \cdot \sigma$

$SE\mu = 0.61$
$SE\sigma = 0.50$
$SE\mu\sigma = 0.31$

3. Test-of-fit

$w_i := \exp\left(-\exp\left(-\frac{x_i - \mu}{\sigma}\right)\right)$

$A := \left[-n - \frac{1}{n} \cdot \sum_i (2 \cdot i - 1) \cdot (\ln(w_i) + \ln(1 - w_{n-i+1}))\right] \cdot \left(1 + \frac{0.2}{\sqrt{n}}\right)$

$A = 0.140$

The 5% critical value is 0.757. Since the sample value A is smaller, there is no reason to reject the estimated model at the 5% significance level.

4. Probability plot of data and estimated model

$p_i := \frac{i - 0.3}{n + 0.4} \qquad r_i := -\ln(-\ln(p_i))$ (ordinates for the data)

$z_i := \frac{x_i - \mu}{\sigma_\mu}$ (ordinates for the model when the cdf is substituted for p)

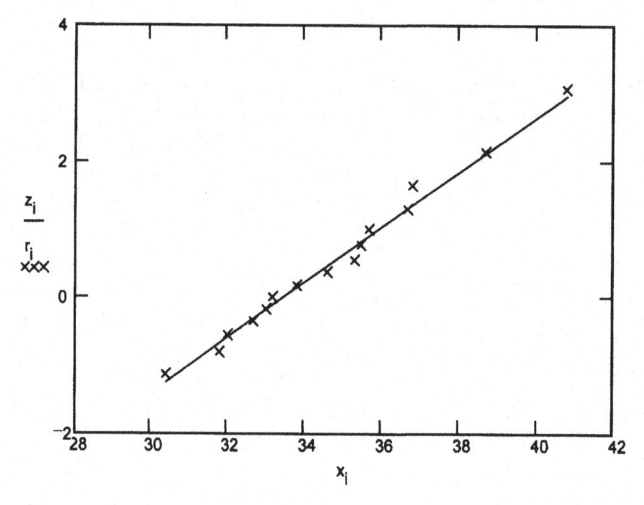

5. pdf plot

$j := 1..200 \qquad t_j := 20 + 0.2 \cdot j$

$$f(x) := \frac{1}{\sigma_u} \cdot \exp\left(-\frac{x-\mu}{\sigma_u} - \exp\left(-\frac{x-\mu}{\sigma_u}\right)\right)$$

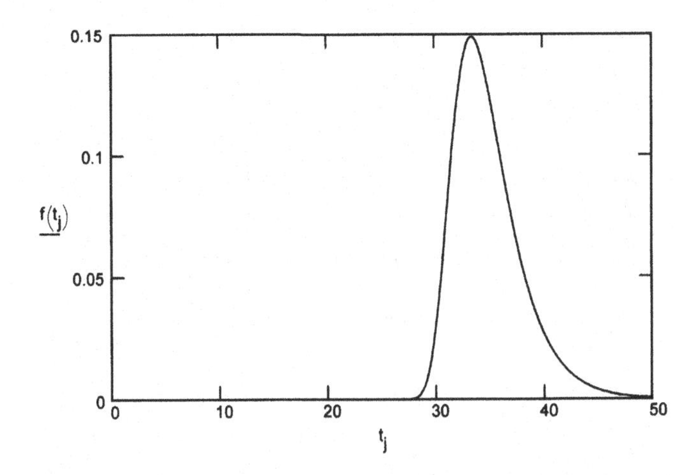

6. 95-percentile

$$c := \ln\left(\ln\left(\frac{1}{0.95}\right)\right)$$

$x95(a, b) := a - b \cdot c$ $\qquad\qquad x95(\mu, \sigma_u) = 40.8$

standard error: $da := 1 \qquad db := -c$

$\text{var} := (da \cdot \text{SE}\mu)^2 + (db \cdot \text{SE}\sigma)^2$

$\qquad +2 \cdot da \cdot db \cdot \text{SE}\mu\sigma^2$

$\text{se}X95 := \sqrt{\text{var}}$ $\qquad\qquad \text{se}X95 = 1.8$

EXAMPLE 15.3

Based on experience with similar designs, an engineer judges the average maximum design load during the mission of the design prototype to be 3,500 kg. He further estimates that the coefficient of variation for these maximum loads would be 0.3. Estimate a reasonable Gumbel model for these loads. Based on the given information, what is the probability that the maximum load exceeds 4,500 kg?

$\text{xbar} := 3500 \qquad \text{cv} := 0.3 \qquad S := X\text{bar} \cdot \text{cv} \qquad S = 1050$

$\mu := X\text{bar} - 0.45 \cdot S$ $\qquad\qquad \mu = 3028$

$\sigma := 0.7797 \cdot S$ $\qquad\qquad \sigma = 819$

The estimated Gumbel cdf is therefore

$$F(x) := \exp\left(-\exp\left(-\frac{x-\mu}{\sigma}\right)\right)$$

The probability that X exceeds 4,500 kg is $\quad 1 - F(4500) = 0.15$

EXAMPLE 15.4

For the data of Example 15.1, calculate the 90% confidence intervals on the parameters and on the 95-percentile.

From Example 15.2 the ML estimates and their standard errors are

$\mu := 33.5 \qquad \sigma_u := 2.47 \qquad \sigma := 2.241 \qquad x95 := 40.81$

$\mathrm{SE}\mu := 0.61 \qquad \mathrm{SE}\sigma := 0.50 \qquad \mathrm{se}X95 := 1.77$

1. Parameters

i) Normal approximation: Standard Normal 95-percentile: $z95 := 1.645$

$L\sigma := \sigma_u - z95 \cdot \mathrm{SE}\sigma \qquad L\sigma = 1.65$

$U\sigma := \sigma_u + z95 \cdot \mathrm{SE}\sigma \qquad U\sigma = 3.29$

$L\mu := \mu - z95 \cdot \mathrm{SE}\mu \qquad L\mu = 32.5$

$U\mu := \mu + z95 \cdot \mathrm{SE}\mu \qquad U\mu = 34.5$

ii) Likelihood ratio method: Chi-squared 90-percentile at $v = 1$: $K90 := 2.71$

$n := 15 \qquad i := 1..n$

$x_i := \qquad xb := \frac{1}{n} \cdot \sum_i x_i$

x_i
32.7
30.4
31.8
33.2
33.8
35.3
34.6
33.0
32.0
35.7
35.5
36.8
40.8
38.7
36.7

log of likelihood function:

$$LL(\mu,\sigma) := -n \cdot \ln(\sigma) - \frac{n \cdot xb}{\sigma} + \frac{n \cdot \mu}{\sigma} - \sum_i \exp\left(-\frac{x_i - \mu}{\sigma}\right)$$

scale parameter:

$$m(a) := -a \cdot \ln\left(\frac{1}{n} \cdot \sum_i \exp\left(-\frac{x_i}{a}\right)\right)$$

$$LR(a) := 2 \cdot LL(\mu,\sigma) - 2 \cdot LL(m(a), a)$$

$a := \sigma - 1 \qquad \sigma l := \mathrm{root}(LR(a) - K90, a) \qquad \sigma l = 1.65$

$a := \sigma + 1 \qquad \sigma u := \mathrm{root}(LR(a) - K90, a) \qquad \sigma u = 3.21$

location parameter:

$$a := \sigma \qquad s(b) := \text{root}\left(a + b + \frac{1}{n} \cdot \sum_i x_i \cdot \exp\left(-\frac{x_i - b}{a}\right) - \frac{b}{n} \cdot \sum_i \exp\left(-\frac{x_i - b}{a}\right) - xb,\ a\right)$$

$$LR(b) := 2 \cdot LL(\mu, \sigma) - 2 \cdot LL(b, s(b))$$

$b := \mu - 1 \qquad \mu l := \text{root}(LR(b) - K90, b) \qquad \mu l = 32.4$

$b := \mu + 1 \qquad \mu u := \text{root}(LR(b) - K90, b) \qquad \mu u = 34.6$

iii) simulated results:

From Table 17.1, at $n = 15$ and $\gamma = 0.05$ $l1 := 0.770$
at $\gamma = 0.95$ $l2 := 1.564$

Thus, from (15.56):

$L\sigma 1 := \sigma \cdot l1 \qquad L\sigma 1 = 1.73$

$L\sigma 2 := \sigma \cdot l2 \qquad L\sigma 2 = 3.50$

From Table 17.2, at $n = 15$ and $\gamma = 0.05$ $l1 := -0.509$
at $\gamma = 0.95$ $l2 := 0.499$

Thus, from (15.57):

$L\mu 1 := \mu + \sigma \cdot l1 \qquad L\mu 1 = 32.4$

$L\mu 2 := \mu + \sigma \cdot l2 \qquad L\mu 2 = 34.6$

2. Percentile

$q := 0.95$

i) Normal approximation:

$xl := X95 - z95 \cdot \text{se}X95 \qquad xl = 37.9$

$xu := X95 + z95 \cdot \text{se}X95 \qquad xu = 43.7$

ii) Likelihood ratio method (denote the quantile by Q):

$$ML(a, Q) := \frac{n}{a} \cdot \left(\frac{xb - Q}{a} + \frac{\ln(q)}{n} \cdot \sum_i \exp\left(-\frac{x_i - Q}{a}\right) \cdot \frac{x_i - Q}{a} - 1\right)$$

$$a := \sigma - 1 \qquad s1(Q) := \text{root}(ML(a, Q), a)$$

$$m(Q) := Q + \ln\left(\ln\left(\frac{1}{q}\right)\right) \cdot s1(Q)$$

$$LR(Q) := 2 \cdot LL(\mu, \sigma) - 2 \cdot LL(m, (Q), s1(Q))$$

$Q := X95 - 5 \qquad Ql := \text{root}(LR(Q) - K90, Q) \qquad Ql = 37.9$

$Q := X95 + 3 \qquad Qu := \text{root}(LR(Q) - K90, Q) \qquad Qu = 43.5$

EXAMPLE 15.5

Consider the data of Example 15.1. The designer of a cooling tower near the weather station needs to know the 90-percentile of the annual maximum temperature in terms of a 70% upper confidence limit. He would also like to know the most likely maximum annual temperature for a 50-year period.

From Example 15.2:

$\mu := 33.5 \qquad \mathrm{SE}_\mu := 0.61$

$\sigma_u := 2.47 \qquad \mathrm{SE}_\sigma := 0.50 \qquad \mathrm{SE}_{\mu\sigma} := 0.31$

1. 90-percentile

$c := \ln\left(\ln\left(\frac{1}{0.9}\right)\right) \qquad X90 := \mu - \sigma_u \cdot c \qquad\qquad X90 = 39.1$

derivatives of $X90$: $\quad dm := 1 \qquad ds := -c$

std. error of $X90$:

$\mathrm{SE}_x := \sqrt{(dm \cdot \mathrm{SE}_\mu)^2 + (ds \cdot \mathrm{SE}_\sigma)^2 + 2 \cdot dm \cdot ds \cdot \mathrm{SE}_{\mu\sigma}{}^2} \qquad\qquad \mathrm{SE}_x = 1.4$

Assuming a Normal sampling distribution for the ML estimate $X90$, the upper 70% confidence limit is obtained as follows:

std. Normal 70-percentile:

$s := 1 \qquad Z70 := \mathrm{root}(\mathrm{cnorm}(s) - 0.7, s) \qquad Z70 = 0.524$

The required confidence limit on $X90$ is therefore

$CL_x := X90 + Z70 \cdot \mathrm{SE}_x \qquad\qquad CL_x = 39.8$

2. 50-year maximum

From the reproductive property of the Gumbel model, the distribution of the 50-year maximum is also Gumbel, with location parameter shifted to

$\mu_{50} := \mu + \sigma_u \cdot \ln(50)$

The mode is equal to the location parameter, hence the mode is $\mu_{50} = 43.2$

derivatives of μ_{50} : $\quad Dm := 1 \qquad Ds := \ln(50)$

std. error of this estimate:

$\mathrm{SE}_{50} = \sqrt{(Dm \cdot \mathrm{SE}_\mu)^2 + (Ds \cdot \mathrm{SE}_\sigma)^2 + 2 \cdot Dm \cdot Ds \cdot \mathrm{SE}_{\mu\sigma}{}^2} \qquad\qquad \mathrm{SE}_{50} = 2.2$

3. pdfs

In order to clarify the relative positions of the several models used here, the designer is provided with the following graph of pdfs:

$$k := 1..200$$

$$x_k := 28 + 0.1 \cdot k \quad G_k := \frac{1}{\sigma_u} \cdot \exp\left[-\left(\frac{x_k - \mu}{\sigma_u}\right) - \exp\left[-\left(\frac{x_k - \mu}{\sigma_u}\right)\right]\right]$$

(measurement model for max. temp.)

$$s_k := 30 + 0.1 \cdot k \quad N_k := \frac{1}{SE_x \cdot \sqrt{2 \cdot \pi}} \cdot \exp\left[-\frac{1}{2} \cdot \left(\frac{s_k - X90}{SE_x}\right)^2\right]$$

(sampling model for $X90$)

$$p_k := 37 + 0.1 \cdot k \quad P_k := \frac{1}{\sigma_u} \cdot \exp\left(\frac{p_k - \mu_{50}}{\sigma_u} - \exp\left(\frac{p_k - \mu_{50}}{\sigma_\mu}\right)\right)$$

(derived model for 50-year max.)

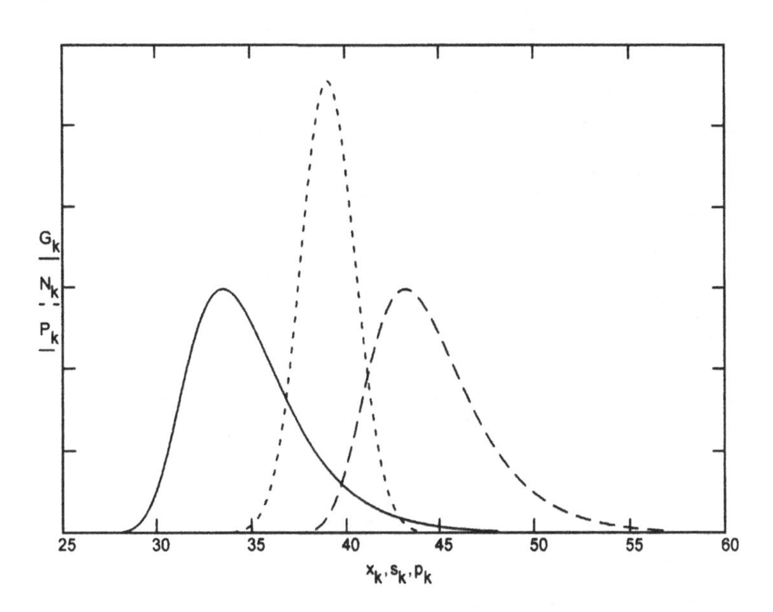

EXAMPLE 15.6

A highway culvert is to be designed to accommodate annual floods of a mountain creek at a risk of exceedance of 8% for a service life of 30 years. From records of the past 18 years the Gumbel parameters were estimated as $\mu = 2.3$ and $\sigma_u = 0.8$ (m^3/s). What is the design flow rate, the return period of exceedances, and the probability that an exceedance occurs sooner than the return period?

1. Design flow rate

$\alpha := 0.08 \qquad N := 30 \qquad c := -\ln\left(-\frac{1}{N}\cdot \ln(1-\alpha)\right)$

$\mu := 2.3 \qquad \sigma_u := 0.8 \qquad c = 5.9$

$X_D := \mu + c\cdot\sigma_u \qquad\qquad X_D = 7.01$

standard error: $n := 18 \qquad b_n := 1 + \frac{2.2}{n^{1.13}} \qquad \sigma := \frac{\sigma_u}{b_n}$

$$SE_\mu := 1.053\cdot\frac{\sigma}{\sqrt{n}} \qquad SE_\sigma := 0.789\cdot\frac{\sigma_u}{\sqrt{n}}$$

$$SE_{\mu\sigma} := 0.507\cdot\sqrt{\frac{b_n}{n}}\cdot\sigma$$

$$SEd := \sqrt{SE_\mu^2 + (c\cdot SE_\sigma)^2 + 2\cdot c\cdot SE_{\mu\sigma}^2} \qquad SEd = 0.95$$

2. Return period

$$t := \left[1-(1-\alpha)^{\frac{1}{N}}\right]^{-1} \qquad t = 360$$

3. Exceedance probability

The annual exceedance probability is

$$p := 1 - \exp\left(-\exp\left(-\frac{X_D-\mu}{\sigma_u}\right)\right) \qquad p = 0.00278$$

From the Geometric distribution, the probability of occurrence of an exceedance event is

$$P := 1-(1-p)^{t-1} \qquad p = 0.63$$

EXAMPLE 15.7

The critical dimension of a tension link in a clam-shell fastener for a spacecraft has been designed to a cross-sectional area $A = 0.38$ cm^2. The ultimate tensile stress (N/cm^2) of the material is Log-Normal with parameters $\mu_y = 11.3$ and $\sigma_y = 0.079$. The maximum load (N) on the link occurs during launch and has been postulated to be Gumbel distributed with parameters $\mu_L = 20{,}000$ and $\sigma_L = 3{,}000$. What is the reliability of the design? How should the link be dimensioned to give a reliability of 0.99?

1. The strength pdf is obtained from the tensile stress pdf by a scale transformation from stress Y to strength S:

   $Y = S/A$ with Jacobian $1/A$

$A := 0.38$
$\mu_y := 11.3$

$\sigma_y := 0.079 \quad f_s(s, A) := \dfrac{1}{\frac{s}{A} \cdot \sigma_y \cdot \sqrt{2 \cdot \pi}} \cdot \exp\left[-0.5 \cdot \left(\dfrac{\ln\left(\frac{s}{A}\right) - \mu_y}{\sigma_y}\right)^2\right] \cdot \dfrac{1}{A}$

(Log-Normal pdf)

The maximum load cdf is

$\mu_L := 20000 \qquad \sigma_L := 3000 \qquad F_L(L) := \exp\left(-\exp\left(-\dfrac{L - \mu_L}{\sigma_L}\right)\right)$

(Gumbel cdf)

The reliability integral (15.64) is evaluated as

$R(A) := \displaystyle\int_{25,000}^{60,000} F_L(t) \cdot f_s(t, A)\, dt \qquad R(A) = 0.96$

(Note: the limits must be chosen carefully to bracket the integrand!)

2. For a specified reliability value $R = 0.99$, the required dimension Ad is

$a := 0.45 \qquad A_d := \text{root}(R(a) - 0.99, a) \qquad A_d = 0.432$

The load and strength pdfs plot as follows:

$i := 1..200 \qquad s_i := 25000 + 120 \cdot i \qquad L_i := 10000 + 150 \cdot i$

$f_L(L) := \dfrac{1}{\sigma_L} \cdot \exp\left(-\dfrac{L - \mu_L}{\sigma_L} - \exp\left(-\dfrac{L - \mu_L}{\sigma_L}\right)\right)$

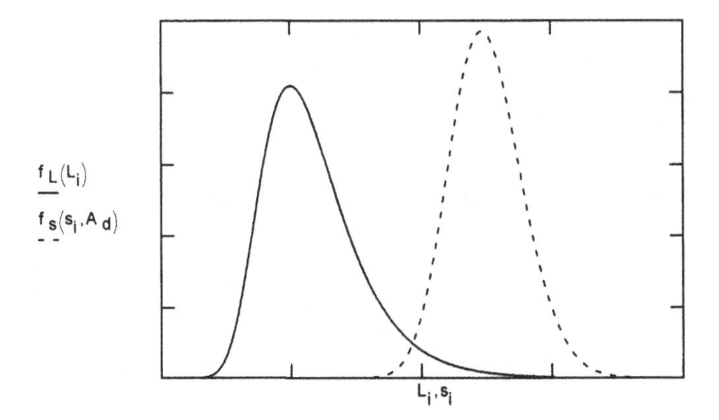

EXAMPLE 15.8

A cable has been designed to tether an experimental weather balloon. Measurements on gust speeds during weather disturbances at the design altitude are available and have been modeled by a Gamma distribution. When transformed to cable tension L (N), the

Gamma parameters become $\sigma_L = 840$ and $\lambda_L = 4.34$. Atmospheric disturbances occur according to a Poisson process with estimated rate 1.7 per week. The ultimate tensile stress Y (N/cm^2) of the cable material is modeled as a Weibull variable with parameters $\sigma_Y = 83{,}600$ and $\lambda_Y = 16.4$. The effective cable cross-sectional area is $A = 0.196$ cm^2. The experiment mission is 12 months. What is the cable reliability?

1. Loads

Initial load pdf: $\sigma_L := 840 \qquad \lambda_L := 4.34$

$$f(x) := \frac{1}{\sigma_L \cdot \Gamma(\lambda_L)} \cdot \left(\frac{x}{\sigma_L}\right)^{\lambda_L - 1} \cdot \exp\left(-\frac{x}{\sigma_L}\right) \qquad \text{(see Gamma chapter)}$$

initial load cdf: $F(x) = \int_0^x f(t)dt$

maximum load pdf (15.67): $\alpha t := 1.7 \cdot 52 \quad f_L(x) := \alpha t \cdot f(x) \cdot \exp(-\alpha t \cdot (1 - F(x)))$

2. Strength

The strenght pdf is obtained from the given stress pdf by a scale transformation $\sigma_s = \sigma_{y \cdot A}$, where A is the given cross-sectional cable area.

$A := 0.196 \qquad \sigma_y := 83600 \qquad \lambda_y := 16.4 \qquad \sigma_s := \sigma_y \cdot A \qquad \lambda_s := \lambda_y$

strength cdf: $F_s(s) := 1 - \exp\left[-\left(\frac{s}{\sigma_s}\right)^{\lambda_s}\right]$ (see Weibull chapter)

strength pdf: $f_s(s) := \frac{\lambda_s}{\sigma_s^{\lambda_s}} \cdot s^{\lambda_s - 1} \cdot \exp\left[-\left(\frac{s}{\sigma_s}\right)^{\lambda_s}\right]$

The initial and maximum load pdfs and the strength pdf plot as follows:

$i := 1 .. 100 \qquad x_i := i \cdot 160 \qquad xx_i := 5000 + i \cdot 150 \qquad s_i := 5000 + i \cdot 200$

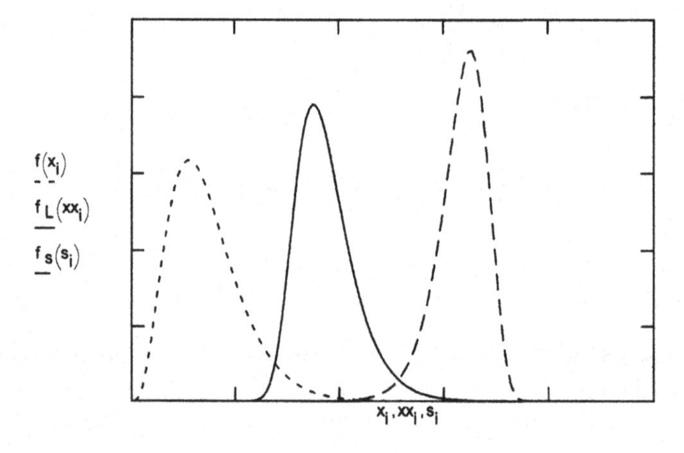

3. Reliability evaluation

$$R := 1 - \int_{8000}^{19000} F_s(t) \cdot f_L(t)dt \qquad R = 0.997$$

(Note: The limits must be chosen carefully to bracket the integrand!)

CHAPTER SIXTEEN

Frechet Distributions

INTRODUCTION

16.1 Definition

A continuous random variable X has a Frechet distribution if its pdf has the form

$$f(x; \sigma, \lambda) = \frac{\lambda}{\sigma}\left(\frac{\sigma}{x}\right)^{\lambda+1} \exp\left\{-\left(\frac{\sigma}{x}\right)^{\lambda}\right\}; \quad x \geq 0; \quad \sigma, \lambda > 0. \tag{16.1}$$

This model arises as the limiting form of the distribution (2.6) (see Section 2.5) of the largest order statistic in a sample of size n from an initial distribution for which not all moments exist. In this context the Frechet distribution is termed a *type II extreme value* distribution of *maxima* (see Section 15.2).

For example, the order statistic $X_{(n)}$ has the limiting form (16.1) if X has the *Pareto* cdf

$$F(x) = 1 - \left(\frac{\sigma}{x}\right)^{\lambda},$$

for which only moments of order less than λ exist, or if X has the *Cauchy* cdf

$$F(x) = \frac{1}{2} + \frac{1}{\pi}\tan^{-1}\left(\frac{x}{\sigma}\right),$$

for which moments of any order do not exist.

This extreme value distribution is becoming increasingly important in engineering statistics as a suitable model to represent phenomena with unusually large maximum observations. In engineering circles, this distribution is often called the *Frechet model*, in honor of one of the pioneers of extreme value statistics.

PROPERTIES

16.2 Distribution Function

A Frechet variable X, as defined by (16.1), has the cdf

$$F(x; \sigma, \lambda) = \exp\left\{-\left(\frac{\sigma}{x}\right)^{\lambda}\right\}. \tag{16.2}$$

This model has scale structure (see Section 9.2), with σ a scale parameter and λ a shape parameter. The pdf of the reduced variable $Z = X/\sigma$ is therefore a function of λ only:

$$f(z; \lambda) = \lambda z^{-\lambda-1} \exp\{-z^{-\lambda}\}. \tag{16.3}$$

Figure 16.1 shows this pdf for several values of λ, indicating the shape flexibility of the Frechet model.

The Frechet variable X is related to a Gumbel variable Y (see Chapter 15) by

$$X = \exp\{Y\}, \tag{16.4}$$

with the Frechet parameters σ and λ expressed in terms of Gumbel parameters μ_G and σ_G as

$$\sigma = \exp\{\mu_G\} \quad \text{and} \quad \lambda = \frac{1}{\sigma_G}. \tag{16.5}$$

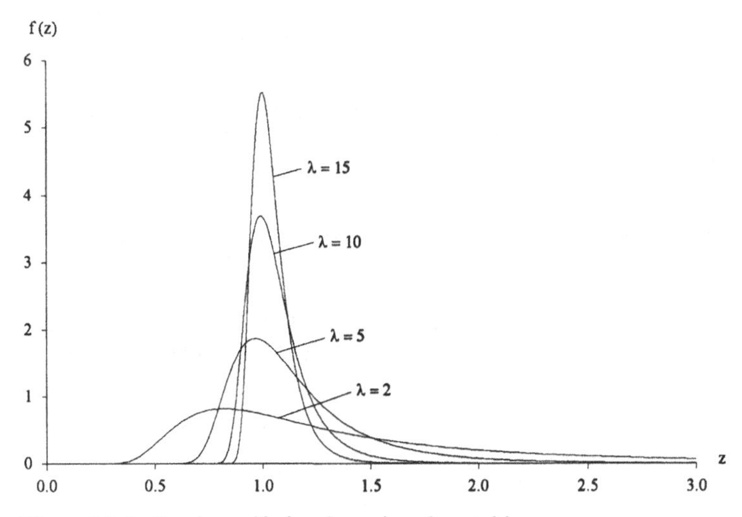

Figure 16.1. Frechet pdfs for the reduced variable z.

16.3 Frechet Variable

The rth moment of X about the origin is directly evaluated from the definition of expectations (see Section 1.7) as

$$\mu_r'(X) = \sigma^r \Gamma\left(1 - \frac{r}{\lambda}\right), \tag{16.6}$$

where Γ is the *gamma function* (see Section 13.1). Clearly, only moments of order $r < \lambda$ exist. In engineering applications λ is usually greater than 2.

The expected value of X is

$$\mu_1'(X) = \sigma\Gamma\left(1 - \frac{1}{\lambda}\right), \qquad \lambda > 1. \tag{16.7}$$

The variance of X is

$$\mu_2(X) = \sigma^2\left[\Gamma\left(1 - \frac{2}{\lambda}\right) - \Gamma^2\left(1 - \frac{1}{\lambda}\right)\right], \qquad \lambda > 2. \tag{16.8}$$

The coefficient of variation of X is therefore a function of λ only:

$$cv(X) = \sqrt{\frac{\Gamma(1 - \frac{2}{\lambda})}{\Gamma^2(1 - \frac{1}{\lambda})} - 1}, \qquad \lambda > 2 \tag{16.9}$$

(see Figure 16.2). The mode value is given by

$$x_m = \sigma\left(\frac{\lambda}{1 + \lambda}\right)^{\frac{1}{\lambda}}. \tag{16.10}$$

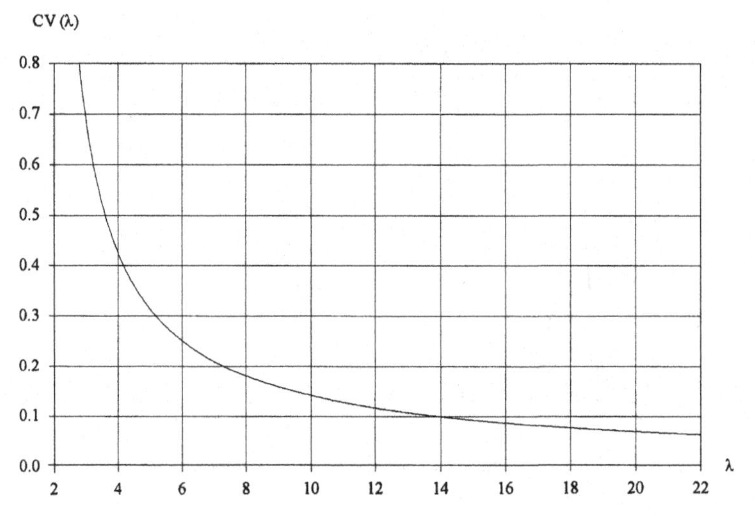

Figure 16.2. The coefficient of variation as a function of the shape parameter.

The quantile of order q is

$$x_q = \sigma \left[\ln \left(\frac{1}{q} \right) \right]^{-\frac{1}{\lambda}}. \tag{16.11}$$

The first shape factor (see Section 1.10) is

$$\gamma_1 = \frac{\Gamma\left(1 - \frac{3}{\lambda}\right) - 3\Gamma\left(1 - \frac{1}{\lambda}\right)\Gamma\left(1 - \frac{2}{\lambda}\right) + 2\Gamma^3\left(1 - \frac{1}{\lambda}\right)}{\left[\Gamma\left(1 - \frac{2}{\lambda}\right) - \Gamma^2\left(1 - \frac{1}{\lambda}\right)\right]^{1.5}}, \quad \lambda > 3 \tag{16.12}$$

(see Fig. 16.3) which shows that the pdf is always positively skewed, with decreasing skew as λ increases. The second shape factor is

$$\gamma_2 = \frac{\Gamma\left(1 - \frac{4}{\lambda}\right) - 4\Gamma\left(1 - \frac{1}{\lambda}\right)\Gamma\left(1 - \frac{3}{\lambda}\right) + 6\Gamma^2\left(1 - \frac{1}{\lambda}\right)\Gamma\left(1 - \frac{2}{\lambda}\right) - 3\Gamma^4\left(1 - \frac{1}{\lambda}\right)}{\left[\Gamma\left(1 - \frac{2}{\lambda}\right) - \Gamma^2\left(1 - \frac{1}{\lambda}\right)\right]^2}, \quad \lambda > 4 \tag{16.13}$$

(see Fig. 16.4). These two figures suggest that, even as λ becomes quite large, the Normal shape values $\gamma_1 = 0$ and $\gamma_2 = 3$ are not approached, and the Frechet pdf retians its distinct shape, different from a Normal pdf.

The matrix of minimum-variance-bounds (see Section 3.2) for estimators of σ and λ is

$$\begin{bmatrix} V_{\sigma\sigma} & V_{\sigma\lambda} \\ V_{\sigma\lambda} & V_{\lambda\lambda} \end{bmatrix} = \frac{1}{n} \begin{bmatrix} 1.10867\frac{\sigma^2}{\lambda^2} & 0.25702\sigma \\ 0.25702\sigma & 0.60793\lambda^2 \end{bmatrix}, \tag{16.14}$$

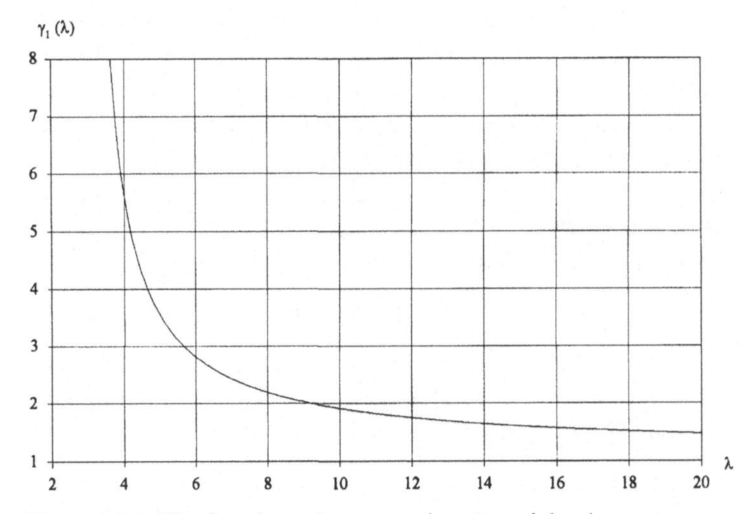

Figure 16.3. The first shape factor as a function of the shape parameter.

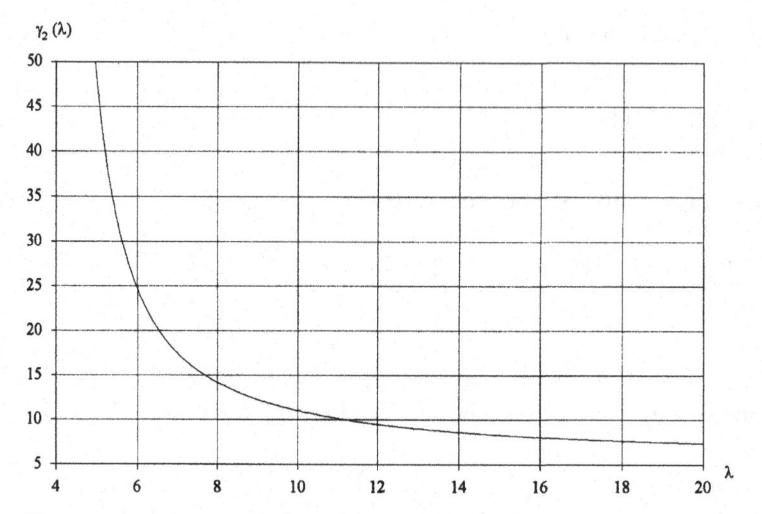

Figure 16.4. The second shape factor as a function of the shape parameter.

so that the MVBs are

$$MVB_\sigma = 1.10867\frac{\sigma^2}{n\lambda^2}, \tag{16.15}$$

$$MVB_\lambda = 0.60793\frac{\lambda^2}{n}, \tag{16.16}$$

$$MVB_{\sigma\lambda} = 0.25702\frac{\sigma}{n}. \tag{16.17}$$

16.4 Reproductive Properties

The Frechet distribution features a reproductive property for its maximum extreme. That is, the distribution of $X_{(n)}$ is again Frechet, with the same shape parameter but with the scale parameter increased to $\sigma n^{1/\lambda}$. This feature is easily verified by substituting (16.2) into (2.7). Thus, the pdf of $X_{(n)}$ has the same shape as that of X but is rescaled as given above.

In addition, a power transformation of X yields a Frechet variable. That is, if the transformed variable is $T = bX^a$, then T is also Frechet, but with scale parameter $b\sigma^a$ and shape parameter λ/a.

16.5 Simulation

Random observations from a Frechet process with known parameter values σ and λ are simulated by inversion (see Section 2.8) as

$$x_i = \sigma\left[\ln\left(\frac{1}{u_i}\right)\right]^{-\frac{1}{\lambda}}, \tag{16.18}$$

where u_i is a Uniform random number on $(0, 1)$ (see Section 14.6). It is advisable to check the adequacy of a simulated sample by comparing at least its first two moments with those of the given model.

PROBABILITY PLOT

16.6 Plotting Procedure

A probability plot of ordered data (see Section 3.13) is routinely used to check the distributional assumption of the Frechet model. Taking natural logs twice of the cdf (16.2) gives a linear relation in $\ln(x_{(i)})$:

$$-\ln[-\ln\{F(x_{(i)})\}] = -\lambda \ln(\sigma) + \lambda \ln(x_{(i)}).$$

Replacing $F(x_{(i)})$ by a plotting position such as the approximate median value, of $F(x_{(i)})$

$$p_i = \frac{i - 0.3}{n + 0.4},$$

gives

$$-\ln[-\ln\{p_i\}] = a + b\ln(x_{(i)}). \tag{16.19}$$

Thus, plotting $-\ln[-\ln\{p_i\}]$ versus the logs of ordered data $x_{(i)}$ results in approximately a straight line, if the data came from a Frechet process. A noticeable departure of the plot from a straight line would put the Frechet postulate in question (see Section 3.14). The parameters σ and λ could be estimated from the intercept a and the slope b of the plot. Because the statistical properties of the resulting estimates are not known, this procedure is not recommended. See Example 16.1 for a Mathcad plot of some Frechet data.

POINT ESTIMATES

16.7 Maximum Likelihood Estimates

The likelihood function of a sample of n independent observations on a Frechet variable X is

$$L(\sigma, \lambda) = \lambda^n \sigma^{n\lambda} \prod_{i=1}^{n} x_i^{-(\lambda+1)} \exp\left\{-\sigma^\lambda \sum_{i=1}^{n} x_i^{-\lambda}\right\}. \tag{16.20}$$

The maximum likelihood equations (see Section 3.6) are

$$\widehat{\sigma} = \left(\frac{1}{n}\sum_{i=1}^{n} x_i^{-\widehat{\lambda}}\right)^{-\frac{1}{\widehat{\lambda}}} \tag{16.21}$$

and

$$\frac{n}{\widehat{\lambda}} + n\ln(\widehat{\sigma}) - \sum_{i=1}^{n} \ln(x_i) - \sum_{i=1}^{n} \left(\frac{\widehat{\sigma}}{x_i}\right)^{\widehat{\lambda}} \ln\left(\frac{\widehat{\sigma}}{x_i}\right) = 0. \tag{16.22}$$

Substituting the first equation into the second gives an equation in $\widehat{\lambda}$ only:

$$\frac{1}{\widehat{\lambda}} + \frac{\sum_{i=1}^{n} x_i^{-\widehat{\lambda}} \ln(x_i)}{\sum_{i=1}^{n} x_i^{-\widehat{\lambda}}} = \frac{1}{n} \sum_{i=1}^{n} \ln(x_i). \tag{16.23}$$

The solution value $\widehat{\lambda}$ is readily obtained computationally with an equation solver. The ML estimate $\widehat{\sigma}$ is then computed from (16.21). See Example 16.2 for an illustration of the computations.

The estimator $\widehat{\lambda}$ is biased upward. Using a convenient bias-correction formula of

$$b_n = 1 + \frac{2.2}{n^{1.13}}, \tag{16.24}$$

we get an unbiased estimate of the shape parameter:

$$\widehat{\lambda}_{\mathrm{u}} = \frac{\widehat{\lambda}}{b_n}. \tag{16.25}$$

The *MVBs* (16.15–16.17) are adjusted for b_n, where appropriate, to give the following standard errors:

$$\mathrm{se}(\widehat{\sigma}) = 1.05293 \frac{\sigma}{\lambda\sqrt{n}}, \tag{16.26}$$

$$\mathrm{se}(\widehat{\lambda}_{\mathrm{u}}) = 0.77970 \frac{\lambda}{b_n\sqrt{n}}, \tag{16.27}$$

$$\mathrm{se}(\widehat{\sigma}, \widehat{\lambda}_{\mathrm{u}}) = 0.50697 \sqrt{\frac{\sigma}{n b_n}}. \tag{16.28}$$

These standard errors are evaluated at the (biased) estimate $\widehat{\lambda}$. The above ML parameter and error estimates are recommended, whenever a sample of observations is available on a Frechet variable.

The standard errors of parameter functions $g(\sigma, \lambda)$ are approximated from the error propagation formula (see Section 3.3) using the above standard errors:

$$\mathrm{Var}(g) = \left(\frac{\partial g}{\partial \sigma}\mathrm{se}_{\sigma}\right)^2 + \left(\frac{\partial g}{\partial \lambda}\mathrm{se}_{\lambda}\right)^2 + 2\left(\frac{\partial g}{\partial \sigma}\right)\left(\frac{\partial g}{\partial \lambda}\right)\mathrm{se}^2_{\sigma\lambda}. \tag{16.29}$$

See Example 16.2 where a percentile function is estimated.

For a statistical test-of-fit the *Anderson–Darling* statistic is recommended:

$$A = -n - \frac{1}{n}\sum_{i=1}^{n}(2i-1)[\ln(w_i) + \ln(1 - w_{n-i+1})]. \tag{16.30}$$

Here w_i is the Frechet cdf (16.2), evaluated at the order statistic $x_{(i)}$ and the ML estimates $\widehat{\sigma}$ and $\widehat{\lambda}$ (biased). The value A is then modified to account for sample size:

$$A_{\mathrm{m}} = A\left(1 + \frac{0.2}{\sqrt{n}}\right). \tag{16.31}$$

Critical values[1] for A_{m} are:

α	0.100	0.050	0.025	0.010
A_c	0.637	0.757	0.877	1.038

Example 16.2 demonstrates the estimation process.

16.8 Moment Estimates

Occasionally the engineer has available to him only prior information on the average value $\bar{x}$ of a variable X, and its coefficient of dispersion cv, rather than a set of observations on X. A load process for a structural design is an example. A model of X can be deduced from the given information by calculating the *moment* estimates of its parameters. The moment estimate $\widetilde{\lambda}$ is obtained by inverting expression (16.9) with an equation solver. Equation (16.7) then provides a moment estimate of σ:

$$\widetilde{\sigma} = \frac{\bar{x}}{\Gamma\left(1 - \frac{1}{\widetilde{\lambda}}\right)}. \tag{16.32}$$

Example 16.3 provides an illustration. Because the efficiencies of these estimators are relatively low, they should not be used when a sample of Frechet observations is available.

INTERVAL ESTIMATES AND TESTS

16.9 Normal Approximation

For large samples, the asymptotic Normal sampling distributions of ML estimators can be used to construct approximate confidence intervals (see Section 3.7).

[1] Extracted from R. B. D'Agostino, M. A. Stephens, *Goodness-of-Fit Techniques*, Marcel Dekker Inc., New York, 1986, by courtesy of the publisher.

That is, the sampling pdf of the ML estimate $\widehat{\theta}$ is asymptotically Normal, with mean value θ and variance MVB_θ:

$$f_N(\widehat{\theta}; \theta, \sqrt{MVB_\theta}), \tag{16.33}$$

where θ stands for σ or λ. The $(1-\alpha)$-level confidence interval on θ is therefore

$$(l_1, l_2) = \theta \pm z_{\frac{\alpha}{2}} \sqrt{MVB_\theta}, \tag{16.34}$$

such that

$$Pr(l_1 \leq \theta \leq l_2) = 1 - \alpha.$$

See Example 16.4 for an illustration.

For parameter functions $g(\theta)$, the invariance property of ML estimators is used (see Section 3.7). That is, the sampling pdf of $\widehat{g} = g(\widehat{\theta})$ is asymptotically Normal, with mean value g and variance approximated by the error propagation formula (16.29). The unbiased estimate $\widehat{\lambda}_u$ and its bias-corrected variance are recommended for these calculations.

16.10 Likelihood-Ratio Approximation

The preceding Normal approximation becomes less accurate as the sample size decreases, particularly for the lower confidence limit. The likelihood-ratio method (see Section 3.9) tends to give superior results.

Recall that the statistic

$$LR(\theta) = 2\ln[L(\widehat{\theta})] - 2\ln[L(\theta)] \tag{16.35}$$

is approximately Chi-squared distributed with $\nu = 1$ degree of freedom. Thus, a $(1-\alpha)$-level confidence interval on θ comprises those values θ for which $LR(\theta) \leq \chi^2_{1,1-\alpha}$. For a two-parameter model, one parameter in (16.35) must be expressed in terms of the other parameter by its ML equation. Thus, a confidence interval on λ is obtained from

$$LR(\lambda) = 2\ln[L(\widehat{\sigma}, \widehat{\lambda})] - 2\ln[L(\sigma\{\lambda\}, \lambda)], \tag{16.36}$$

where $\sigma\{\lambda\}$ is defined by the ML equation (16.21). Similarly, a confidence interval on σ is obtained from

$$LR(\sigma) = 2\ln[L(\widehat{\sigma}, \widehat{\lambda})] - 2\ln[L(\sigma, \lambda\{\sigma\})], \tag{16.37}$$

where $\lambda\{\sigma\}$ is defined by the ML equation (16.22).

A confidence interval on a parameter function $g(\theta)$ is more difficult to obtain since the relation among parameters, implied by $g(\theta)$, constrains the solution process. Consider, as an illustration, the quantile $x_q = \sigma[\ln(1/q)]^{-1/\lambda}$, so that

$$\sigma\{x_q, \lambda\} = x_q\left[\ln\left(\frac{1}{q}\right)\right]^{\frac{1}{\lambda}}. \tag{16.38}$$

Substituting (16.38) into the likelihood function (16.20) and differentiating with respect to λ gives the constraint relation

$$ML(\lambda) = \frac{n}{\lambda} + n\ln(x_q) - \sum_{i=1}^{n}\ln(x_i) - \ln\left(\frac{1}{q}\right)x_q^{\lambda}\ln(x_q)\sum_{i=1}^{n}x_i^{-\lambda} + \ln\left(\frac{1}{q}\right)x_q^{\lambda}\sum_{i=1}^{n}x_i^{-\lambda}\ln(x_i). \quad (16.39)$$

A confidence interval is then obtained from

$$LR(x_q) = 2\ln[L(\widehat{\sigma},\widehat{\lambda})] - 2\ln[L(\sigma\{x_q,\lambda\{x_q\}\},\lambda\{x_q\})], \quad (16.40)$$

where $\sigma\{x_q,\lambda\}$ is given by (16.38) and $\lambda\{x_q\}$ is the solution of $ML(\lambda) = 0$ according to (16.39). Note that the *biased* estimate $\widehat{\lambda}$ is used in expressions (16.36) to (16.40). See Example 16.4 for an illustration.

A test on the null hypothesis $H_o\!:\lambda = \lambda_o$ is based on the test statistic $LR(\lambda_o)$, given by (16.36). Similary, a test on $H_o\!:\sigma = \sigma_o$ is based on the test statistic $LR(\sigma_o)$, given by (16.37). Critical values are obtained as the α-quantiles of the Chi-squared distribution with $\nu = 1$.

16.11 Simulation Results

For small samples, the simplest approach to obtain confidence intervals on Frechet parameters is based on the Monte Carlo simulation[2] of samples from the standard Exponential model ($\mu_E = 0, \mu_E = 1$). This approach is based on the following results.

If X is a Frechet variable with parameters σ and λ, then the transformed variable

$$Y = \left(\frac{\sigma}{X}\right)^{\lambda}$$

is a *standard* Exponential variable. By transforming the ML equations (16.23) and (16.21) from X to Y, the Weibull ML equations (17.42) and (17.40) are obtained with $T_1 = \lambda/\widehat{\lambda}$ replacing $\widehat{\lambda}_w$ and $T_2 = (\widehat{\sigma}/\sigma)^{\widehat{\lambda}}$ replacing $\widehat{\sigma}_w^{\widehat{\lambda}_w}$. Thus, the sampling distributions of T_1 and T_2 are independent of σ and λ, and they can be generated from the Weibull estimates $\widehat{\sigma}_w$ and $\widehat{\lambda}_w$ of standard Exponential samples. Tables 17.1 and 17.2 present the simulated results. These are used for Frechet parameters as follows.

[2] Thoman, D. R., Bain, L. J., Antle, C. E., "Inferences on the Parameters of the Weibull Distribution," *Technometrics*, Vol. 11, No. 3, pp. 445–460, 1969.

For selected confidence levels $(1-\alpha)$ and sample sizes n, Table 17.1 gives the values l_1 and l_2 such that the confidence interval on λ is obtained from

$$Pr\left(\frac{\widehat{\lambda}}{l_2} \leq \lambda \leq \frac{\widehat{\lambda}}{l_1}\right) = 1 - \alpha. \tag{16.41}$$

Table 17.2 gives the values l_1 and l_2 such that the confidence interval on σ is obtained from

$$Pr\left(\widehat{\sigma}\exp\left\{\frac{l_1}{\widehat{\lambda}}\right\} \leq \sigma \leq \widehat{\sigma}\exp\left\{\frac{l_1}{\widehat{\lambda}}\right\}\right) = 1 - \alpha. \tag{16.42}$$

The probability level γ of the tables relates to the above α values as $\gamma = \alpha/2$ for the lower limit and $\gamma = 1 - \alpha/2$ for the upper limit. Example 16.4 illustrates this. Note that the tables are based on the (biased) ML estimate $\widehat{\lambda}$, which should therefore be used in (16.41) and (16.42).

APPLICATIONS

16.12 General

The engineering design of structures and devices requires information on the maximum loads the structure or device can be expected to experience during its service mission (see Sections 15.17 and 15.18). These loads are often caused by natural phenomena such as floods, snow accumulation, wave forces, earthquakes, wind pressure, and so forth. Data on the maxima of such phenomena occassionally feature larger values in the upper tail than one would expect from a Gumbel postulate. If there is no reason to suspect these values to be *outliers*, the Frechet distribution often leads to a superior data fit due to its inherently longer upper tail. Thus, the Frechet distribution finds application as an alternative to the Gumbel distribution, modeling maximum extreme value phenomena.

EXAMPLE 16.1

The maximum yearly snow load (coded data) on a proposed stadium roof has been calculated on the basis of a 16-year precipitation record:

13.1, 14.8, 17.0, 18.2, 15.3, 14.9, 14.0, 16.8,
32.1, 22.8, 21.1, 15.3, 15.7, 21.0, 28.8, 21.6

In view of several unusually high values, a Frechet model has been postulated for this load. Check graphically if this postulate is reasonable.

$n := 16 \qquad i := 1..n$

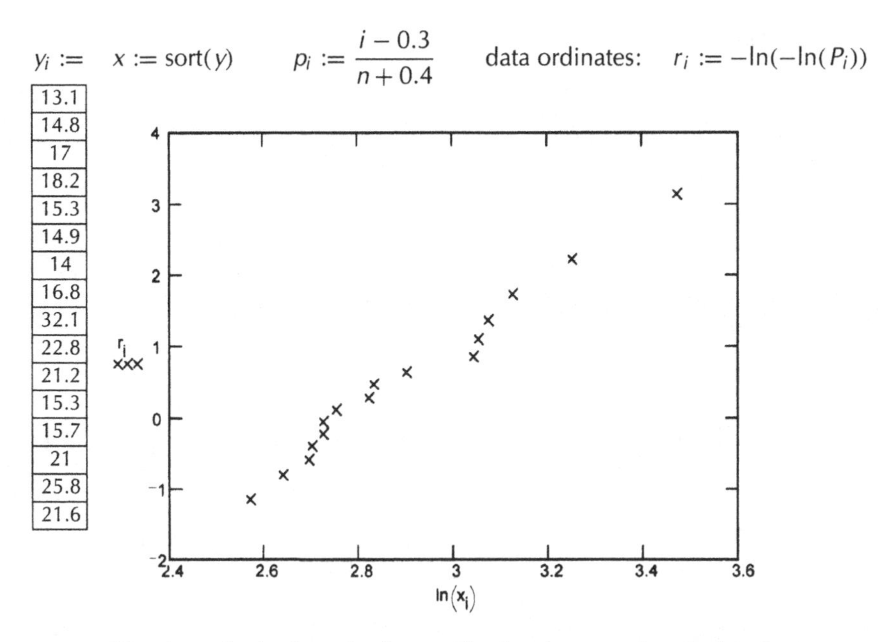

The data plot looks quite linear. The Frechet postulate is therefore not unreasonable.

EXAMPLE 16.2

Given the data on snow loads of Example 16.1, estimate the parameters of the assumed Frechet model and their standard errors. Test the estimated model for fit at the 5% level of significance, and produce a probability plot of the data and the ML model. For a stadium roof service life of 30 years, estimate the 80-percentile of a maximum 30 year snow load and its standard error.

$n := 16 \qquad i := 1..n$

$y_i := \qquad x := \mathrm{sort}(y)$

y_i
13.1
14.8
17
18.2
15.3
14.9
14
16.8
32.1
22.8
21.2
15.3
15.7
21
25.8
21.6

1. Estimation

$$s1 := \sum_i \ln(x_i) \qquad s2(\lambda) := \sum_i (x_i)^{-\lambda} \cdot \ln(x_i) \qquad s3(\lambda) := \sum_i (x_i)^{-\lambda}$$

$$L := 5 \qquad \lambda := \mathrm{root}\left(\frac{1}{L} + \frac{s2(L)}{s3(L)} - \frac{s1}{n}, L\right) \qquad \lambda = 5.48$$

$$\sigma := \left(\frac{s3(\lambda)}{n}\right)^{-\frac{1}{\lambda}} \qquad \sigma = 16.27$$

$$bn := 1 + \frac{2.2}{n^{1.13}} \qquad \lambda_u := \frac{\lambda}{bn} \qquad \lambda_u = 5.00$$

2. Standard errors

$$SE\sigma := 1.05293 \cdot \frac{\sigma}{\lambda \cdot \sqrt{n}} \qquad SE\sigma = 0.78$$

$$SE\lambda := 0.7797 \cdot \frac{\lambda}{bn \cdot \sqrt{n}} \qquad SE\lambda = 0.98$$

$$SE\sigma\lambda := 0.50697 \cdot \sqrt{\frac{\sigma}{bn \cdot n}} \qquad SE\sigma\lambda = 0.49$$

3. Test-of-fit

$$w_i := \exp\left[-\left(\frac{\sigma}{x_i}\right)^{\lambda}\right]$$

$$A := \left[-n - \frac{1}{n} \cdot \sum_i (2 \cdot i - 1) \cdot (\ln(w_i) + \ln(1 - w_{n-i+1}))\right] \cdot \left(1 + \frac{0.2}{\sqrt{n}}\right)$$

$$A = 0.340$$

The 5% critical value is 0.757. Because the sample value A is smaller, there is no reason to reject the estimated model at the 5% significance level.

4. Probability plot of data and estimated model

$$p_i := \frac{i - 0.3}{n + 0.4} \qquad r_i := -\ln(-\ln(p_i)) \quad \text{(ordinates for the data)}$$

$$z_i := \lambda_u \cdot (\ln(x_i) - \ln(\sigma)) \quad \text{(ordinates for the model when the cdf is substituted for } p\text{)}$$

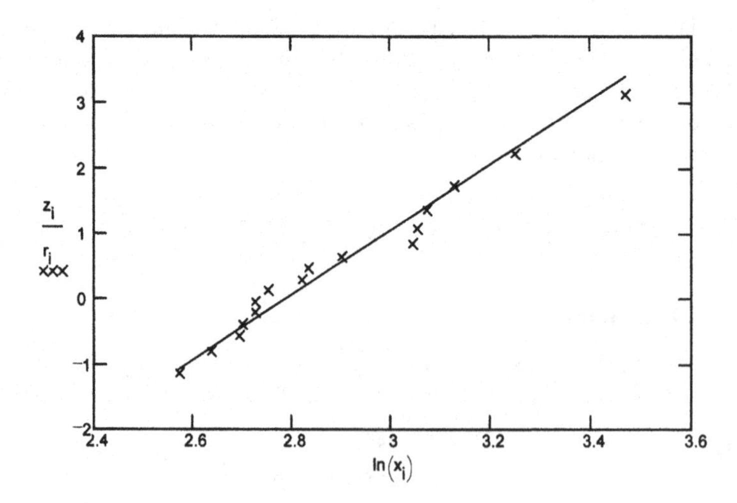

5. Percentile

From the reproductive property of the Frechet model, the maximum over 30 years is also Frechet distributed, with scale parameter

$$\sigma_m(a, b) := a \cdot 30^{\frac{1}{b}} \qquad \sigma_m(\sigma, \lambda_u) = 32.11$$

80-percentile: $X80(a, b) := \sigma_m(a, b) \cdot \ln\left(\frac{1}{0.8}\right)^{\left(-\frac{1}{b}\right)}$ $\qquad X80(\sigma, \lambda_u) = 43.3$

standard error: $da := \frac{d}{d\sigma} X80(\sigma, \lambda_u) \qquad db := \frac{d}{d\lambda_u} X80(\sigma, \lambda_u)$

$\text{var} := (da \cdot SE\sigma)^2 + (db \cdot SE\lambda)^2 + 2 \cdot da \cdot db \cdot SE\sigma\lambda^2 \qquad seX80 := \sqrt{\text{var}}$

$seX80 = 7.9$

6. pdf plot

$j := 1..200 \qquad t_j := 10 + 0.2 \cdot j \qquad f(x) := \frac{\lambda_u}{\sigma} \cdot \left(\frac{\sigma}{x}\right)^{\lambda_u + 1} \cdot \exp\left[-\left(\frac{\sigma}{x}\right)^{\lambda_u}\right]$

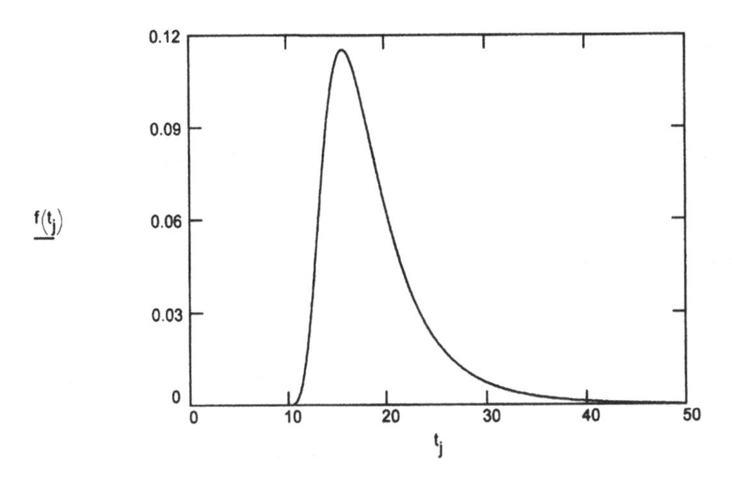

EXAMPLE 16.3

An offshore structure is to be designed in an area noted for its violent winter storms. One of the design parameters is the force on the structure due to sea waves. No data on wave heights are available for the proposed location. However, on the basis of weather maps, scaling of available measurements from a distant location gave an average maximum storm wave of 7 m for the service duration of the structure. The coefficient of variation was judged to be 0.5. The 95-percentile is to be used for design. Estimate this design value, based on the Frechet assumption.

$X\text{bar} := 7 \qquad \text{cv} := 0.5$

$CV(a) := \sqrt{\frac{\Gamma\left(1-\frac{2}{a}\right)}{\Gamma\left(1-\frac{1}{a}\right)^2} - 1} \qquad a := 3 \qquad \lambda := \text{root}(CV(a) - \text{cv}, a)$

$\lambda = 3.59$

$$\sigma := \frac{X\text{bar}}{\Gamma(1 - \frac{1}{\lambda})} \qquad \sigma = 5.53$$

95-percentile: $X95 := \sigma \cdot \ln\left(\frac{1}{0.95}\right)^{\left(-\frac{1}{\lambda}\right)}$ $\qquad X95 = 13$

EXAMPLE 16.4

For the data of Example 16.1, calculate the 90% confidence intervals on the estimated distribution parameters. Estimate the 95-percentile of yearly maximum snow load and its 90% upper confidence limit.

From Example 16.2 the ML estimates and their standard errors are

$\sigma := 16.27 \qquad \lambda_u := 5.00 \qquad \lambda := 5.48$

$SE\sigma := 0.78 \qquad SE\lambda := 0.98 \qquad SE\sigma\lambda := 0.49$

1. Parameters

i) Normal approximation: The standard Normal 95-percentile is $z95 := 1.645$

The required intervals are

$L\lambda := \lambda_u - z95 \cdot SE\lambda \qquad L\lambda = 3.39$

$U\lambda := \lambda_u + z95 \cdot SE\lambda \qquad U\lambda = 6.61$

$L\sigma := \sigma - z95 \cdot SE\sigma \qquad L\sigma = 15.0$

$U\sigma := \sigma + z95 \cdot SE\sigma \qquad U\sigma = 17.6$

ii) Likelihood ratio method: Chi-squared 90-percentile at $v = 1$ is $K90 := 2.71$

$x_i :=$ $\qquad n := 16 \qquad i := 1..n$

x_i
13.1
14.8
17
18.2
15.3
14.9
14
16.8
32.1
22.8
21.2
15.3
15.7
21
25.8
21.6

log of likelihood function:

$$s1 := \sum_i \ln(x_i) \qquad s3(\lambda) := \sum_i (x_i)^{-\lambda}$$

$$LL(\sigma, \lambda) := n \cdot \ln(\lambda) + n \cdot \lambda \cdot \ln(\sigma) - (\lambda + 1) \cdot s1 - \sigma^{\lambda} \cdot s3(\lambda)$$

shape parameter:

$$s(a) := \left[\frac{1}{n} \cdot \sum_i (x_i)^{-a}\right]^{-\frac{1}{a}} \qquad LR(a) := 2 \cdot LL(\sigma, \lambda) - 2 \cdot LL(s(a), a)$$

$a := \lambda - 1 \qquad \lambda l := \text{root}(LR(a) - K90, a) \qquad \lambda l = 3.81$

$a := \lambda + 1 \qquad \lambda u := \text{root}(LR(a) - K90, a) \qquad \lambda u = 7.50$

scale parameter:

$$l(b) := \text{root}\left[\left(\frac{n}{a} + n\cdot\ln(b)\right) - \sum_i \ln(x_i) - \sum_i \left(\frac{b}{x_i}\right)^a \cdot \ln\left(\frac{b}{x_i}\right), a\right]$$

$$LR(b) := 2\cdot LL(\sigma,\lambda) - 2\cdot LL(b, l(b))$$

$a := \lambda \qquad b := \sigma - 1 \qquad \sigma l := \text{root}(LR(b) - K90, b) \qquad \sigma l = 15.0$

$\qquad\qquad b := \sigma + 1 \qquad \sigma u := \text{root}(LR(b) - K90, b) \qquad \sigma u = 17.7$

iii) Simulated results:

From Table 17.1, at $n = 16$ and $\gamma = 0.05$ $\quad l1 := 0.775$

at $\gamma = 0.95$ $\quad l1 := 1.535$

Thus, from (16.41)

$$L\lambda := \frac{\lambda}{l2} \qquad L\lambda = 3.57$$

$$U\lambda := \frac{\lambda}{l1} \qquad U\lambda = 7.07$$

From Table 17.2 at $n = 16$ and $\gamma = 0.05$ $\quad l1 := -0.489$

at $\gamma = 0.95$ $\quad l2 := 0.480$

Thus, from (16.42)

$$L\sigma := \sigma\cdot\exp\cdot\left(\frac{l1}{\lambda}\right) \qquad L\sigma = 14.9$$

$$U\sigma := \sigma\cdot\exp\cdot\left(\frac{l2}{\lambda}\right) \qquad U\sigma = 17.8$$

2. 95-Percentile

$c := \ln\left(\frac{1}{0.95}\right) \qquad x(a,b) := a\cdot c^{-\frac{1}{b}} \qquad x(\sigma,\lambda_u) = 29.5$

standard error: $da := \frac{d}{d\sigma}x(\sigma,\lambda_u) \quad db := \frac{d}{d\lambda_u}x(\sigma,\lambda_u)$

$\text{var} := (da\cdot SE\sigma)^2 + (db\cdot SE\lambda)^2 + 2\cdot da\cdot db\cdot SE\sigma\lambda^2 \quad seX := \sqrt{\text{var}}$

$seX = 3.3$

90% upper confidence limit:

i) Normal approximation:

$z90 := 1.28$

$XU := x(\sigma,\lambda_u) + z90\cdot seX \qquad XU = 33.7$

ii) LR method:

$K80 := 1.642 \qquad c := \ln\left(\frac{1}{0.95}\right) \qquad s(Q,\lambda) := Q\cdot c^{\frac{1}{\lambda}}$

$$s2(\lambda) := \sum_i (x_i)^{-\lambda} \cdot \ln(x_i)$$

$$LL(\sigma, \lambda) := n \cdot \ln(\lambda) + n \cdot \lambda \cdot \ln(\sigma) - (\lambda + 1) \cdot s1 - \sigma^{\lambda} \cdot s3(\lambda)$$

$$ML(Q, b) := \frac{n}{b} + n \cdot \ln(Q) - s1 - c \cdot Q^{b} \cdot \ln(Q) \cdot s3(b) + c \cdot Q^{b} \cdot s2(b)$$

$b := \lambda \qquad l(Q) := \mathrm{root}(ML(Q, b), b)$

$$LR(Q) := 2 \cdot LL(\sigma, \lambda) - 2 \cdot LL(s(Q, l(Q)), l(Q))$$

$Q := 34 \qquad Qu := \mathrm{root}(LR(Q) - K80, Q) \qquad\qquad Qu = 34.3$

CHAPTER SEVENTEEN

Weibull Distributions

INTRODUCTION

17.1 Definition

A continuous random variable X has a Weibull distribution if its pdf has the form

$$f(x; \sigma, \lambda) = \frac{\lambda}{\sigma}\left(\frac{x}{\sigma}\right)^{\lambda-1} \exp\left\{-\left(\frac{x}{\sigma}\right)^{\lambda}\right\}; \quad x \geq 0; \quad \sigma, \lambda > 0. \tag{17.1}$$

This model arises as the limiting form of the distribution (2.4) of the smallest order statistic in a sample of size n from an initial distribution with a bounded tail. In this context the Weibull distribution is termed a *type III extreme value distribution* of *minima* (see Section 15.2).

For example, consider a length of chain comprising many links. The weakest link in the chain determines the chain's strength. Although the strength distribution of the links may not be known, the chain strength approaches a Weibull distribution, with increasing chain length. Engineering components can often be regarded as made up of many series-connected critical elements whose statistical properties cannot be determined. A measurable statistical component variable such as strength or life length then often closely follows the Weibull (extreme value) distribution, which is a most important statistical model in engineering.

This distribution was originally derived by Fisher and Tippet in 1928 as an asymptotic extreme value distribution. In 1939 the Swedish physicist Weibull derived the same distribution on the basis of practical requirements in the analysis of material breaking strength. It was not until 1951, however, when one of Weibull's articles received wide circulation among engineers concerned with modeling the statistical variation of their data, that this distribution became prominent in the engineering community. Weilbull's name has since been associated with this distribution.

PROPERTIES: TWO-PARAMETER MODEL

17.2 Distribution Function

A Weibull variable X, as defined by (17.1), has the cdf

$$F(x; \sigma, \lambda) = 1 - \exp\left\{-\left(\frac{x}{\sigma}\right)^{\lambda}\right\}. \tag{17.2}$$

This model has scale structure (see Section 9.2), that is, σ is a scale parameter, while λ is a shape parameter. The pdf of the reduced variable $Z = X/\sigma$ is therefore a function of λ only:

$$f(z; \lambda) = \lambda z^{\lambda-1} \exp\{-z^{\lambda}\}, \quad z \geq 0. \tag{17.3}$$

Figure 17.1 shows this pdf for several values of λ, indicating the wide shape flexibility of the Weibull model.

The Weibull variable X is related to a Gumbel variable Y (see Chapter 15) by

$$X = \exp\{-Y\}, \tag{17.4}$$

with the Weibull parameters σ and λ expressed in terms of Gumbel parameters μ_G and σ_G as

$$\sigma = \exp\{-\mu_G\} \quad \text{and} \quad \lambda = \frac{1}{\sigma_G}. \tag{17.5}$$

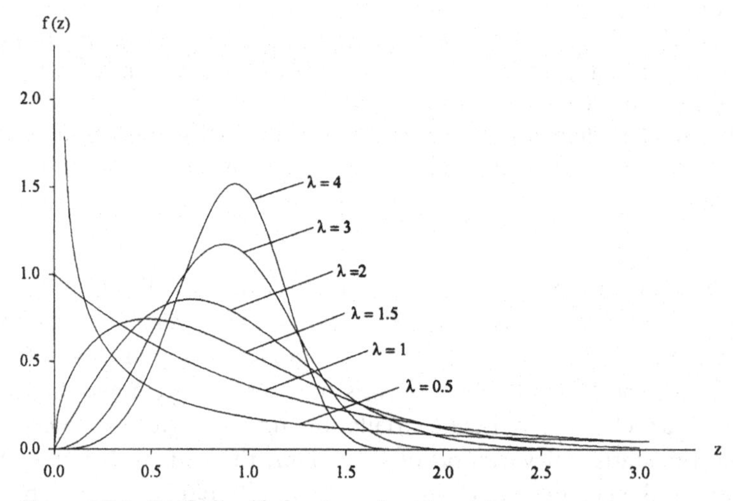

Figure 17.1. Weibull pdfs for the reduced variable z.

17.3 Weibull Variable

The rth moment of X about the origin is directly evaluated from the definition of expectations (see Section 1.7) as

$$\mu_r'(X) = \sigma^r \Gamma\left(1 + \frac{r}{\lambda}\right), \tag{17.6}$$

where Γ is the *gamma function* (see Section 13.1).
Thus, the expected value of X is

$$\mu_1'(X) = \sigma \Gamma\left(1 + \frac{1}{\lambda}\right). \tag{17.7}$$

The variance of X is

$$\mu_2(X) = \sigma^2 \left[\Gamma\left(1 + \frac{2}{\lambda}\right) - \Gamma^2\left(1 + \frac{1}{\lambda}\right)\right]. \tag{17.8}$$

The coefficient of variation of X is therefore a function of λ only:

$$cv(X) = \sqrt{\frac{\Gamma\left(1 + \frac{2}{\lambda}\right)}{\Gamma^2\left(1 + \frac{1}{\lambda}\right)} - 1} \tag{17.9}$$

(see Fig. 17.2). The mode value is given by

$$x_m = \sigma\left(\frac{\lambda - 1}{\lambda}\right)^{\frac{1}{\lambda}}, \quad \lambda \geq 1. \tag{17.10}$$

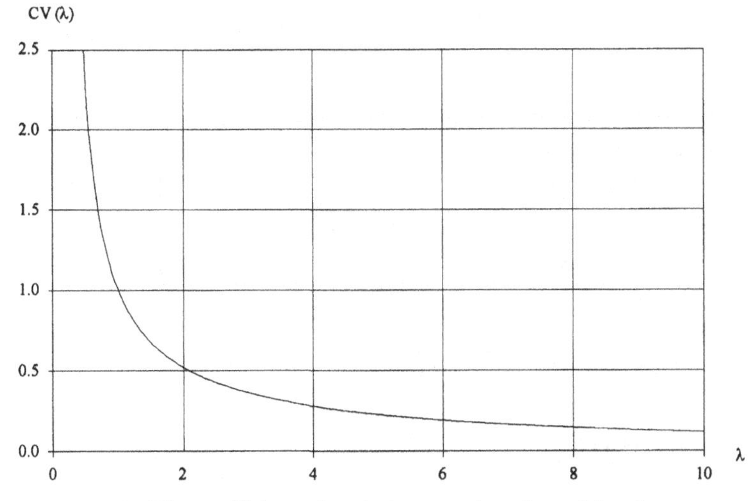

Figure 17.2. The coefficient of variation as a function of the shape parameter.

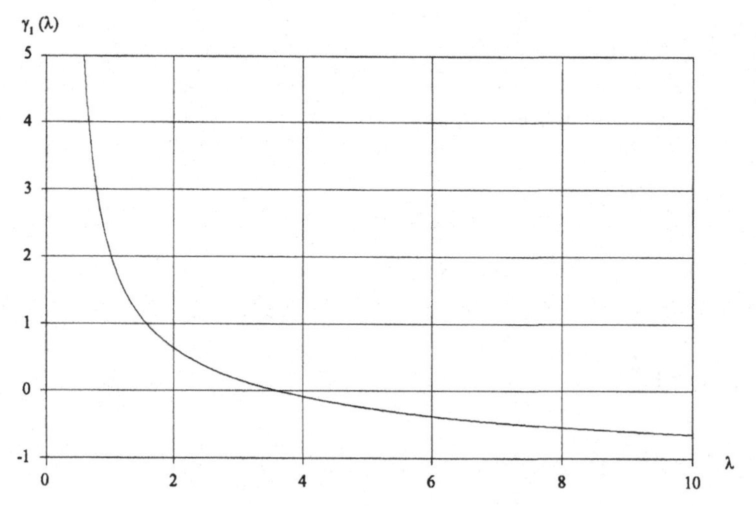

Figure 17.3. The first shape factor as a function of the shape parameter.

The quantile of order q is

$$x_q = \sigma \left[\ln\left(\frac{1}{1-q} \right) \right]^{\frac{1}{\lambda}}. \tag{17.11}$$

The first shape factor (see Section 1.10) is

$$\gamma_1 = \frac{\Gamma\left(1+\frac{3}{\lambda}\right) - 3\Gamma\left(1+\frac{1}{\lambda}\right)\Gamma\left(1+\frac{2}{\lambda}\right) + 2\Gamma^3\left(1+\frac{1}{\lambda}\right)}{\left[\Gamma\left(1+\frac{2}{\lambda}\right) - \Gamma^2\left(1+\frac{1}{\lambda}\right)\right]^{1.5}} \tag{17.12}$$

(see Fig. 17.3), which shows that the pdf is skewed to the left for $\lambda > 3.6$. The second shape factor is

$$\gamma 2 = \frac{\Gamma\left(1+\frac{4}{\lambda}\right) - 4\Gamma\left(1+\frac{1}{\lambda}\right)\Gamma\left(1+\frac{3}{\lambda}\right) + 6\Gamma^2\left(1+\frac{1}{\lambda}\right)\Gamma\left(1+\frac{2}{\lambda}\right) - 3\Gamma^4\left(1+\frac{1}{\lambda}\right)}{\left[\Gamma\left(1+\frac{2}{\lambda}\right) - \Gamma^2\left(1+\frac{1}{\lambda}\right)\right]^2} \tag{17.13}$$

(see Fig. 17.4), indicating that near the value $\lambda \doteq 2.2$ the Weibull pdf looks somewhat like the Normal pdf.

The matrix of minimum-variance-bounds for estimators of σ and λ is

$$\begin{bmatrix} V_{\sigma\sigma} & V_{\sigma\lambda} \\ V_{\sigma\lambda} & V_{\lambda\lambda} \end{bmatrix} = \frac{1}{n} \begin{bmatrix} 1.10867\,\frac{\sigma^2}{\lambda^2} & 0.25702\,\sigma \\ 0.25702\,\sigma & 0.60793\,\lambda^2 \end{bmatrix}, \tag{17.14}$$

so that the MVBs are

$$MVB_\sigma = 1.10867\,\frac{\sigma^2}{n\lambda^2}, \tag{17.15}$$

$$MVB_\lambda = 0.60793\,\frac{\lambda^2}{n}, \tag{17.16}$$

$$MVB_{\sigma\lambda} = 0.25702\,\frac{\sigma}{n}. \tag{17.17}$$

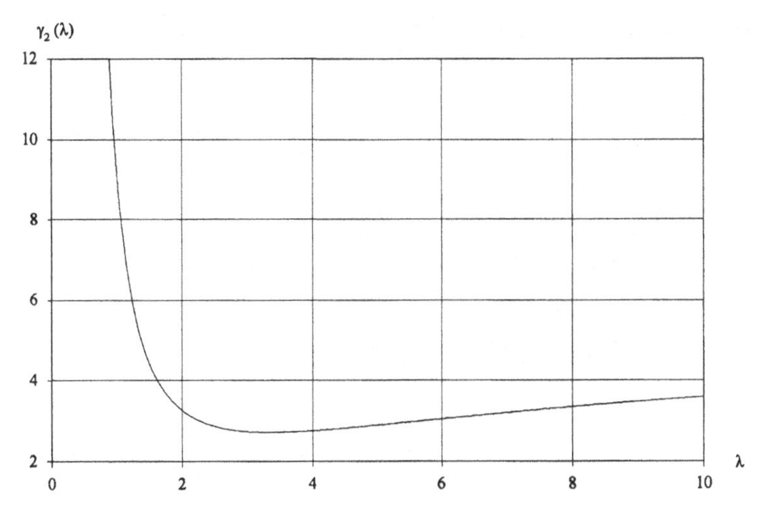

Figure 17.4. The second shape factor as a function of the shape parameter.

17.4 Reproductive Properties

The Weibull distribution features a reproductive property for its own minimum extreme. That is, the distribution of $X_{(1)}$ is again Weibull, with the same shape parameter but with the scale parameter decreased to $\sigma n^{-1/\lambda}$. This feature is easily verified by substituting the cdf (17.2) into (2.5). Thus, the pdf of $X_{(1)}$ has the same shape as that of X but is rescaled as given above.

Furthermore, a power transformation of X yields another Weibull model. That is, if the transformed variable is $T = bX^a$, then T is also Weibull, but with scale parameter $b\sigma^a$ and shape parameter λ/a.

When the shape parameter is of known value λ_o, the Weibull variable X can be transformed to an Exponential variable $Y = X^{\lambda_o}$, with Exponential scale parameter σ^{λ_o}. This transformation is useful in certain life-testing situations where design changes on a component affect only the value of the scale parameter σ. In that case the methods associated with the Exponential model (see Chapter 12) can be used directly. Section 17.28 elaborates this application.

17.5 Simulation

Random observations from a Weibull process with known parameter values are easily simulated by inversion (see Section 2.8) as

$$x_i = \sigma \left[\ln\left(\frac{1}{u_i} \right) \right]^{\frac{1}{\lambda}}, \tag{17.18}$$

where u_i is a Uniform random number on (0, 1) (see Section 14.6). Note that Mathcad 6+ has a built-in function that generates random Weibull variates z_i from $F(z; \sigma = 1, \lambda)$; a general variate is then $x_i = \sigma z_i$. It is advisable to check the adequacy of a simulated sample by comparing at least its first two moments with those of the given model.

PROPERTIES: THREE-PARAMETER MODEL

17.6 Definition and Properties

For some engineering applications a location parameter μ is introduced to the model. The pdf of the resulting three-parameter Weibull model is

$$f(x;\mu,\sigma,\lambda)=\frac{\lambda}{\sigma}\left(\frac{x-\mu}{\sigma}\right)^{\lambda-1}\exp\left\{-\left(\frac{x-\mu}{\sigma}\right)^{\lambda}\right\};\quad x\geq\mu;\quad \sigma,\lambda>0; \tag{17.19}$$

with cdf

$$F(x;\mu,\sigma,\lambda)=1-\exp\left\{-\left(\frac{x-\mu}{\sigma}\right)^{\lambda}\right\}. \tag{17.20}$$

The location parameter μ serves as a *threshold value* below which the variable X is not realized. For example, the strength of many manufactured engineering materials has not been observed to be zero or close to zero. Hence, some positive minimum value μ can often be assumed to exist for such strength variables.

For the three-parameter model all location measures are displaced to the right by μ with respect to their two-parameter equivalents, while central moments remain the same. Thus, the expected value of the three-parameter variable X is

$$\mu_1'(X)=\mu+\sigma\Gamma\left(1+\frac{1}{\lambda}\right). \tag{17.21}$$

The mode value is

$$x_{\text{m}}=\mu+\sigma\left(\frac{\lambda-1}{\lambda}\right)^{\frac{1}{\lambda}},\quad \lambda>1. \tag{17.22}$$

The quantile of order q is

$$x_q=\mu+\sigma\left[\ln\left(\frac{1}{1-q}\right)\right]^{\frac{1}{\lambda}}. \tag{17.23}$$

The variance and shape factors are unchanged, but the coefficient of variation is now

$$cv(X)=\frac{\sqrt{\Gamma\left(1+\frac{2}{\lambda}\right)-\Gamma^2\left(1+\frac{1}{\lambda}\right)}}{\frac{\mu}{\sigma}+\Gamma\left(1+\frac{1}{\lambda}\right)}. \tag{17.24}$$

For the three-parameter model the minimum-variance-bounds for estimators of μ,σ, and λ are obtained from the inverse of the *expected* information matrix (see Section 3.1). Provided that $\lambda > 2$, this symmetric matrix is (in the order μ, σ, λ)

$$\begin{bmatrix} \frac{n}{\sigma^2}(\lambda-1)^2\Gamma(1-\frac{2}{\lambda}) & \frac{n}{\sigma^2}\lambda(\lambda-1)\Gamma(1-\frac{1}{\lambda}) & -\frac{n}{\sigma\lambda}(\lambda-1)\Gamma(1-\frac{1}{\lambda})\left[1+\psi(1-\frac{1}{\lambda})\right] \\ & n\left(\frac{\lambda}{\sigma}\right)^2 & -\frac{n}{\sigma}\,0.42278 \\ & & \frac{n}{\lambda^2}\,1.82368 \end{bmatrix}, \tag{17.25}$$

where ψ is the *digamma* function (see Section 13.2). Alternatively, the *MVBs* can be estimated from the inverse of the symmetric *local* information matrix (see Section 3.7), provided $\lambda > 1$:

$$\begin{bmatrix} I_{\mu\mu} = \frac{\lambda-1}{\sigma^2}(s_2 + s_5\lambda) & I_{\mu\sigma} = \frac{\lambda(\lambda-1)}{\sigma^2}s_1 & I_{\mu\lambda} = \frac{s_1}{\sigma\lambda} - \frac{\lambda}{\sigma}s_6 \\ & I_{\sigma\sigma} = n\left(\frac{\lambda}{\sigma}\right)^2 & I_{\sigma\lambda} = -\frac{\lambda}{\sigma}\left(s_3 + \frac{n}{\lambda}\right) \\ & & I_{\lambda\lambda} = \frac{n}{\lambda^2} + s_4 \end{bmatrix}. \tag{17.26}$$

The sums in the above matrix elements are defined for $z_i = (x_i - \mu)/\sigma$ as

$$s_1 = \sum_i z_i^{-1}, \qquad s_2 = \sum_i z_i^{-2}, \qquad s_3 = \sum_i \ln(z_i),$$
$$s_4 = \sum_i z_i^{\lambda}\ln^2(z_i), \qquad s_5 = \sum_i z_i^{\lambda-2}, \qquad s_6 = \sum_i z_i^{\lambda-1}\ln(z_i).$$

SPECIAL CASES: EXPONENTIAL AND RAYLEIGH MODELS

17.7 Definition

When the shape parameter is $\lambda = 1$, the Weibull distribution specializes to the Exponential distribution; see Chapter 12 for details. When $\lambda = 2$ and the Weibull scale parameter σ is replaced by $\sigma_R\sqrt{2}$, the Rayleigh model results, with pdf

$$f_R(x; \sigma_R) = \frac{x}{\sigma_R^2}\exp\left\{-\frac{1}{2}\left(\frac{x}{\sigma_R}\right)^2\right\}; \qquad x \geq 0, \quad \sigma > 0; \tag{17.27}$$

and cdf

$$F_R(x; \sigma_R) = 1 - \exp\left\{-\frac{1}{2}\left(\frac{x}{\sigma_R}\right)^2\right\}. \tag{17.28}$$

Figure 17.5 plots the pdf for several values of the scale parameter σ_R. The pdf of the reduced variable $Z = X/\sigma_R$ is parameter free:

$$f_R(z) = z\exp\left\{-\frac{z^2}{2}\right\}. \tag{17.29}$$

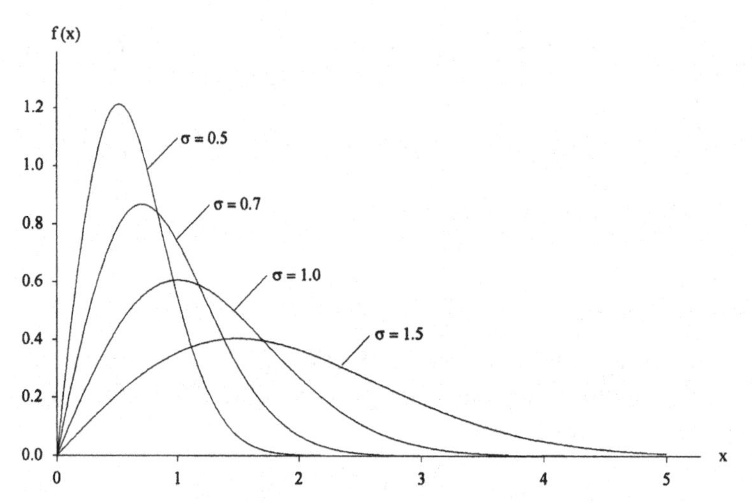

Figure 17.5. Rayleigh pdfs for several values of the scale parameter.

17.8 Properties

The properties of the Rayleigh distribution follow directly from those of the Weibull distribution, with $\lambda = 2$ and $\sigma = \sigma_R\sqrt{2}$:

$$\mu_1'(X) = 1.2533\,\sigma_R, \tag{17.30}$$

$$\mu_2(X) = 0.4292\,\sigma_R^2 \tag{17.31}$$

$$x_m = \sigma_R, \tag{17.32}$$

$$x_q = \sigma_R\sqrt{2\ln\left(\frac{1}{1-q}\right)}, \tag{17.33}$$

and

$$cv(X) = 0.5227. \tag{17.34}$$

The shape factors are constant: $\gamma_1 = 0.631$ and $\gamma_2 = 3.245$. The smallest observation $X_{(1)}$ of a Rayleigh sample is again Rayleigh, with scale parameter $\sigma_R\sqrt{2/n}$. The power transformation of a Rayleigh variable X to $T = bX^a$ results in a Weibull variable T with scale parameter $b(\sigma_R\sqrt{2})^a$ and shape parameter $2/a$.

17.9 Application

The Rayleigh model is chiefly applied to the sum of squares of Normal variables X_1 and X_2 with zero means and common variance σ^2.

For example, the rms value of two Normal error sources X_1 and X_2 is $S_m = \sqrt{\frac{1}{2}(X_1^2 + X_2^2)}$, with distribution $F_R(s_m; \sigma/\sqrt{2})$. Similarly, the resultant of two orthogonal Normal errors is $S_r = \sqrt{X_1^2 + X_2^2}$, with distribution $F_R(s_r; \sigma)$. An example is the radial targeting error S_r resulting from independent Cartesian error components X_1 and X_2. See Example 17.1 for an illustration.

PROBABILITY PLOT

17.10 Complete Samples

A probability plot of ordered observations (see Section 3.13) is routinely used to check the distributional assumption of the Weibull model. Taking natural logs twice of the cdf (17.2) gives a linear relation in terms of $\ln(x_{(i)})$:

$$\ln\left[-\ln\{1 - F\left(x_{(i)}\right)\}\right] = -\lambda \ln(\sigma) + \lambda \ln\left(x_{(i)}\right). \tag{17.35}$$

Replacing $F(x_{(i)})$ by a plotting position such as the approximate median value of $F(x_{(i)})$:

$$p_i = \frac{i - 0.3}{n + 0.4} \tag{17.36}$$

gives

$$\ln[-\ln\{1 - p_i\}] = a + b \ln\left(x_{(i)}\right). \tag{17.37}$$

Thus, for *complete* or *right-tail censored* samples, plotting $\ln[-\ln\{1 - p_i\}]$ versus the logs of ordered data $x_{(i)}$ results in approximately a straight line, if the data came from a two-parameter Weibull process. A noticeable departure of the plot from a straight line would put the Weibull postulate in question (see Section 3.14). The parameters σ and λ could be estimated from the intercept a and the slope b of the plot. Since the statistical properties of the resulting estimates are not known, this procedure is not recommended. See Example 17.2 for a Mathcad plot of some two-parameter Weibull data.

When the appearance of the data plot suggests that a location parameter μ may be present (the lower end of the data plot curves downward), a *preliminary* estimate of its value may be obtained by subtracting trial values μ_k from the observation $x_{(i)}$ until the corresponding plot has straightened.

For data $x_{(i)}$ from a Rayleigh process it is easiest to plot the square of the data $y_{(i)} = x_{(i)}^2$ as Exponential values (see Section 12.6), since y is Exponentially distributed with scale parameter $2\sigma_R^2$.

17.11 Censored Samples

Many engineering problems center around the operating time-to-failure (TTF) of an engineered device. The Weibull distribution is undoubtedly the most prominent life model in use today to represent TTF. See Chapter 12 and Sections 17.22 to 17.28 for a discussion of some special problems associated with life modeling in engineering.

One of the peculiarities of this application area is that the database is often multiply time-censored (see Section 12.2). This requires a modified procedure for data plotting and parameter estimation.

For example, consider a critical component (e.g., the transmission) of an ore-hauling truck in an open-pit mining operation. There is a fixed number of units of the component in the fleet of trucks, and each unit has its own date of installation. At any given time there may have been r failures, with corresponding TTFs known from records. The remainder of the sample (including overhauled and replaced units) is still operating, each unit with its own accumulated operating time. The database consists of the total number of units installed (sample size n), r failure times, and $m = n - r$ current survival times. The problem is to estimate the life model of the component, utilizing *all* of the database information. Note that the current survival times depend on when these units were put into service. The nature of that selection process is usually unknown.

Consider a multiply censored sample of size n, with r failure times x_i and $m = n - r$ survival times c_j. Label the combined and *ordered* set of failure and survival data as $\{t_{(k)}\}_n$. Thus, some of the $t_{(k)}$ are failure data x_i and some are survival data c_j.

The plotting position (17.36) needs to be modified to accommodate the information implicit in the survival times c_j. A suitable modification is based on the distribution-free product limit estimate of the reliability function (see Section 3.15). The modified median plotting position is then

$$p_k = 1 - \frac{n + 0.7}{n + 0.4} \prod_{q=1}^{k} \frac{n - q + 0.7}{n - q + 1.7}, \tag{17.38}$$

where $q \leq k$ is the subscript of a *failure* time in the combined, ordered set $\{t_{(k)}\}_n$. When there are no censored data, (17.38) reduces to the complete-sample median plotting position (17.36).

Plotting $\ln[-\ln\{1 - p_k\}]$ versus the logs of the failure data in the combined set $\{t_{(k)}\}_n$ will produce an approximate straight line if the data came from a two-parameter Weibull process. This plot incorporates the information of the survival data c_j. Example 17.3 illustrates this plotting procedure.

POINT ESTIMATES: TWO-PARAMETER WEIBULL MODEL

17.12 Maximum Likelihood Estimates

The likelihood function of a *complete* sample of n independent observations on a Weibull variable X is

$$L(\sigma, \lambda) = \lambda^n \sigma^{-n\lambda} \prod_{i=1}^{n} x_i^{\lambda-1} \exp\left\{-\sigma^{-\lambda} \sum_{i=1}^{n} x_i^{\lambda}\right\}. \tag{17.39}$$

The ML equations (see Section 3.6) are

$$\widehat{\sigma} = \left(\frac{1}{n}\sum_{i=1}^{n} x_i^{\widehat{\lambda}}\right)^{\frac{1}{\widehat{\lambda}}} \tag{17.40}$$

and

$$\frac{n}{\widehat{\lambda}} - n\ln(\widehat{\sigma}) + \sum_{i=1}^{n}\ln(x_i) - \sum_{i=1}^{n}\left(\frac{x_i}{\widehat{\sigma}}\right)^{\widehat{\lambda}}\ln\left(\frac{x_i}{\widehat{\sigma}}\right) = 0. \tag{17.41}$$

Substituting the first equation into the second gives an equation in only $\widehat{\lambda}$:

$$\frac{1}{\widehat{\lambda}} - \frac{\sum_{i=1}^{n} x_i^{\widehat{\lambda}}\ln(x_i)}{\sum_{i=1}^{n} x_i^{\widehat{\lambda}}} + \frac{1}{n}\sum_{i=1}^{n}\ln(x_i) = 0. \tag{17.42}$$

The solution value $\widehat{\lambda}$ is readily obtained with an equation solver. The ML estimate $\widehat{\sigma}$ is then computed from (17.40).

The estimate $\widehat{\lambda}$ is biased upward. Using a convenient bias-correction formula of

$$b_n = 1 + \frac{2.2}{n^{1.13}}, \tag{17.43}$$

the unbiased estimate of the shape parameter becomes

$$\widehat{\lambda}_{\mathrm{u}} = \frac{\widehat{\lambda}}{b_n}. \tag{17.44}$$

The *MVB*s (17.15) to (17.17) are corrected for b_n, where appropriate, to give the following standard errors:

$$\mathrm{se}(\widehat{\sigma}) = 1.05293\frac{\sigma}{\lambda\sqrt{n}}, \tag{17.45}$$

$$\mathrm{se}(\widehat{\lambda}_{\mathrm{u}}) = 0.77970\frac{\lambda}{b_n\sqrt{n}}, \tag{17.46}$$

$$\mathrm{se}(\widehat{\sigma}, \widehat{\lambda}_{\mathrm{u}}) = 0.50697\sqrt{\frac{\sigma}{nb_n}}. \tag{17.47}$$

These standard errors are evaluated at the (biased) estimate $\widehat{\lambda}$. The above ML parameter estimates $\widehat{\sigma}$ and $\widehat{\lambda}_{\mathrm{u}}$ and their standard errors are recommended, whenever observations on the Weibull variable are available.

The standard errors of parameter functions $g(\sigma, \lambda)$ are approximated from the error propagation formula (see Section 3.3). For the two-parameter case of a complete sample this formula is

$$\mathrm{Var}(g) \doteq \left(\frac{\partial g}{\partial \sigma}\mathrm{se}_\sigma\right)^2 + \left(\frac{\partial g}{\partial \lambda}\mathrm{se}_\lambda\right)^2 + 2\left(\frac{\partial g}{\partial \sigma}\right)\left(\frac{\partial g}{\partial \lambda}\right)\mathrm{se}^2_{\sigma\lambda}, \tag{17.48}$$

using the above standard errors. See Example 17.4, where a percentile function is estimated.

For a statistical test-of-fit the *Anderson–Darling* statistic is recommended:

$$A = -n - \frac{1}{n}\sum_{i=1}^{n}(2i-1)[\ln(w_i) + \ln(1 - w_{n-i+1})]. \tag{17.49}$$

Here w_i is the Weibull cdf (17.2), evaluated at the order statistic $x_{(i)}$ and the MLEs $\widehat{\sigma}$ and $\widehat{\lambda}$ (biased). The value A is then modified to account for sample size:

$$A_\mathrm{m} = A\left(1 + \frac{0.2}{\sqrt{n}}\right). \tag{17.50}$$

Critical values[1] for A_m are:

α	0.100	0.050	0.025	0.010
A_c	0.637	0.757	0.877	1.038

Example 17.4 demonstrates the test procedure.

17.13 Moment Estimates

Sometimes the engineer does not have available a sample of observations on a variable X, but he may have prior information of its average value $\bar{x}$ and its coefficient of variation cv. A load process for a structural design is a typical example. A model of X can be deduced from the given information by calculating moment estimates of the parameters λ and σ. The moment estimate $\widetilde{\lambda}$ can be obtained by inverting expression (17.9) with an equation solver. A good approximation for $\lambda \geq 1$ is

$$\widetilde{\lambda} \doteq cv^{-1.0852}. \tag{17.51}$$

Expression (17.7) then provides a moment estimate of σ:

$$\widetilde{\sigma} = \frac{\bar{x}}{\Gamma(1 + \frac{1}{\widetilde{\lambda}})}. \tag{17.52}$$

See Example 17.5 for an illustration. Because the efficiencies of these estimators can be expected to be low, they should not be used when Weibull data are available.

[1] Extracted from R. B. D'Agostino, M. A. Stephens, *Goodness-of-Fit Techniques*, Marcel Dekker Inc., New York, 1986, by courtesy of the publisher.

17.14 Censored Samples

The likelihood function of a multiply censored sample, defined in Section 17.11, from a two-parameter Weibull process is

$$L(\sigma, \lambda) = \lambda^r \sigma^{-r\lambda} \prod_{i=1}^{r} x_i^{\lambda-1} \exp\left\{ -\sigma^{-\lambda} \sum_{k=1}^{n} t_k^{\lambda} \right\}. \tag{17.53}$$

This function has the same structure as expression (17.39) for a complete sample. The maximum likelihood equations are therefore similar as well:

$$\widehat{\sigma} = \left(\frac{1}{r} \sum_{k=1}^{n} t_k^{\widehat{\lambda}} \right)^{\frac{1}{\widehat{\lambda}}} \tag{17.54}$$

and

$$\frac{r}{\widehat{\lambda}} - r \ln(\widehat{\sigma}) + \sum_{i=1}^{r} \ln(x_i) - \sum_{k=1}^{n} \left(\frac{t_k}{\widehat{\sigma}} \right)^{\widehat{\lambda}} \ln\left(\frac{t_k}{\widehat{\sigma}} \right) = 0. \tag{17.55}$$

Substituting the first equation into the second gives an equation in only $\widehat{\lambda}$:

$$\frac{1}{\widehat{\lambda}} - \frac{\sum_{k=1}^{n} t_k^{\widehat{\lambda}} \ln(t_k)}{\sum_{k=1}^{n} t_k^{\widehat{\lambda}}} + \frac{1}{r} \sum_{i=1}^{r} \ln(x_i) = 0. \tag{17.56}$$

The solution value $\widehat{\lambda}$ is readily obtained with an equation solver. The ML estimate $\widehat{\sigma}$ is then computed from (17.54). It is seen that the complete-sample estimating equations (17.40) and (17.42) are special cases of (17.54) and (17.56). See Example 17.6 for an illustration.

The estimator $\widehat{\lambda}$ is biased, but a bias-correction factor for different sample sizes and levels of censoring is not available. The factor (17.43) is recommended, with r replacing n. Inferences are then based on the (approximately) unbiased estimate $\widehat{\lambda}_u$.

In the general case of multiply censored samples it is difficult to calculate the *expected* information matrix for the ML estimates. Instead, the *local* information matrix (see Section 3.7) can be computed from the observed values of the log-likelihood function. For the two-parameter Weibull model that matrix simplifies to

$$\begin{bmatrix} I_{\sigma\sigma} = r\left(\frac{\lambda}{\sigma}\right)^2 & I_{\sigma\lambda} = \frac{\lambda}{\sigma}\left(r \ln(\sigma) - \frac{r}{\lambda} - \sum_{i=1}^{r} \ln(x_i)\right) \\ I_{\sigma\lambda} & I_{\lambda\lambda} = \frac{r}{\lambda^2} + \sum_{i=1}^{n} \left(\frac{t_k}{\sigma}\right)^{\lambda} \ln^2\left(\frac{t_k}{\sigma}\right) \end{bmatrix}. \tag{17.57}$$

The covariance matrix of the parameter estimates $\widehat{\sigma}, \widehat{\lambda}$ is the inverse of matrix I, evaluated at these estimates. Example 17.6 illustrates the computations.

Convenient statistical tests-of-fit are not available for multiply censored Weibull samples. Probability plotting is recommended as a reliable check on the distributional postulate.

INTERVAL ESTIMATES AND TESTS: TWO-PARAMETER WEIBULL MODEL

17.15 Normal Approximation

For large samples, the asymptotic sampling distributions of ML estimators are convenient for constructing approximate confidence intervals on parameters and parameter functions. Let θ stand for σ or λ, and recall from Section 3.7 that the sampling pdf of the MLE $\widehat{\theta}$ is asymptotically Normal, with mean value θ and variance MVB_θ:

$$f_{\mathrm{N}}(\widehat{\theta}; \theta, \sqrt{MVB_\theta}). \tag{17.58}$$

Thus, the $(1-\alpha)$-level confidence interval on θ is

$$(l_1, l_2) = \theta \pm z_{\frac{\alpha}{2}} \sqrt{MVB_\theta}, \tag{17.59}$$

such that

$$Pr(l_1 \leq \theta \leq l_2) = 1 - \alpha.$$

See Example 17.7 for an illustration.

For parameter functions $g(\theta)$, the invariance property of ML estimators is used (see Section 3.7). That is, the sampling pdf of $\widehat{g} = g(\widehat{\theta})$ is asymptotically Normal, with mean value g and variance approximated by the error propagation formula (17.48). The unbiased estimate $\widehat{\lambda}_{\mathrm{u}}$ and the bias-corrected variances are recommended for these calculations.

17.16 Likelihood Ratio Approximation

The preceding Normal approximation loses accuracy as the sample size decreases, particularly for highly censored samples. The likelihood-ratio method (see Section 3.9) tends to give more accurate results for such samples.

Recall that the statistic

$$LR(\theta) = 2\ln[L(\widehat{\theta})] - 2\ln[L(\theta)] \tag{17.60}$$

is approximately Chi-squared distributed with $\nu = 1$ degree of freedom. Thus, a $(1-\alpha)$-level confidence interval on θ comprises those values θ for which $LR(\theta) \leq \chi^2_{1,1-\alpha}$. For the two-parameter model, one parameter in (17.60) must be expressed in terms of the other parameter by its ML equation. Thus, a confidence interval on λ is obtained from

$$LR(\lambda) = 2\ln[L(\widehat{\sigma}, \widehat{\lambda})] - 2\ln[L(\sigma\{\lambda\}, \lambda)], \tag{17.61}$$

where $\sigma\{\lambda\}$ is defined by the ML equation (17.40) or (17.54), depending on the nature of the sample.

Similarly, a confidence interval on σ is obtained from

$$LR(\sigma) = 2\ln[L(\widehat{\sigma}, \widehat{\lambda})] - 2\ln[L(\sigma, \lambda\{\sigma\})], \tag{17.62}$$

where $\lambda\{\sigma\}$ is defined by the ML equation (17.41) or (17.55). See Example 17.7 for an illustration.

A confidence interval on a parameter function $g(\theta)$ is more difficult to obtain since the relation among parameters, implied by $g(\theta)$, constrains the solution process. As an illustration, consider the quantile $x_q = \sigma[-\ln(1-q)]^{1/\lambda}$, so that

$$\sigma\{x_q, \lambda\} = x_q[-\ln(1-q)]^{-\frac{l}{\lambda}}. \tag{17.63}$$

Substituting (17.63) into the likelihood function (17.53) and differentiating with respect to λ gives the constraint relation for censored samples:

$$ML(\lambda) = \frac{r}{\lambda} - r\ln(x_q) + \sum_{i=1}^{r}\ln(x_i) + \ln(1-q)\sum_{k=1}^{n}\left(\frac{t_k}{x_q}\right)^{\lambda}\ln\left(\frac{t_k}{x_q}\right). \tag{17.64}$$

For complete samples, n replaces r and all t_k become x_i. A confidence interval on x_q is then obtained from

$$LR(x_q) = 2\ln[L(\widehat{\sigma}, \widehat{\lambda})] - 2\ln[L(\sigma\{x_q, \lambda\{x_q\}\}, \lambda\{x_q\})], \tag{17.65}$$

where $\sigma\{x_q, \lambda\}$ is given by (17.63) and $\lambda\{x_q\}$ is the solution of $ML(\lambda) = 0$ according to (17.64). See Example 17.7 for an illustration of the computations. Note that the *biased* estimate $\widehat{\lambda}$ is used in formulas (17.61) to (17.65).

A test on the null hypothesis $H_o : \lambda = \lambda_o$ against the alternative $H_a : \lambda \neq \lambda_a$ is based of the test statistic $LR(\lambda_o)$, given by (17.61). Similarly, a test on $H_o : \sigma = \sigma_o$ is based on the test statistic $LR(\sigma_o)$, given by (17.62). Critical values are obtained as the α-quantiles of the Chi-squared distribution with $\nu = 1$.

17.17 Simulation Results

For small samples, one of the simplest approaches for obtaining confidence intervals for the *two-parameter* Weibull model is based on the Monte Carlo simulation[2] of samples from the standard Exponential model $\mu_E = 0, \sigma_E = 1$. This approach is based on the following results.

If X is Weibull distributed with parameters σ and λ, then the transformed variable $Y = (X/\sigma)^{\lambda}$ is distributed as a *standard* Exponential variable. Transforming the ML equations (17.42) and (17.40) from X to Y, we see that the

[2] Thoman, D. R., Bain, L. J., Antle, C. E., "Inferences on the Parameters of the Weibull Distribution," *Technometrics*, Vol. 11, No. 3, pp. 445–460, 1969.

quantities $\widehat{\lambda}/\lambda$ and $(\widehat{\sigma}/\sigma)^{\widehat{\lambda}}$ have sampling distributions that can be generated from samples of Y and that are independent of σ and λ. Thus, by simulating standard Exponential samples and computing for each the Weibull estimates $\widehat{\lambda}$ and $\widehat{\sigma}$, the required sampling distributions are generated. Tables 17.1 and 17.2 present the simulated results, which are used for Weibull parameters as follows.

For selected confidence levels $1 - \alpha$ and sample sizes n, Table 17.1 gives the values l_1 and l_2 such that the confidence interval on λ is obtained from

$$Pr\left(\frac{\widehat{\lambda}}{l_2} \leq \lambda \leq \frac{\widehat{\lambda}}{l_1}\right) = 1 - \alpha. \tag{17.66}$$

Table 17.2 gives values l_1 and l_2 such that the confidence interval on σ is obtained from

$$Pr\left(\widehat{\sigma}\exp\left\{-\frac{l_2}{\widehat{\lambda}}\right\} \leq \sigma \leq \widehat{\sigma}\exp\left\{-\frac{l_1}{\widehat{\lambda}}\right\}\right) = 1 - \alpha. \tag{17.67}$$

The probability level γ of the tables relates to the above α values as $\gamma = \alpha/2$ for the lower limit and $\gamma = 1 - \alpha/2$ for the upper limit. See Example 17.7 for an illustration. Note that the tables are based on the (biased) MLE $\widehat{\lambda}$, which should therefore be used in (17.66) and (17.67).

POINT ESTIMATES: THREE-PARAMETER WEIBULL MODEL

17.18 Complete Samples

For a complete sample of n independent Weibull observations, the likelihood function is

$$L(\mu, \sigma, \lambda) = \lambda^n \sigma^{-n\lambda} \prod_{i=1}^{n} (x_i - \mu)^{\lambda-1} \exp\left\{-\sigma^{-\lambda} \sum_{i=1}^{n} (x_i - \mu)^{\lambda}\right\}. \tag{17.68}$$

The maximum likelihood equations (see Section 3.6) are then

$$\widehat{\sigma}^{\widehat{\lambda}} \cdot \frac{\widehat{\lambda} - 1}{\widehat{\lambda}} = \frac{\sum_{i=1}^{n} (x_i - \widehat{\mu})^{(\widehat{\lambda}-1)}}{\sum_{i=1}^{n} (x_i - \widehat{\mu})^{-1}}, \tag{17.69}$$

$$\widehat{\sigma} = \left(\frac{1}{n} \sum_{i=1}^{n} (x_i - \widehat{\mu})^{\widehat{\lambda}}\right)^{\frac{1}{\widehat{\lambda}}}, \tag{17.70}$$

and

$$\frac{n}{\widehat{\lambda}} - n\ln(\widehat{\sigma}) + \sum_{i=1}^{n} \ln(x_i - \widehat{\mu}) - \sum_{i=1}^{n} \left(\frac{x_i - \widehat{\mu}}{\widehat{\sigma}}\right)^{\widehat{\lambda}} \ln\left(\frac{x_i - \widehat{\mu}}{\widehat{\sigma}}\right) = 0. \tag{17.71}$$

Table 17.1.[a] *Percentage points l_γ such that $Pr(\widehat{\lambda}/\lambda < l_\gamma) = \gamma$*

	Probability Level γ					
n	0.02	0.05	0.10	0.90	0.95	0.98
5	0.604	0.683	0.766	2.277	2.779	3.518
6	0.623	0.697	0.778	2.030	2.436	3.067
7	0.639	0.709	0.785	1.861	2.183	2.640
8	0.653	0.720	0.792	1.747	2.015	2.377
9	0.665	0.729	0.797	1.665	1.896	2.199
10	0.676	0.738	0.802	1.602	1.807	2.070
11	0.686	0.745	0.807	1.553	1.738	1.972
12	0.695	0.752	0.811	1.513	1.682	1.894
13	0.703	0.759	0.815	1.480	1.636	1.830
14	0.710	0.764	0.819	1.452	1.597	1.777
15	0.716	0.770	0.823	1.427	1.564	1.732
16	0.723	0.775	0.826	1.406	1.535	1.693
17	0.728	0.779	0.829	1.388	1.510	1.660
18	0.734	0.784	0.832	1.371	1.487	1.630
19	0.739	0.788	0.835	1.356	1.467	1.603
20	0.743	0.791	0.838	1.343	1.449	1.579
22	0.752	0.798	0.843	1.320	1.418	1.538
24	0.759	0.805	0.848	1.301	1.392	1.504
26	0.766	0.810	0.852	1.284	1.370	1.475
28	0.772	0.815	0.856	1.269	1.351	1.450
30	0.778	0.820	0.860	1.257	1.334	1.429
32	0.783	0.824	0.863	1.246	1.319	1.409
34	0.788	0.828	0.866	1.236	1.306	1.392
36	0.793	0.832	0.869	1.227	1.294	1.377
38	0.797	0.835	0.872	1.219	1.283	1.363
40	0.801	0.839	0.875	1.211	1.273	1.351
42	0.804	0.842	0.877	1.204	1.265	1.339
44	0.808	0.845	0.880	1.198	1.256	1.329
46	0.811	0.847	0.882	1.192	1.249	1.319
48	0.814	0.850	0.884	1.187	1.242	1.310
50	0.817	0.852	0.886	1.182	1.235	1.301
52	0.820	0.854	0.888	1.177	1.229	1.294
54	0.822	0.857	0.890	1.173	1.224	1.286
56	0.825	0.859	0.891	1.169	1.218	1.280
58	0.827	0.861	0.893	1.165	1.213	1.273
60	0.830	0.863	0.894	1.162	1.208	1.267
70	0.840	0.871	0.901	1.146	1.188	1.242
80	0.848	0.878	0.907	1.134	1.173	1.222
90	0.855	0.883	0.912	1.124	1.160	1.206
100	0.861	0.888	0.916	1.116	1.150	1.192
110	0.866	0.893	0.920	1.110	1.141	1.181
120	0.871	0.897	0.923	1.104	1.133	1.171

[a] Extracted by permission of the publisher from Thoman, D. R., Bain, L. J., Antle, C. E., "Inferences on the Parameters of the Weibull Distribution," *Technometrics*, Vol. 11, No. 3, pp. 445–460, 1969.

Table 17.2.[a] *Percentage points l_γ such that $Pr(\hat{\lambda}\ln(\hat{\sigma}/\sigma) < l_\gamma) = \gamma$.*

	Probability Level γ					
n	0.02	0.05	0.10	0.90	0.95	0.98
5	−1.631	−1.247	−0.888	0.772	1.107	1.582
6	−1.396	−1.007	−0.740	0.666	0.939	1.291
7	−1.196	−0.874	−0.652	0.598	0.829	1.120
8	−1.056	−0.784	−0.591	0.547	0.751	1.003
9	−0.954	−0.717	−0.544	0.507	0.691	0.917
10	−0.876	−0.665	−0.507	0.475	0.644	0.851
11	−0.813	−0.622	−0.477	0.448	0.605	0.797
12	−0.762	−0.587	−0.451	0.425	0.572	0.752
13	−0.719	−0.557	−0.429	0.406	0.544	0.714
14	−0.683	−0.532	−0.410	0.389	0.520	0.681
15	−0.651	−0.509	−0.393	0.374	0.499	0.653
16	−0.624	−0.489	−0.379	0.360	0.480	0.627
17	−0.599	−0.471	−0.365	0.348	0.463	0.605
18	−0.578	−0.455	−0.353	0.338	0.447	0.584
19	−0.558	−0.441	−0.342	0.328	0.433	0.566
20	−0.540	−0.428	−0.332	0.318	0.421	0.549
22	−0.509	−0.404	−0.314	0.302	0.398	0.519
24	−0.483	−0.384	−0.299	0.288	0.379	0.494
26	−0.460	−0.367	−0.286	0.276	0.362	0.472
28	−0.441	−0.352	−0.274	0.265	0.347	0.453
30	−0.423	−0.338	−0.264	0.256	0.334	0.435
32	−0.408	−0.326	−0.254	0.247	0.323	0.420
34	−0.394	−0.315	−0.246	0.239	0.312	0.406
36	−0.382	−0.305	−0.238	0.232	0.302	0.393
38	−0.370	−0.296	−0.231	0.226	0.293	0.382
40	−0.360	−0.288	−0.224	0.220	0.285	0.371
42	−0.350	−0.280	−0.218	0.214	0.278	0.361
44	−0.341	−0.273	−0.213	0.209	0.271	0.352
46	−0.333	−0.266	−0.208	0.204	0.264	0.344
48	−0.325	−0.260	−0.203	0.199	0.258	0.336
50	−0.318	−0.254	−0.198	0.195	0.253	0.328
52	−0.312	−0.249	−0.194	0.191	0.247	0.321
54	−0.305	−0.244	−0.190	0.187	0.243	0.315
56	−0.299	−0.239	−0.186	0.184	0.238	0.309
58	−0.294	−0.234	−0.183	0.181	0.233	0.303
60	−0.289	−0.230	−0.179	0.177	0.229	0.297
70	−0.266	−0.211	−0.165	0.164	0.211	0.274
80	−0.248	−0.197	−0.153	0.153	0.197	0.255
90	−0.234	−0.184	−0.144	0.143	0.185	0.239
100	−0.221	−0.174	−0.136	0.136	0.175	0.226
110	−0.211	−0.165	−0.129	0.129	0.166	0.215
120	−0.202	−0.158	−0.123	0.123	0.159	0.205

[a] Extracted by permission of the publisher from Thoman, D. R., Bain, L. J., Antle, C. E., "Inferences on the Parameters of the Weibull Distribution," *Technometrics*, Vol. 11, No. 3, pp. 445–460, 1969.

Substituting (17.70) into (17.69) and (17.71) gives two expressions in $\widehat{\mu}$ and $\widehat{\lambda}$ only:

$$\frac{\widehat{\lambda}}{\widehat{\lambda}-1} = \frac{\sum_{i=1}^{n}(x_i - \widehat{\mu})^{\widehat{\lambda}} \sum_{i=1}^{n}(x_i - \widehat{\mu})^{-1}}{n\sum_{i=1}^{n}(x_i - \widehat{\mu})^{\widehat{\lambda}-1}} \tag{17.72}$$

and

$$\frac{1}{\widehat{\lambda}} - \frac{\sum_{i=1}^{n}(x_i - \widehat{\mu})^{\widehat{\lambda}}\ln(x_i - \widehat{\mu})}{\sum_{i=1}^{n}(x_i - \widehat{\mu})^{\widehat{\lambda}}} + \frac{1}{n}\sum_{i=1}^{n}\ln(x_i - \widehat{\mu}) = 0. \tag{17.73}$$

In principle, the solution $\widehat{\mu}, \widehat{\lambda}$ is obtained with an equation solver, and $\widehat{\sigma}$ is then obtained from (17.70). However, a solution may not exist. This occurs, for example, when $\lambda < 1$ (a rare case in engineering applications of the Weibull model). When a solution does exist, it is a local maximum and is accompanied by a second solution, which is a saddle point. This second solution corresponds to the two-parameter Exponential model:

$$\widehat{\mu} = x_{(1)}, \qquad \widehat{\sigma} = \bar{x} - x_{(1)}, \qquad \widehat{\lambda} = 1. \tag{17.74}$$

Occasionally, this solution produces a larger likelihood value (17.68) than the local maximum. In this case, or when a local maximum cannot be found, (17.74) may be taken as the solution. However, this model should always be checked graphically against the data, since it often produces a poor fit. When (17.74) fits poorly, an acceptable solution can be obtained from the two-parameter Weibull model at $\widehat{\mu} = x_{(1)}$ and fit to the $n - 1$ data $(x_{(i)} - \widehat{\mu}), i = 2, \ldots, n$. Subsequent inferences are then conditional on $\widehat{\mu}$.

The recommended approach to finding a local maximum likelihood solution is to display the log-likelihood function over chosen parameter ranges for μ and λ in matrix form (or as a contour plot). One can then conveniently search for the local maximum to obtain starting values of μ and λ for the equation solver. Substituting Equation (17.70) into the logarithm of (17.68) gives the desired expression for plotting:

$$LLF(\mu, \lambda) = n\ln(\lambda) - n\ln\left(\sum_{i=1}^{n}(x_i - \mu)^{\lambda}\right) + (\lambda - 1)\sum_{i=1}^{n}\ln(x_i - \mu) + C. \tag{17.75}$$

The constant C is chosen to give small positive values of the function near the local maximum. Example 17.8 illustrates the procedure.

Approximate standard errors of the estimates are then found from the inverse of the information matrix (17.25) or (17.26). The unbiased shape parameter $\widehat{\lambda}_u$ from (17.44) is recommended for inferences. A statistical test-of-fit for the three-parameter Weibull model is not available.

17.19 Censored Samples

For a multiply censored sample, defined in Section 17.11, from a three-parameter Weibull process, the likelihood function is

$$L(\mu, \sigma, \lambda) = \lambda^r \sigma^{-r\lambda} \prod_{i=1}^{r} (x_i - \mu)^{\lambda-1} \exp\left\{-\sigma^{-\lambda} \sum_{k=1}^{n} (t_k - \mu)^\lambda\right\}. \qquad (17.76)$$

The three likelihood equations are

$$\widehat{\sigma}^{\widehat{\lambda}} \cdot \frac{\widehat{\lambda} - 1}{\widehat{\lambda}} = \frac{\sum_{k=1}^{n} (t_k - \widehat{\mu})^{\widehat{\lambda}-1}}{\sum_{i=1}^{r} (x_i - \widehat{\mu})^{-1}}, \qquad (17.77)$$

$$\widehat{\sigma} = \left(\frac{1}{r} \sum_{k=1}^{n} (t_k - \widehat{\mu})^{\widehat{\lambda}}\right)^{\frac{1}{\widehat{\lambda}}}, \qquad (17.78)$$

and

$$\frac{r}{\widehat{\lambda}} - r \ln(\widehat{\sigma}) + \sum_{i=1}^{r} \ln(x_i - \widehat{\mu}) - \sum_{k=1}^{n} \left(\frac{t_k - \widehat{\mu}}{\widehat{\sigma}}\right)^{\widehat{\lambda}} \ln\left(\frac{t_k - \widehat{\mu}}{\widehat{\sigma}}\right) = 0. \qquad (17.79)$$

Substituting (17.78) into (17.77) and (17.79) gives two expressions in $\widehat{\mu}$ and $\widehat{\lambda}$ only:

$$\frac{\widehat{\lambda}}{\widehat{\lambda} - 1} = \frac{\sum_{k=1}^{n} (t_k - \widehat{\mu})^{\widehat{\lambda}} \sum_{i=1}^{r} (x_i - \widehat{\mu})^{-1}}{r \sum_{k=1}^{n} (t_k - \widehat{\mu})^{\widehat{\lambda}-1}} \qquad (17.80)$$

and

$$\frac{1}{\widehat{\lambda}} - \frac{\sum_{k=1}^{n} (t_k - \widehat{\mu})^{\widehat{\lambda}} \ln(t_k - \widehat{\mu})}{\sum_{k=1}^{n} (t_k - \widehat{\mu})^{\widehat{\lambda}}} + \frac{1}{r} \sum_{i=1}^{r} \ln(x_i - \widehat{\mu}) = 0. \qquad (17.81)$$

The discussion of the solution process in the preceding section applies here as well.

The covariance matrix of the estimates $\widehat{\mu}, \widehat{\sigma}, \widehat{\lambda}$ is obtained from the inverse of the *local* information matrix for $\lambda > 1$. This symmetric matrix is similar to the complete-sample case (17.26):

$$\begin{bmatrix} I_{\mu\mu} = \frac{\lambda-1}{\sigma^2}(s_2 + s_5\lambda) & I_{\mu\sigma} = \frac{\lambda(\lambda-1)}{\sigma^2} s_1 & I_{\mu\lambda} = \frac{s_1}{\sigma\lambda} - \frac{\lambda}{\sigma} s_6 \\ & I_{\sigma\sigma} = R\left(\frac{\lambda}{\sigma}\right)^2 & I_{\sigma\lambda} = -\frac{\lambda}{\sigma}\left(s_3 + \frac{r}{\lambda}\right) \\ & & I_{\lambda\lambda} = \frac{r}{\lambda^2} + s_4 \end{bmatrix}. \qquad (17.82)$$

The sums in the above matrix elements are defined for $z_i = (x_i - \mu)/\sigma$ and $u_k = (t_k - \mu)/\sigma$ as

$$s_1 = \sum_{i=1}^{r} z_i^{-1}, \qquad s_2 = \sum_{i=1}^{r} z_i^{-2}, \qquad s_3 = \sum_{i=1}^{r} \ln(z_i),$$

$$s_4 = \sum_{k=1}^{n} u_k^{\lambda} \ln^2(u_k), \qquad s_5 = \sum_{k=1}^{n} u_k^{\lambda-2}, \qquad s_6 = \sum_{k=1}^{n} u_k^{\lambda-1} \ln(u_k).$$

This matrix is, of course, evaluated at the ML estimates.

For three-parameter Weibull models, convenient small-sample methods for confidence intervals are not available. Although the asymptotic formula (17.59) can be used for this case, caution is advised, particularly with heavily censored samples, since the sampling distributions are non-Normal unless the sample size is quite large.

When the threshold value μ is not of primary interest, inferences on σ and λ can be constructed by conditioning on the estimated value $\widehat{\mu}$: Transform the data to $y_i = x_i - \widehat{\mu}$, and then apply two-parameter methods. Often such conditional inferences are insensitive to $\widehat{\mu}$, since the likelihood function does not appear to hold much information on μ. See Example 17.9 for an illustration.

ESTIMATES: RAYLEIGH MODEL

17.20 Point and Interval Estimates

The likelihood function of a complete sample of n independent Rayleigh observations is

$$L(\sigma_R) = \sigma_R^{-2n} \prod_{i=1}^{n} x_i \exp\left\{ -\frac{1}{2\sigma_R^2} \sum_{i=1}^{n} x_i^2 \right\}, \tag{17.83}$$

from which the maximum likelihood estimator of the parameter σ_R is obtained (see Section 3.6) as

$$\widehat{\sigma}_R = \sqrt{\frac{1}{2n} \sum_{i=1}^{n} x_i^2}. \tag{17.84}$$

The standard error of $\widehat{\sigma}_R$ is

$$\text{se}(\widehat{\sigma}_R) = \frac{\sigma_R}{2\sqrt{n}}. \tag{17.85}$$

A Rayleigh variable X can be transformed to a one-parameter Exponential variable $Y = X^2$ with scale parameter $2\sigma_R^2$. Thus, methods associated with the

Exponential model (see Chapter 12) can be used for the parameter σ_R. In particular, the $(1-\alpha)$-level confidence interval on σ_R is obtained from (12.31) as

$$\left(\sqrt{\frac{\sum_i x_i^2}{\chi^2_{2n,1-\frac{\alpha}{2}}}}, \sqrt{\frac{\sum_i x_i^2}{\chi^2_{2n,\frac{\alpha}{2}}}}\right), \tag{17.86}$$

where $\chi^2_{2n,q}$ is the q-quantile of the χ^2 distribution, with $\nu = 2n$ degrees of freedom. See Example 17.10 for an illustration.

APPLICATIONS

17.21 General

The Weibull model is an appropriate *postulate* for variables that arise from the minimum extreme of an underlying random process whose characteristics have not been measured. Many engineering variables are the result of such minimum extreme value phenomena. This would account for the prominence of the Weibull model in engineering statistics.

Most engineering applications of the Weibull model fall into the related areas of material strength and component life. For *material strength*, the usual argument is that strength-degrading flaws are of random size and distributed randomly through the volume of a test specimen. The specimen can therefore be regarded as made up of many small volumes of varying strength, the weakest of which governs the measurable strength of the test specimen.

For example, the tensile strength of a lumber board is governed by the minimum strength of element-volumes containing knots and other irregularities in the wood grain. Although these irregularities are not strictly random in size and location, the Weibull distribution often models wood strength very well.

For *component life*, it can often be similarly argued that a given component is made up of many small elements (whose lives cannot be measured) that are critical to the component's survival. Hence the life of the weakest element determines the measurable component life. This line of reasoning applies widely, perhaps explaining why the Weibull distribution has become a most prominent life model in engineering. Beginning with the next section, the remainder of this chapter considers the role of the Weibull model in some basic aspects of life-data analysis.

When one considers the life of a *device* or *system*, which is made up of independent critical components, the above "chain-link" model also applies. If any number of identical Weibull-distributed components make up the device, then the reproductive property of the Weibull model (see Section 17.4) ensures that the device is also Weibull distributed. If, however, many critical non-Weibull components make up the device, the Weibull distribution can be postulated as an asymptotic model, regardless of what the component distributions might be.

For example, the life of a high-speed turbine disk is identical to the life of the weakest blade assembled to the disk. If the number of blades per disk is large (say 50 or more), then the disk life will be closely modeled by a Weibull distribution, regardless of the distribution of blade life.

There are many situations, spanning the spectrum of engineering work, where not enough is known about the underlying physical process that produces a measurable random variable, so that it is not possible to postulate a specific distribution model. For such cases the Weibull distribution remains a good model choice, because of its inherent flexibility to fit data. In engineering practice one therefore often finds the Weibull model applied successfully, even when an extreme-value argument cannot be made. As a data-fit model, the Weibull distribution clearly competes with the Log-Normal and Gamma distributions (see Chapters 11 and 13, respectively).

LIFE TESTING

17.22 Hazard Function

Many engineering applications of statistical methods arise from the need to model the operating time-to-failure of equipment. Chapter 12 on Exponential distributions details some of the basic concepts central to the analysis of life data. Recall from Section 12.4 that the Exponential distribution features the memoryless property implying that its hazard function is constant regardless of the device's life history. For nonelectronic devices that property is difficult to justify in view of the wear most products experience during their uses. The Weibull model is one of the distributions that extend to the modeling of life phenomena to nonconstant hazard functions, thus generalizing the Exponential model. See the curve in Figure 12.2, called a "bathtub curve," for the typical appearance of a hazard function for a mechanical device.

Expression (12.62) defines the hazard function as

$$h(X) = \frac{f(X)}{1 - F(X)}.$$

For the Weibull model this reduces to a power function:

$$h(x; \sigma, \lambda) = \frac{\lambda}{\sigma^{\lambda}} x^{\lambda - 1}. \tag{17.87}$$

Figure 17.6 shows this function for $\sigma = 1$ and several values of λ, indicating the flexibility of the Weibull distribution to model the various hazard regimes of

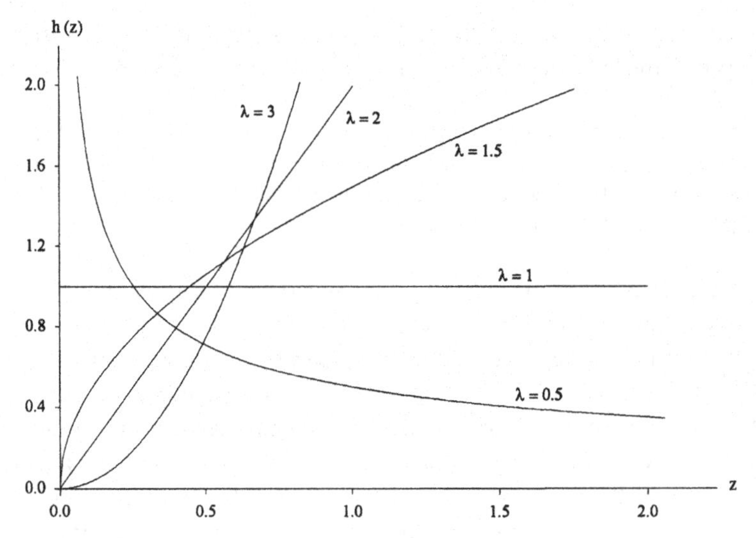

Figure 17.6. Weibull hazard functions for the reduced variable z.

life phenomena. Note that for the Rayleigh model ($\lambda = 2$) the hazard function increases linearly.

Expression (12.64) of Section 12.16 defines the empirical version of the hazard function. When the sample size is large, this function can be used to check the Weibull model graphically. That is, the Weibull hazard plot follows a linear trend when the log of the hazard function is plotted versus the log of the ordered data.

17.23 Multiple Failure Modes

Many engineered devices are subject to several independent hazards $h_i(x)$, each associated with a different failure mode. To survive its mission, the device must survive each of these, say k, failure-mode hazards, so that the device reliability $R(x) = 1 - F(x)$ is related to failure-mode reliabilities $R_i(x)$ as

$$R(x) = \prod_{i=1}^{k} R_i(x).$$

Taking logarithms and differentiating with respect to x gives

$$\frac{\partial \ln[R(x)]}{\partial x} = \sum_{i=1}^{k} \frac{\partial \ln[R_i(x)]}{\partial x} \quad \text{or} \quad \frac{f(x)}{1 - F(x)} = \sum_{i=1}^{k} \frac{f_i(x)}{1 - F_i(x)}$$

or

$$h(x) = \sum_{i=1}^{k} h_i(x). \tag{17.88}$$

Hence, the device hazard equals the sum of failure-mode hazards. From the definition of the hazard function, the device cdf can easily be expressed as a function of the failure-mode hazards as

$$F(x) = 1 - \exp\left\{-\int_0^x \sum_{i=1}^{k} h_i(x)\,dx\right\}. \tag{17.89}$$

Here, $F(x)$ represents the life model of devices that are subject to multiple independent failure modes of comparable severity. Although this cdf is not usually one of the recognized statistical models, it forms the basis for inferences on the device when separate failure-mode hazards $h_i(x)$ are available. In practice, however, one particular failure mode often dominates all others. The other modes may therefore be ignored until the product has been improved with respect to the dominant mode to the point where they begin to influence the failure process. Until then, single-failure-mode analysis is appropriate.

17.24 Reliability Function

One of the decision functions of central practical interest in reliability analysis is the *reliability value*, which must be estimated accurately (i.e., without bias). This value is defined as the probability of a component, or a system of components, to survive a stated operating time x: $R \mid x = 1 - F(x)$. This measure is conditional on the chosen model F, the estimated parameters θ, and the specified operating conditions of the device. When considered as a function of x, it is the *reliability function*. For the Weibull model that function is

$$R(x) = \exp\left\{-\left(\frac{x}{\sigma}\right)^{\lambda}\right\}. \tag{17.90}$$

Figure 17.7 shows that function for $\sigma = 1$ and several values of λ.

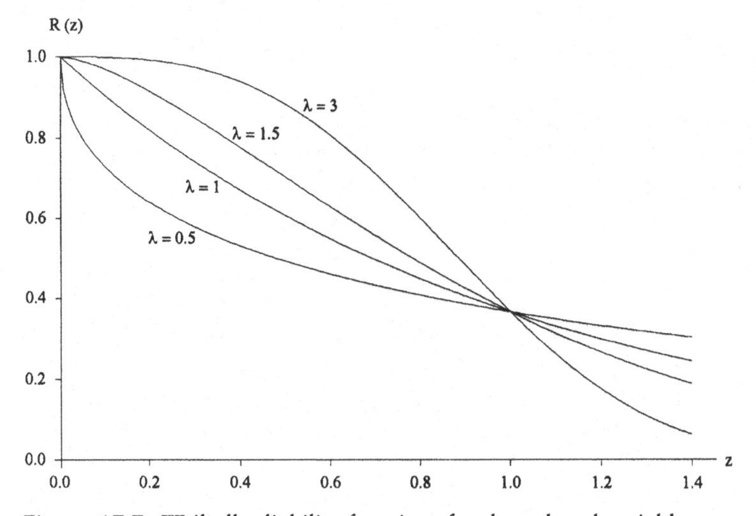

Figure 17.7. Weibull reliability functions for the reduced variable z.

The Weibull reliability value $R \mid x$ can be estimated by maximum likelihood as

$$\widehat{R} \mid x = \exp\left\{ -\left(\frac{x}{\widehat{\sigma}} \right)^{\widehat{\lambda}} \right\}. \tag{17.91}$$

In the high-reliability region of interest in engineering ($R > 0.95$, say), this estimator is virtually unbiased and highly efficient, even for small samples. Note that when a distributional assumption, such as the Weibull postulate, cannot be made, the reliability value can be estimated in a distribution-free manner by the product-limit estimate (see Section 3.15).

An accurate approximation to the variance of the estimate (17.91) is obtained directly from the error propagation formula (17.48) as

$$\mathrm{Var}(\widehat{R} \mid x) \doteq (R \ln[R])^2 \left\{ \left(\frac{\lambda}{\sigma} \right)^2 \mathrm{Var}(\sigma) + \left(\frac{\ln[-\ln(R)]}{\lambda} \right)^2 \mathrm{Var}(\lambda) - \frac{2 \ln[-\ln(R)]}{\sigma} \mathrm{Cov}(\sigma, \lambda) \right\}, \tag{17.92}$$

where the parameter variance terms represent the *MVB*s corresponding to the particular sample on hand. For complete samples, for instance, the variance terms (17.15–17.17) are substituted in (17.92), resulting in

$$\mathrm{Var}(\widehat{R} \mid x) \doteq \frac{(R \ln[R])^2}{n} \{1.109 + 0.608 \ln^2[-\ln(R)] - 0.514 \ln[-\ln(R)]\}. \tag{17.93}$$

The *biased* estimate $\widehat{\lambda}$ and its standard error are recommended for (17.91) and (17.92).

17.25 Confidence Limit on Reliability

Instead of the point estimate $\widehat{R} \mid x$ and its standard error $\sqrt{\mathrm{Var}(\widehat{R} \mid x)}$, it is often required to obtain a *lower* $(1-\alpha)$-level confidence limit $L_{1-\alpha} \mid x$ on the reliability value. Asymptotically, that limit is given by

$$L_{1-\alpha} \mid x = \widehat{R} \mid x + z_\alpha \sqrt{\mathrm{Var}(\widehat{R} \mid x)}, \tag{17.94}$$

where z_α is the standard Normal quantile of order α. However, the sampling distribution of $\widehat{R} \mid x$ is *not* Normal, and (17.94) gives results that can be significantly higher than true values, even for large samples. For confidence levels of 95% or higher, the asymptotic estimate (17.94) can be improved significantly by a single recursion:

$$L'_{1-\alpha} \mid x = \widehat{R} \mid x + z_\alpha \sqrt{\mathrm{Var}(L_{1-\alpha} \mid x)}. \tag{17.95}$$

The resulting value is in close agreement with tabulated results[3] of a simulation study for complete samples. See Example 17.11.

An alternate approach to calculating a lower confidence limit on R is to use the likelihood-ratio method (see Section 17.16). For a given value x, the scale parameter σ can be expressed, from (17.90), as

$$\sigma\{R, \lambda\} = x[-\ln(R)]^{-\frac{1}{\lambda}}. \tag{17.96}$$

Substituting this quantity into the likelihood function (17.53), and differentiating with respect to λ, gives a constraint relation for censored samples:

$$ML(\lambda) = \frac{r}{\lambda} - r\ln(x) + \sum_{i=1}^{r}\ln(x_i) + \ln(R)\sum_{k=1}^{n}\left(\frac{t_k}{x}\right)^{\lambda}\ln\left(\frac{t_k}{x}\right). \tag{17.97}$$

For complete sample, n replaces r and all t_k become x_i. A lower confidence limit on R is then obtained from

$$LR(R) = 2\ln[L(\widehat{\sigma}, \widehat{\lambda})] - 2\ln[L(\sigma\{R, \lambda\{R\}\}, \lambda\{R\})], \tag{17.98}$$

where $\sigma\{R, \lambda\}$ is given by (17.96) and $\lambda\{R\}$ is the solution of $ML(\lambda) = 0$ according to (17.97). See Example 17.11 for an illustration of the computations. Note that the *biased* estimate $\widehat{\lambda}$ is used in formulas (17.96) to (17.98).

17.26 Expected Number of Failures

When dealing with a population of components that are subject to failure, one wants to predict the number of failures over a given time period.

For example, each transmission in a fleet of trucks has accumulated a current operating time t_i. Assuming that life data for a particular failure mode of the transmission are available, the life model $F(t)$ can be estimated. An important decision quantity is then the expected number of transmission failures at present or over a specified future period. That estimate allows the operations engineer to assess the economic impact of the dominant failure mode in question and to optimize the stock level of backup units.

Given the Weibull life model $F(t)$, and a record of the current operating time t_i (since installation or overhaul) of the ith unit, the *current* expected number of failures among currently operating units is

$$\mathrm{E}\{N\}_{\mathrm{cur}} = \sum_{i=1}^{n} F(t_i) = n - \sum_{i=1}^{n}\exp\left\{-\left(\frac{t_i}{\sigma}\right)^{\lambda}\right\}, \tag{17.99}$$

where n is the number of units operating.

[3] Thoman, D. R., Bain, L. B., Antle, C. E., "Maximum Likelihood Estimation, Exact Confidence Intervals for Reliability, and Tolerance Limits in the Weibull Distribution," *Technometrics*, Vol. 12, No. 2, pp. 363–371, 1970.

When the time horizon extends over some *future* period T, and failed units are *not* replaced, the expected number of failures is

$$\mathrm{E}\{N\}_{\mathrm{fut}} = \sum_{i=1}^{n} \frac{F(t_i + u_i) - F(t_i)}{1 - F(t_i)} = n - \sum_{i=1}^{n} \exp\left\{\left(\frac{t_i}{\sigma}\right)^{\lambda} - \left(\frac{t_i + u_i}{\sigma}\right)^{\lambda}\right\}, \tag{17.100}$$

where u_i is the anticipated usage time of the ith unit during the decision period T. Example 17.12 provides an illustration.

> Continuing with the preceding example, we suppose that a decision has been made to replace failed transmissions with an improved design. That is, the current design is to be phased out. Expression (17.100) estimates the expected number of failures for the *current* design, for input to the replacement stock level of the *improved* units for the period T.

When failed units are replaced by new or overhauled units of the same type, failure prediction requires *renewal theory* or simulation. However, if the usages u_i are sufficiently small so that $F(u_i)$ is also small, (17.100) will give a close approximation to the expected number of failures.

17.27 Age Replacement

When the replacement cost C_f of a failed unit is higher than the cost C_r for the planned replacement of an aged but functioning unit, the cost-optimal replacement time t^* can be determined. The relevant decision function to minimize is the average cost per unit operating time for a component:

$$Ca(t) = \frac{C_f F(t) + C_r[1 - F(t)]}{\int_0^t [1 - F(x)]\, dx}. \tag{17.101}$$

Minimizing (17.101) with respect to t results in the solution equation:

$$h(t^*) \int_0^{t^*} R(x)\, dx + R(t^*) - \frac{C_f}{C_f - C_r} = 0, \tag{17.102}$$

where $R(x) = 1 - F(x)$, and $h(t) = f(t)/R(t)$ is the hazard function. When only failed units are replaced (i.e., there are no preventive replacements), the average cost per time unit for a component is

$$C = \frac{C_f}{\mathrm{E}\{t\}}. \tag{17.103}$$

The cost advantage of the age-replacement policy, per unit time and per component, is obtained as the difference between (17.103) and (17.101).

For the two-parameter Weibull life model, the above expressions become

$$Ca(t) = \frac{C_f - (C_f - C_r)\exp\{-(\frac{t}{\sigma})^\lambda\}}{\int_0^t \exp\{-(\frac{x}{\sigma})^\lambda\}\,dx}, \tag{17.104}$$

$$\frac{\lambda}{\sigma}\left(\frac{t^*}{\sigma}\right)^{\lambda-1}\int_0^{t^*}\exp\left\{-\left(\frac{x}{\sigma}\right)^\lambda\right\}dx + \exp\left\{-\left(\frac{t^*}{\sigma}\right)^\lambda\right\} - \frac{C_f}{C_f - C_r} = 0, \tag{17.105}$$

and

$$C = \frac{C_f}{\sigma\Gamma\left(1+\frac{1}{\lambda}\right)}. \tag{17.106}$$

Equation (17.105) is easily solved for the cost-optimal replacement time t^* with an equation solver. See Example 17.13.

17.28 Acceptance Tests

In Sections 12.20 to 12.22, some details of acceptance test procedures are given. These test procedures specify life-test parameters, such as sample size and test termination rules, to decide whether or not a specified life criterion has been met. The relevant life criterion could be the product's reliability value R at a given operating time t or, equivalently, a given percentile value t_R of time-to-failure:

$$R(t_R) = \exp\left\{-\left(\frac{t_R}{\sigma}\right)^\lambda\right\} \quad \text{or} \quad t_R = \sigma\left[\ln\left(\frac{1}{R}\right)\right]^{\frac{1}{\lambda}}. \tag{17.107}$$

For the general case when both σ and λ are unknown, limited results for Weibull test plans are available.[4] In many practical situations, however, the objective of an acceptance test is to decide if a modification to an existing design has improved its life characteristics with respect to a specific failure mode. This usually implies that failure data for the old design are available to estimate its Weibull parameters. Often the value of the shape parameter λ is associated with the particular failure mode in question and does not change appreciably through design modifications. An improvement of the design then manifests as an increase in the scale parameter σ alone. When the shape value $\tilde{\lambda}$ can be assumed known, the two-parameter Weibull model of the variable X can be transformed to a one-parameter Exponential model of the variable Y by $Y = X^{\tilde{\lambda}}$. The Exponential test plans of Chapter 12 can then be used on the parameter $\sigma^{\tilde{\lambda}}$. However,

[4] Fertig, K. W., Mann, N. R., "Life-Test Sampling Plans for Two-Parameter Weibull Populations," *Technometrics*, Vol. 22, No. 2, pp. 165–177, 1980.

for this limited decision frame, it is worthwhile to present directly some simple results.

Perhaps the simplest case, called a *substantiation requirement*, is a significance test (see Section 3.10) based on a zero-failure decision rule. That is, a test is to be conducted to demonstrate that a component under development has improved significantly with respect to a particular failure mode. The null hypothesis under test is that the original scale parameter remained unchanged, $H_0 \mid \tilde{\lambda} : \sigma = \sigma_0$, implying no improvement. The decision rule is to reject H_0, implying that the component has indeed improved, if none of the n units put on test fails during the test period T. The motivation for this test is to control the probability α that an underdeveloped component passes the acceptance test. If n units are available for testing (the usual case in practice), the test plan is defined by the test duration T for each unit:

$$\alpha = Pr(\text{no failures among } n \text{ items during } T \mid H_0 \text{ is true}) = \exp\left\{-n\left(\frac{T}{\sigma_0}\right)^{\tilde{\lambda}}\right\},$$

or

$$T = \sigma_0 \left[-\frac{1}{n}\ln(\alpha)\right]^{\frac{1}{\tilde{\lambda}}}. \tag{17.108}$$

Conversely, if the test duration T needs to be limited, the test plan is defined by the required number n of units to be tested:

$$n = -\ln(\alpha)\left(\frac{T}{\sigma_0}\right)^{-\tilde{\lambda}} \tag{17.109}$$

It is instructive to check the probability β that the test rejects a fully developed component (i.e., the scale parameter is in fact $\sigma_a > \sigma_0$):

$$\beta = 1 - \exp\left\{-n\left(\frac{T}{\sigma_a}\right)^{\tilde{\lambda}}\right\}. \tag{17.110}$$

Clearly, β should be small. See Example 17.14 for an illustration.

A more general test plan controls both probabilities α and β and allows for r failures in the decision rule. The result is a hypothesis test (conditional on the assumed value $\tilde{\lambda}$) that discriminates between the null hypothesis $H_0 \mid \tilde{\lambda} : \sigma = \sigma_0$ and the alternative hypothesis $H_a \mid \tilde{\lambda} : \sigma \geq \sigma_a$, where $\sigma_a > \sigma_0$ (see Section 3.11). When failed test units are not replaced, the test process conforms to a *Bernoulli scheme*, with the number of failed units modeled by a Binomial distribution (see

Section 6.1). The equations defining this test plan are

$$\alpha = Pr(r < r_c \mid \sigma_0) = \sum_{r=0}^{r_c-1} \binom{n}{r} p_0^r (1 - p_0)^{n-r} \tag{17.111}$$

and

$$\beta \leq Pr(r \geq r_c \mid \sigma_a) = 1 - \sum_{r=0}^{r_c-1} \binom{n}{r} p_a^r (1 - p_a)^{n-r}, \tag{17.112}$$

where $p_x = Pr$ *(a unit fails during* $T \mid \sigma_x) = 1 - \exp\{-(T/\sigma_x)^{\tilde{\lambda}}$ and r_c is the critical number of failures that terminates the test and precipitates the decision H_0. If fewer than r_c units failed during T, the decision is H_a. The above expressions contain three unknowns: The number n of units to be tested, the critical number r_c of failures, and the test duration T for each unit tested. The decision parameters α and β are given. Two of the three unknowns are determined when the third value is given. It is usually the case that the number n of available test units is fixed, so that r_c and T constitute the test design. Example 17.15 provides an illustration. Note that this simple test plan ignores the information from the r failure times x_i. For more powerful test plans, see Sections 12.21 and 12.22.

EXAMPLE 17.1

The error in the x and y directions of an accurate medical targeting device is Normally distributed with standard deviation 0.012 mm. What is the 90-percentile of the error resultant?

The distribution of the error resultant is Rayleigh with scale parameter $\sigma := 0.012$
The required 90-percentile resultant is therfore

$$E90 := \sigma \cdot \sqrt{2 \cdot \ln\left(\frac{1}{1 - 0.9}\right)} \qquad E90 = 0.026$$

EXAMPLE 17.2

Twenty specimens of a new engineering material were tested for rupture strength measured in ($N/cm^2 \times 10{,}000$):

62.3, 42.0, 47.3, 54.8, 48.5, 61.3, 43.1, 55.6, 51.3, 46.3,
61.8, 66.1, 53.5, 40.7, 46.0, 54.0, 57.0, 57.0, 45.1, 59.1.

Check graphically the Weibull postulate for the rupture strength of this material.

$n := 20 \qquad i := 1..n$

$y_i :=$ $\quad x := \text{sort}(y) \quad p_i := \dfrac{1 - 0.3}{n + 0.4} \quad$ data ordinates: $r_i := \ln(-\ln(1 - p_i))$

y_i
62.3
42
47.3
54.8
48.5
61.3
43.1
55.6
51.3
46.3
61.8
66.1
53.5
40.7
46
54
57
57
45.1
59.1

Except for the first two points, this data plot looks fairly linear. The Weibull postulate is therefore not unreasonable. However, the downward sloping trend on the lower left indicates that there may be a strength threshhold.

EXAMPLE 17.3

Bearings for a 6-ton turret mount were subjected to an accelerated life test. A number of bearings, marked *, were removed during the test for inspection purposes. The ordered data are:

211, 397, 461, 500*, 576, 640*, 812, 854, 943, 1000*,
1000*, 1211, 1482, 1487, 1500*, 1822, 2103, 2585.

A two-parameter Weibull distribution is postulated for the accelerated time-to-failure. Does this postulate appear resonable?

$n := 18 \qquad r := 13$
$k := 1..n \qquad i := 1..r$

$t_k := \quad x_i := \quad q_i :=$ (q_i is the order of failures in the total sample)

t_k	x_i	q_i
211	211	1
397	397	2
461	461	3

median plotting position:

$$p_i := 1 - \frac{n + 0.7}{n + 0.4} \cdot \prod_{j=1}^{i} \frac{n - q_j + 0.7}{n - q_j + 1.7}$$

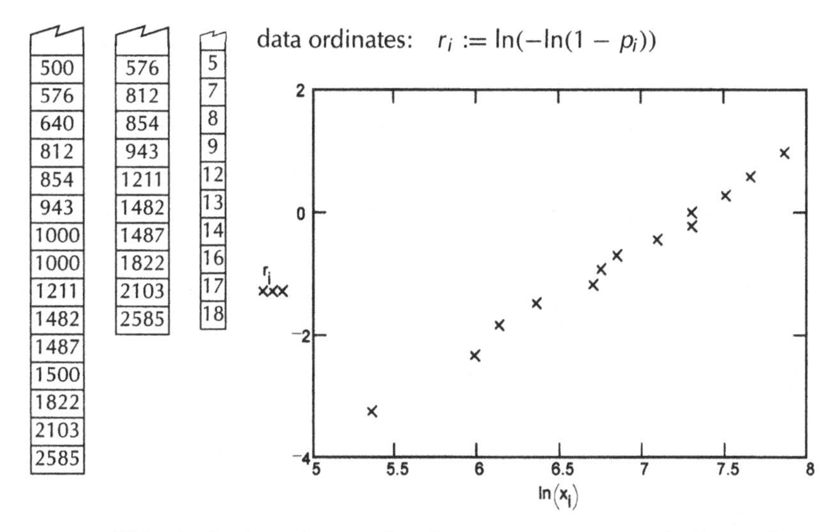

This plot looks quite straight. The two-parameter Weibull postulate is therefore not unreasonable.

EXAMPLE 17.4

Given the strength data of Example 17.2, estimate the Weibull parameters and their standard errors. Test the estimated model for fit at the 5% level of significance. Estimate the 5-percentile and its standard error. Generate a probability plot of the data and the estimated model.

$n := 20 \qquad i := 1\,..\,n$

$y_i :=$ $\qquad x := \mathrm{sort}(y)$

y_i
62.3
42
47.3
54.8
48.5
61.3
43.1
55.6
51.3
46.3
61.8
66.1
53.5
40.7
46
54
57
57
45.1
59.1

1. Estimation

$$s1 := \sum_i \ln(x_i) \qquad s2(\lambda) := \sum_i (x_i)^{\lambda} \cdot \ln(x_i) \qquad s3(\lambda) := \sum_i (x_i)^{\lambda}$$

$$L := 5 \qquad \lambda := \mathrm{root}\left(-\frac{1}{L} + \frac{s2(L)}{s3(L)} - \frac{s1}{n}, L\right) \qquad \lambda = 8.12$$

$$\sigma := \left(\frac{s3(\lambda)}{n}\right)^{\frac{1}{\lambda}} \qquad \sigma = 55.84$$

$$bn := 1 + \frac{2.2}{n^{1.13}} \qquad \lambda_u := \frac{\lambda}{bn} \qquad \lambda_u = 7.55$$

2. Standard errors

$$\mathrm{SE}\sigma := 1.05293 \cdot \frac{\sigma}{\lambda \cdot \sqrt{n}} \qquad \mathrm{SE}\sigma = 1.62$$

$$\mathrm{SE}\lambda := 0.7797 \cdot \frac{\lambda}{bn \cdot \sqrt{n}} \qquad \mathrm{SE}\lambda = 1.32$$

$$\mathrm{SE}\sigma\lambda := 0.50697 \cdot \sqrt{\frac{\sigma}{bn \cdot n}} \qquad \mathrm{SE}\sigma\lambda = 0.82$$

3. Test-of-fit

$$w_i := 1 - \exp\left[-\left(\frac{x_i}{\sigma}\right)^{\lambda}\right]$$

$$A := \left[-n - \frac{1}{n} \cdot \sum_i (2 \cdot i - 1) \cdot (\ln(w_i) + \ln(1 - w_{n-i+1}))\right] \cdot \left(1 + \frac{0.2}{\sqrt{n}}\right)$$

$$A = 0.31$$

The 5% critical value is 0.757. Because the sample value A is smaller, there is no reason to reject the estimated model at the 5% significance level.

4. 5-Percentile

$$c := \ln\left(\frac{1}{1 - 0.05}\right) \qquad X05(a, b) := a \cdot c^{\frac{1}{b}} \qquad X05(\sigma, \lambda_u) = 37.7$$

standard error: $da(a, b) := \frac{d}{da} X05(a, b) \qquad db(a, b) := \frac{d}{db} X05(a, b)$

$$\text{var}(a, b) := (da(a, b) \cdot \text{SE}\sigma)^2 + (db(a, b) \cdot \text{SE}\lambda)^2 + 2 \cdot da(a, b) \cdot db(a, b) \cdot \text{SE}\sigma\lambda^2$$

$$\text{se}X05 := \sqrt{\text{var}(\sigma, \lambda_u)} \qquad \text{se}X05 = 3.1$$

5. Probability plot of data and estimated model

$p_i := \dfrac{i - 0.3}{n + 0.4}$ $\quad r_i := \ln(-\ln(1 - p_i))$ (ordinates for the data)

$z_i := \lambda_u \cdot (\ln(x_i) - \ln(\sigma))$ (ordinates for the model when the cdf is substituted for p)

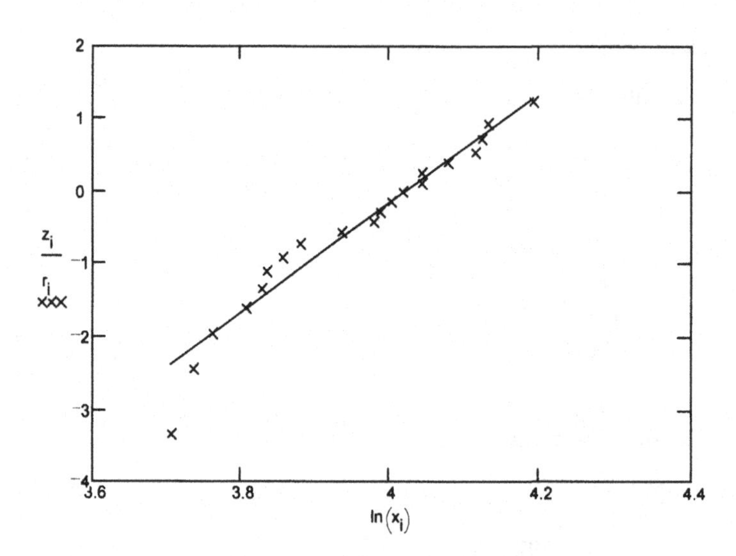

Except, for the first two points, the model fits the data reasonably well. The curved lower end of the data plot suggests that a threshold parameter may be present, so that the 5-percentile estimate is likely to be low. See Example 17.8 for a continuation of the data analysis.

6. pdf plot

$j := 1..200 \qquad t_j := 10 + 04 \cdot j \qquad f(x) := \frac{\lambda_u}{\sigma^{\lambda_u}} \cdot x^{\lambda_u - 1} \cdot \exp\left[-\left(\frac{x}{\sigma}\right)^{\lambda_u}\right]$

EXAMPLE 17.5

The average time-to-failure (TTF) of a newly designed electro mechanical actuator is judged to be 1,200 hours, based on similar designs in service. Assuming a Weibull distribution for TTF and a coefficient of variation of 0.4, estimate the 5-percentile TTF.

$X\text{bar} := 1200 \qquad \text{cv} := 0.4$

$$\text{CV}(a) := \sqrt{\frac{\Gamma\left(1 + \frac{2}{a}\right)}{\Gamma\left(1 + \frac{1}{a}\right)^2} - 1} \qquad a := 3 \qquad \lambda := \text{root}(\text{CV}(a) - \text{cv}, a) \qquad \lambda = 2.7$$

or, from (17.51):

$\lambda_a := \text{cv}^{-1.0852} \qquad \lambda_a = 2.7$

$$\sigma := \frac{X\text{bar}}{\Gamma\left(1 + \frac{1}{\lambda}\right)} \qquad \sigma = 1349$$

$$\text{5-percentile:} \quad X05 := \sigma \cdot \ln\left(\frac{1}{1 - 0.05}\right)^{\frac{1}{\lambda}} \qquad X05 = 448$$

EXAMPLE 17.6

Given the data on bearing TTF from the Example 17.3, estimate the Weibull parameters and their standard errors. Produce a probability plot of the data and the estimated model.

$n := 18 \qquad r := 13$

$k := 1..n \qquad i := 1..r$

$t_k :=$ $\quad x_i :=$ $\quad q_i :=$ (q_i is the order of failures in the total sample)

t_k	x_i	q_i
211	211	1
397	397	2
461	461	3
500	576	5
576	812	7
640	854	8
812	943	9
854	1211	12
943	1482	13
1000	1487	14
1000	1822	16
1211	2103	17
1482	2585	18
1487		
1500		
1822		
2103		
2585		

1. Estimation

$s1 := \sum_i \ln(x_i) \qquad s2(\lambda) := \sum_k (t_k)^\lambda \cdot \ln(t_k)$

$s3(\lambda) := \sum_k (t_k)^\lambda$

$L := 2 \quad \lambda := \text{root}\left(-\frac{1}{L} + \frac{s2(L)}{s3(L)} - \frac{s1}{r}, L\right) \qquad \lambda = 1.90$

$\sigma := \left(\frac{s3(\lambda)}{r}\right)^{\frac{1}{\lambda}} \qquad \sigma = 1468$

$br := 1 + \frac{2.2}{r^{1.13}} \qquad \lambda_u := \frac{\lambda}{br} \qquad \lambda_u = 1.69$

2. Standard errors

$$I := \begin{bmatrix} r \cdot \left(\frac{\lambda}{\sigma}\right)^2 & \frac{\lambda}{\sigma} \cdot \left(r \cdot \ln(\sigma) - \frac{r}{\lambda} - s1\right) \\ \frac{\lambda}{\sigma}\left(r \cdot \ln(\sigma) - \frac{r}{\lambda} - s1\right) & \frac{r}{\lambda^2} + \sum_k \left(\frac{t_k}{\sigma}\right)^\lambda \cdot \ln\left(\frac{t_k}{\sigma}\right)^2 \end{bmatrix} \qquad V := I^{-1}$$

$SE\sigma := \sqrt{V_{0,0}} \qquad SE\sigma = 216$

$\sqrt{V_{1,1}} = 0.405$ bias-correction: $SE\lambda := \frac{1}{br} \cdot \sqrt{V_{1,1}} \qquad SE\lambda = 0.36$

$\sqrt{V_{0,1}} = 2.97 \qquad SE\sigma\lambda := \frac{1}{\sqrt{br}} \cdot \sqrt{V_{0,1}} \qquad SE\sigma\lambda = 2.80$

3. Probability plot of data and estimated model

median plotting position: $p_i := 1 - \frac{n + 0.7}{n + 0.4} \cdot \prod_{j=1}^{i} \frac{n - q_j + 0.7}{n - q_j + 1.7}$

$r_i := \ln(-\ln(1 - p_i))$ (ordinates for the data)

$z_i := \lambda_u \cdot (\ln(x_i) - \ln(\sigma))$ (ordinates for the estimated model when the cdf subsitituted for p)

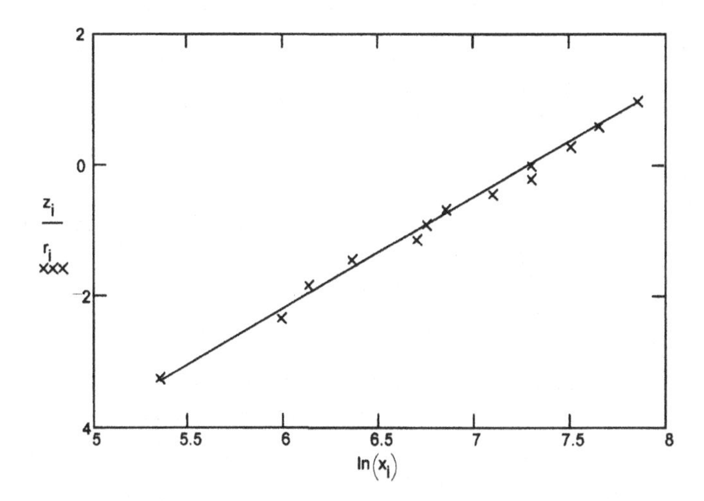

The estimated model fits the data quite well.

4. pdf plot:

$j := 0..200 \qquad t_j := 25 \cdot j \qquad f(x) := \frac{\lambda_u}{\sigma^{\lambda_u}} \cdot x^{\lambda_u - 1} \cdot \exp\left[-\left(\frac{x}{\sigma}\right)^{\lambda_u}\right]$

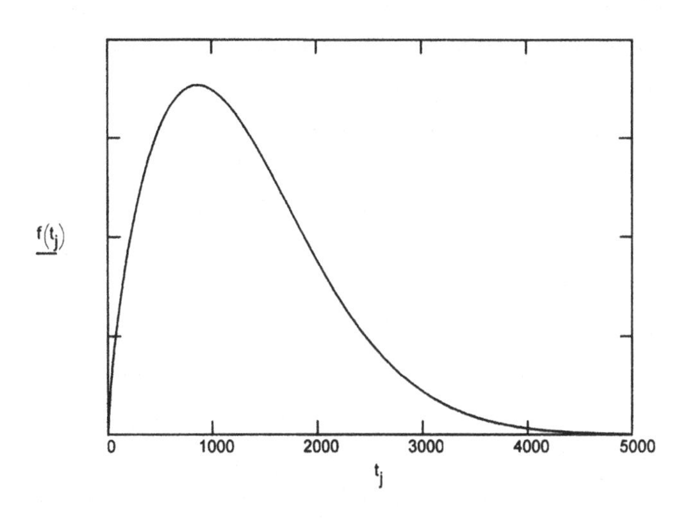

EXAMPLE 17.7

For the data of Example 17.2, calculate 90% confidence intervals on the Weibull parameters and on the 5-percentile.

From Example 17.4:

$n := 20 \qquad \sigma := 55.84 \qquad \lambda u := 7.55 \qquad \lambda := 8.12 \qquad X05 := 37.7$

$i := 1..n \qquad SE\sigma := 1.62 \qquad SE\lambda := 1.32 \qquad seX05 := 3.1$

1. Parameters

i) Normal approximation: Standard Normal 95-percentile is $z95 := 1.645$

$$L\lambda := \lambda u - z95 \cdot SE\lambda \qquad L\lambda = 5.4$$

$$U\lambda := \lambda u + z95 \cdot SE\lambda \qquad U\lambda = 9.7$$

$$L\sigma := \sigma - z95 \cdot SE\sigma \qquad L\sigma = 53.2$$

$$U\sigma := \sigma + z95 \cdot SE\sigma \qquad U\sigma = 58.5$$

ii) Likelihood ratio method: Chi-squared 90-percentile at $v = 1$ is $K90 := 2.71$

$x_i :=$

62.3
42
47.3
54.8
48.5
61.3
43.1
55.6
51.3
46.3
61.8
66.1
53.5
40.7
46
54
57
57
45.1
59.1

log of likelihood function:

$$LL(\sigma, \lambda) := n \cdot \ln(\lambda) - n \cdot \lambda \cdot \ln(\sigma) + (\lambda - 1) \cdot \sum_i \ln(x_i) - \sigma^{-\lambda} \cdot \sum_i (x_i)^{\lambda}$$

shape parameter:

$$s(a) := \left[\frac{1}{n} \cdot \sum_i (x_i)^a\right]^{\frac{1}{a}} \qquad LR(a) := 2 \cdot LL(\alpha, \lambda) - 2 \cdot LL(s(a), a)$$

$$a := \lambda - 1 \qquad \lambda l := \text{root}(LR(a) - K90, a) \qquad \lambda l = 6.0$$

$$a := \lambda + 1 \qquad \lambda u := \text{root}(LR(a) - K90, a) \qquad \lambda u = 10.6$$

scale parameter:

$$a := \lambda$$

$$l(b) := \text{root}\left[\frac{n}{a} - n \cdot \ln(b) + \sum_i \ln(x_i) - \sum_i \left(\frac{x_i}{b}\right)^a \cdot \ln\left(\frac{x_i}{b}\right), a\right]$$

$$LR(b) := 2 \cdot LL(\sigma, \lambda) - 2 \cdot LL(b, l(b))$$

$$b := \sigma - 1 \qquad \sigma l := \text{root}(LR(b) - K90, b) \qquad \sigma l = 53.1$$

$$b := \sigma + 1 \qquad \sigma u := \text{root}(LR(b) - K90, b) \qquad \sigma u = 58.6$$

iii) Simulated results:

From Table 17.1 at $n = 20$ and $\gamma = 0.05$ $\quad L_{11} := 0.791$

and $\gamma = 0.95$ $\quad L_{12} := 1.449$

From Table 17.2 at $n = 20$ and $\gamma = 0.05$ $\quad L_{21} := 0.428$

and $\gamma = 0.95$ $\quad L_{22} := 0.421$

(17.66): $L\lambda := \frac{\lambda}{L_{12}}$ $\qquad L\lambda = 5.6$

$U\lambda := \frac{\lambda}{L_{11}}$ $\qquad U\lambda = 10.3$

(17.67): $L\sigma := \sigma \cdot \exp\left(-\frac{L_{22}}{\lambda}\right)$ $\qquad L\sigma = 53.0$

$U\sigma := \sigma \cdot \exp\left(-\frac{L_{21}}{\lambda}\right)$ $\qquad U\sigma = 58.9$

2. Percentile

$q := 0.05$

i) Normal approximation:

$Xl := X05 - z95 \cdot seX05$ $\qquad Xl = 32.6$

$Xu := X05 + z95 \cdot seX05$ $\qquad Xu = 42.8$

ii) Likelihood ratio method:

Denote the quantile by Q:

$$ML(a, Q) := \frac{n}{a} - n \cdot \ln(Q) + \sum_i \ln(x_i) + \ln(1-q) \cdot \sum_i \left(\frac{x_i}{Q}\right)^a \cdot \ln\left(\frac{x_i}{Q}\right)$$

$a := \lambda \qquad l1(Q) := \text{root}(ML(a, Q), a)$

$s(Q) := Q \cdot (-\ln(1-q))^{-\frac{1}{l1(Q)}}$

$LR(Q) := 2 \cdot LL(\sigma, \lambda) - 2 \cdot LL(s(Q), l1(Q))$

$Q := X05 - 5 \qquad Ql := \text{root}(LR(Q) - K90, Q)$ $\qquad Ql = 33.1$

$Q := X05 + 5 \qquad Qu := \text{root}(LR(Q) - K90, Q)$ $\qquad Qu = 43.2$

EXAMPLE 17.8

Assuming a three-parameter Weibull model for the strength data of Example 17.2, estimate the parameters, the 5-percentile, and their standard errors.

1. Exploration of the log-likelihood function

$n := 20 \qquad i := 1..n \qquad k := 1..10 \qquad p := 1..10$

$m_k := 37.5 + 0.1 \cdot k \qquad l_p := 1.8 + p \cdot 0.05$

$y_i :=$

62.3
42
47.3
54.8
48.5
61.3
43.1
55.6
51.3
46.3
61.8
66.1
53.5
40.7
46
54
57
57
45.1
59.1

$x := \mathrm{sort}(y)$

$$s1_{k,p} := \sum_i (x_i - m_k)^{l_p} \qquad s2_{k,p} := \sum_i \ln(x_i - m_k)$$

$$L_{k,p} := n \cdot \ln(l_p) - n \cdot \ln(s1_{k,p}) + (l_p - 1) \cdot s2_{k,p} + 107.5$$

$$l^T = (1.850 \quad 1.900 \quad 1.950 \quad 2.000 \quad 2.050 \quad 2.100 \quad 2.150 \quad 2.200 \quad 2.250 \quad 2.300)$$

$$L = \begin{bmatrix} -0.38 & -0.266 & -0.173 & -0.098 & -0.041 & -0.002 & 0.021 & 0.028 & 0.02 & -0.002 \\ -0.339 & -0.232 & -0.144 & -0.075 & -0.024 & 0.009 & 0.027 & 0.028 & 0.014 & -0.015 \\ -0.3 & -0.199 & -0.117 & -0.054 & -0.009 & 0.018 & 0.029 & 0.024 & 0.004 & -0.03 \\ -0.263 & -0.168 & -0.092 & -0.036 & 0.003 & 0.024 & 0.029 & 0.018 & -0.008 & -0.049 \\ -0.228 & -0.139 & -0.07 & -0.019 & 0.013 & 0.028 & 0.026 & 0.009 & -0.024 & -0.071 \\ -0.195 & -0.112 & -0.049 & -0.006 & 0.02 & 0.029 & 0.021 & -0.003 & -0.043 & -0.097 \\ -0.164 & -0.088 & -0.032 & 0.005 & 0.025 & 0.026 & 0.012 & -0.019 & -0.065 & -0.126 \\ -0.135 & -0.066 & -0.017 & 0.014 & 0.026 & 0.021 & -0.001 & -0.039 & -0.092 & -0.159 \\ -0.109 & -0.047 & -0.005 & 0.019 & 0.024 & 0.012 & -0.017 & -0.062 & -0.122 & -0.197 \\ -0.085 & -0.031 & 0.004 & 0.02 & 0.018 & -0.001 & -0.037 & -0.089 & -0.157 & -0.239 \end{bmatrix} \qquad m = \begin{bmatrix} 37.6 \\ 37.7 \\ 37.8 \\ 37.9 \\ 38 \\ 38.1 \\ 38.2 \\ 38.3 \\ 38.4 \\ 38.5 \end{bmatrix}$$

$$M_{p,11-k} := L_{k,p}$$

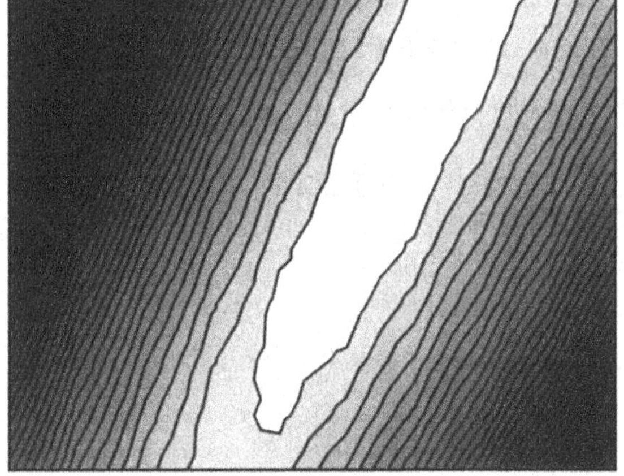

M

Choose starting values: $m := 38.1 \qquad L := 2.1$

2. Maximum likelihood estimates

$$S1(m, L) := \sum_i (x_i - m)^L \qquad S2(m) := \sum_i \ln(x_i - m)$$

$$S3(m) := \sum_i (x_i - m)^{-1}$$

$$S4(m, L) := \sum_i (x_i - m)^{L-1} \qquad S5(m, L) := \sum_i (x_i - m)^L \cdot \ln(x_i - m)$$

GIVEN

$$\frac{L}{L-1} - \frac{S1(m, L) \cdot S3(m)}{n \cdot S4(m, L)} = 0$$

$$-\frac{1}{L}+\frac{S5(m,L)}{S1(m,L)}-\frac{1}{n}\cdot S2(m)=0$$

$$\begin{pmatrix}\mu\\ \lambda\end{pmatrix} := \mathrm{FIND}(m,L) \qquad \lambda = 2.145 \qquad \mu = 37.87$$

$$\sigma := \left(\frac{1}{n}\cdot S1(\mu,\lambda)\right)^{\frac{1}{\lambda}} \qquad \sigma = 16.68$$

$$bn := 1+\frac{2.2}{n^{1.13}} \qquad bn = 1.075 \qquad \lambda_u := \frac{\lambda}{bn} \qquad \lambda_u = 1.996$$

3. Standard errors

$$\mathrm{psi}(a) := \frac{d}{da}\ln(\Gamma(a)) \qquad \text{(digamma function)}$$

expected information matrix:

$$I_{1,1} := \frac{n}{\sigma^2}\cdot(\lambda-1)^2\cdot\Gamma\left(1-\frac{2}{\lambda}\right) \qquad I_{1,2} := \frac{n}{\sigma^2}\cdot\lambda\cdot(\lambda-1)\cdot\Gamma\left(1-\frac{1}{\lambda}\right) \qquad I_{2,1} := I_{1,2}$$

$$I_{1,3} := -\frac{n}{\sigma\cdot\lambda}\cdot(\lambda-1)\cdot\Gamma\left(1-\frac{1}{\lambda}\right)\cdot\left(1+\mathrm{psi}\left(1-\frac{1}{\lambda}\right)\right) \qquad I_{3,1} := I_{1,3} \qquad I_{2,2} := \frac{n}{\sigma^2}\cdot\lambda^2$$

$$I_{2,3} := -\frac{n}{\sigma}\cdot 0.42278 \qquad I_{3,2} := I_{2,3} \qquad I_{3,3} := \frac{n}{\lambda^2}\cdot 1.82368$$

covariance matrix: $V := I^{-1}$

bias correction:

$$B := \begin{bmatrix} V_{1,1} & V_{1,2} & \frac{V_{1,3}}{bn} \\ V_{2,1} & V_{2,2} & \frac{V_{2,3}}{bn} \\ \frac{V_{3,1}}{bn} & \frac{V_{3,2}}{bn} & \frac{V_{3,3}}{bn^2} \end{bmatrix} \qquad \begin{matrix} \mathrm{SE}\mu := \sqrt{B_{1,1}} & \mathrm{SE}\mu = 1.09 \\ \mathrm{SE}\sigma := \sqrt{B_{2,2}} & \mathrm{SE}\sigma = 2.23 \\ \mathrm{SE}\lambda := \sqrt{B_{3,3}} & \mathrm{SE}\lambda = 0.39 \end{matrix}$$

4. 5-Percentile

$$C := \ln\left(\frac{1}{1-0.05}\right) \qquad X05(\mu,\sigma,\lambda) := \mu+\sigma\cdot C^{\frac{1}{\lambda}} \qquad X05(\mu,\sigma,\lambda_u) = 41.6$$

standard error:

$$j := 1..3 \qquad k := 1..3$$

$$d_1 := \frac{d}{d_\mu}X05(\mu,\sigma,\lambda_u) \qquad d_2 := \frac{d}{d_\sigma}X05(\mu,\sigma,\lambda_u) \qquad d_3 := \frac{d}{d\lambda_\mu}X05(\mu,\sigma,\lambda_u)$$

$$\mathrm{var} := \sum_j\sum_k d_j\cdot d_k\cdot V_{j,k} \qquad \mathrm{se}X05 := \sqrt{\mathrm{var}} \qquad \mathrm{se}X05 = 1.3$$

5. Graphical check

$p_i := \dfrac{i - 0.3}{n + 0.4}$ $\quad r_i := \ln(-\ln(1 - p_i))$ (ordinates for the data)

$z_i := \lambda_u \cdot (\ln(x_i - \mu) - \ln(\sigma))$ (ordinates for the model when the cdf is substituted for p)

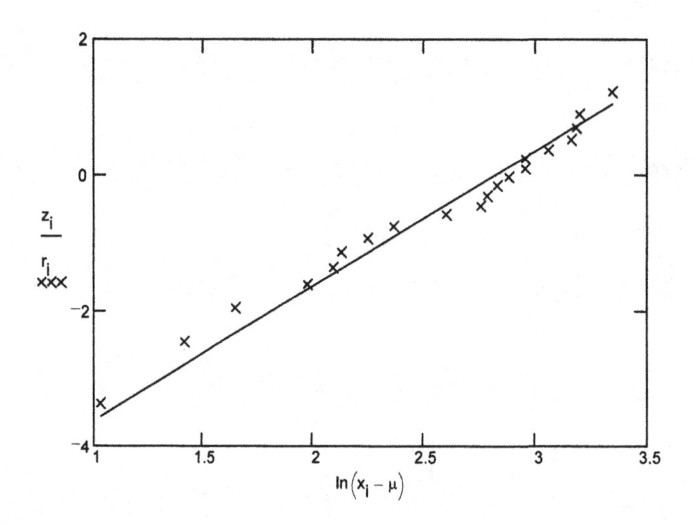

This fit appears superior to the two-parameter fit (Example 17.4). The presence of a threshold value is therefore supported.

6. pdf plot

$j := 1..200 \qquad t_j := 37.5 + 0.4 \cdot j$

$$f(x) := \frac{\lambda_u}{\sigma^{\lambda_u}} \cdot (x - \mu)^{\lambda_u - 1} \cdot \exp\left[-\left(\frac{x-\mu}{\sigma}\right)^{\lambda_u}\right]$$

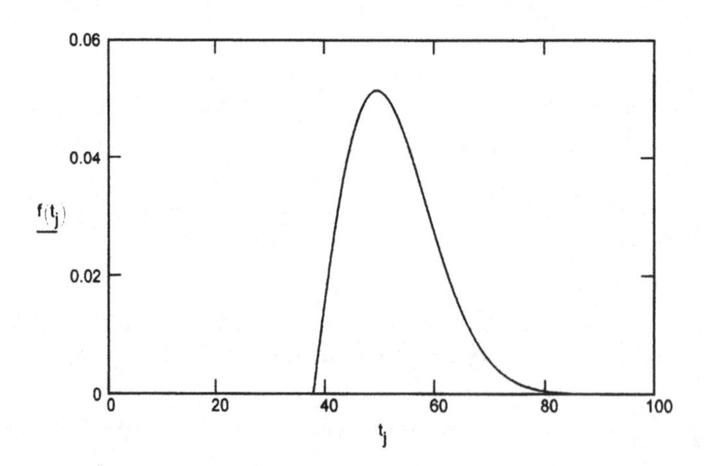

EXAMPLE 17.9

Calculate 90% confidence intervals for the parameters of the Weibull model estimated in Example 17.8 and for the 5-percentile.

From Example 17.8:

$\mu := 37.87$ $\sigma := 16.68$ $\lambda := 1.996$ $X05 := 41.6$

$SE\mu := 1.09$ $SE\sigma := 2.23$ $SE\lambda := 0.39$ $seX05 := 1.3$

Normal approximation

Standard Normal 95-percentile is $z95 := 1.645$.

parameters:

$L\lambda := \lambda - z95 \cdot SE\lambda$ $L\lambda = 1.35$

$U\lambda := \lambda + z95 \cdot SE\lambda$ $U\lambda = 2.64$

$L\sigma := \sigma - z95 \cdot SE\sigma$ $L\sigma = 13.0$

$U\sigma := \sigma + z95 \cdot SE\sigma$ $U\sigma = 20.3$

$L\mu := \mu - z95 \cdot SE\mu$ $L\mu = 36.1$

$U\mu := \mu + z95 \cdot SE\mu$ $U\mu = 39.7$

5-percentile:

$Lx := X05 - z95 \cdot seX05$ $Lx = 39.5$

$Ux := X05 - z95 \cdot seX05$ $Ux = 43.7$

Conditional Inference

Given $\mu = 37.87$:

Transforming the data of Example 17.2 to $(xi - 37.87)$,
and applying the methods of Example 17.4, gives:

$\lambda := 2.00$ $\sigma := 16.68$ $x05 := 3.77$

Thus $X05 := x05 + \mu$ $X05 = 41.6$

Applying the method of Example 17.7 to the transformed data gives the following confidence intervals on the 5-percentile:

i) Normal approximation: $Lx := 39.7$

$Ux := 43.6$

ii) Likelihood ratio method: $Lx := 40.1$

$Ux := 44.3$

EXAMPLE 17.10

A sample of 22 Rayleigh observations on the rms value of noise in a power circuit gave the statistic $\Sigma x^2 = 4491$. Estimate the Rayleigh scale parameter, its standard error, and its 90% confidence interval.

$n := 22 \qquad sx2 := 4491$

1. ML estimate

$$\sigma := \sqrt{\frac{sx2}{2 \cdot n}} \qquad \sigma = 10.10$$

2. Standard error

$$SE\sigma := \frac{\sigma}{2 \cdot \sqrt{n}} \qquad SE\sigma = 1.08$$

3. 90% Confidence interval

For $2n = 44$ degrees of freedom, the χ^2 percentiles are

$K_l := 29.79 \qquad K_u := 60.48$

The lower limit is

$$L_l := \sqrt{\frac{sx^2}{K_u}} \qquad L_l = 8.62$$

The upper limit is

$$L_u := \sqrt{\frac{sx^2}{K_l}} \qquad L_u = 12.28$$

EXAMPLE 17.11

For the bearing data of Example 17.6, estimate the reliability, and its 95% lower confidence limit, at 200 hours of accelerated test time.
From Example 17.6:

$\sigma := 1468 \qquad \lambda := 1.9$ (biased)

$SE\sigma := 216 \qquad SE\lambda := 0.405 \qquad SE\sigma\lambda := 2.97$

Reliability value

$$R := \exp\left[-\left(\frac{200}{\sigma}\right)^{\lambda}\right] \qquad R = 0.978$$

Confidence limit

i) Normal approximation:
Standard Normal 5-percentile is $z := -1.645$

$$\text{var}(R) := (R \cdot \ln(R))^2 \cdot \left[\left(\frac{\lambda}{\sigma}\right)^2 \cdot \text{SE}\sigma^2 + \left(\frac{\ln(-\ln(R))}{\lambda}\right)^2 \cdot \text{SE}\lambda^2 - 2 \cdot \frac{\ln(-\ln(R))}{\sigma} \cdot (\text{SE}\alpha\lambda)^2 \right]$$

$L := R + z \cdot \sqrt{\text{var}(R)}$ $\quad L = 0.946$

95% lower confidence limit: $L1 := R + z \cdot \sqrt{\text{var}(L)},$ $\quad L1 = 0.917$

ii) Likelihood ratio method:

$n := 18 \quad r := 13 \quad to := 200$

$k := 1..n \quad i := 1..r$

Chi-squared 90-percentile at $v = 1$ is $K := 2.71$

$t_k :=$	$x_i :=$
211	211
397	397
461	461
500	576
576	812
640	854
812	943
854	1211
943	1482
1000	1487
1000	1822
1211	2103
1482	2585
1487	
1500	
1822	
2103	
2585	

log of likelihood function:

$$LL(\sigma, \lambda) := r \cdot \ln(\lambda) - r \cdot \lambda \cdot \ln(\sigma) + (\lambda - 1) \cdot \sum_i \ln(x_i) - \sum_k \left(\frac{t_k}{\sigma}\right)^\lambda$$

$$ML(a, R) := \frac{r}{a} - r \cdot \ln(to) + \sum_i \ln(x_i) + \ln(R) \cdot \sum_k \left(\frac{t_k}{to}\right)^a \cdot \ln\left(\frac{t_k}{to}\right)$$

$a := \lambda \qquad \lambda 1(R) := \text{root}(ML(a, R), a)$

$$s(R) := to \cdot (-\ln(R))^{-\frac{1}{\lambda 1(R)}}$$

$$LR(R) := 2 \cdot LL(\alpha, \lambda) - 2 \cdot LL(s(R), \lambda 1(R))$$

$R := 0.9 \qquad Rl := \text{root}(LR(R) - K, R) \qquad Rl = 0.920$

EXAMPLE 17.12

Impeller rupture is the dominant failure mode in a certian type of waste pump. Twenty-eight units are installed in an industrial process, with approximate operating times (hours) as follows:

operating times:	1200	1300	1400	1500	1600	1700
number of units:	6	8	2	4	7	1

Failure data were used to estimate a Weibull life model with $\lambda = 2.6$, $\sigma = 3{,}240$, and variance matrix given below. Usage over the next 3 months is estimated as 500 hours for each unit. A pump with improved impeller design is now available for replacing failed units. Estimate the number of old pump failures expected in the next 3 months.

$i := 1..6 \qquad \lambda := 2.6 \qquad \sigma := 3240 \qquad V := \begin{pmatrix} 1.49 \cdot 10^5 & 12.1 \\ 12.1 & 0.31 \end{pmatrix}$

$t_i :=$ $\quad n_i :=$ $\quad$ (n_i is the number of units is group i)

t_i	n_i
1200	6
1300	8
1400	2
1500	4
1600	7
1700	1

1. Expected number of failures

$$N(\sigma, \lambda) := \sum_i n_i \cdot \left[1 - \exp\left[\left(\frac{t_i}{\sigma}\right)^\lambda - \left(\frac{t_i + 500}{\sigma}\right)^\lambda\right]\right]$$

$$N(\sigma, \lambda) = 3.6$$

2. Standard error

$j := 1..2 \qquad k := 1..2$

$d_1 := \frac{d}{d\sigma} N(\sigma, \lambda) \qquad d_2 := \frac{d}{d\lambda} N(\sigma, \lambda)$

$\text{var} = \sum_j \sum_k d_j \cdot d_k \cdot V_{j,k} \qquad \text{se}N := \sqrt{\text{var}} \qquad \text{se}N = 1.2$

EXAMPLE 17.13

Continuing with Example 17.12, assume the cost of replacing a failed pump during process operation is \$1,500, while a planned replacement during process shutdown costs \$300. What is the optimum replacement time, the cost of this age-replacement policy, and the cost of a failure-replacement policy? Graph the cost as a function of replacement time.

Given information:

$\sigma := 3240 \qquad \text{SE}\sigma := \sqrt{1.49 \cdot 10^5} \qquad Cf = 1500$

$\lambda := 2.6 \qquad \text{SE}\lambda := \sqrt{0.31} \qquad \text{SE}\sigma\lambda := \sqrt{12.1} \qquad Cr = 300$

1. Optimum replacement time

$$G(t, \sigma, \lambda) := \frac{\lambda}{\sigma} \cdot \left(\frac{t}{\sigma}\right)^{\lambda - 1} \cdot \int_0^t \exp\left[-\left(\frac{x}{\sigma}\right)^\lambda\right] dx + \exp\left[-\left(\frac{t}{\sigma}\right)^\lambda\right] - \frac{Cf}{Cf - Cr}$$

$t := 1000 \qquad T(\sigma, \lambda) := \text{root}(G(t, \sigma, \lambda), t) \qquad T(\sigma, \lambda) = 1600$

standard error:

$d\sigma := \frac{d}{d\sigma} T(\sigma, \lambda) \qquad d\lambda := \frac{d}{d\lambda} T(\sigma, \lambda)$

$\text{var} := (d\sigma \cdot \text{SE}\sigma)^2 + (d\lambda \cdot \text{SE}\lambda)^2 + 2 \cdot d\sigma \cdot d\lambda \cdot \text{SE}\sigma\lambda^2 \qquad \text{se}T := \sqrt{\text{var}} \qquad \text{se}T = 193$

2. Age-replacement cost ($ cost/hour/pump)

$$ca := \frac{Cf - (Cf - Cr) \cdot \exp\left[-\left(\frac{T(\sigma,\lambda)}{\alpha}\right)^{\lambda}\right]}{\int_0^{T(\alpha,\lambda)} \exp\left[-\left(\frac{x}{\sigma}\right)^{\lambda}\right] dx} \qquad Ca = 0.311$$

3. Failure-replacement cost ($ cost/hour/pump)

$$C := \frac{Cf}{\sigma \cdot \Gamma\left(1 + \frac{1}{\lambda}\right)} \qquad C = 0.521$$

$ savings/hour/pump: $S := C - Ca$ $\qquad S = 0.21$

4. Cost function

$$i := 1..200 \qquad t_i := 100 + 25 \cdot i \qquad \text{cost}_i := \frac{Cf - (Cf - Cr) \cdot \exp\left[-\left(\frac{t_i}{\sigma}\right)^{\lambda}\right]}{\int_0^{t_i} \exp\left[-\left(\frac{s}{\sigma}\right)^{\lambda}\right] ds}$$

cost/item/unit time

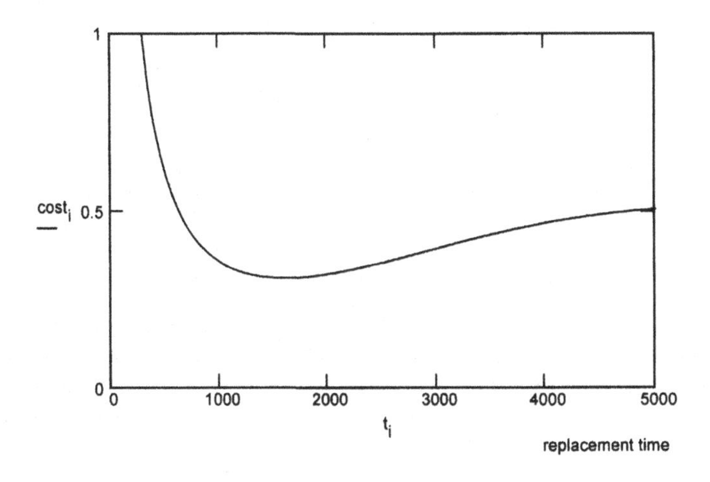

EXAMPLE 17.14

Low-cycle fatigue failures of a support bracket for a hoist drive were analyzed to give a Weibull fit to fatigue life in stress cycles under normal operating conditions. The estimated parameters are $\sigma = 3{,}200$ and $\lambda = 4.7$. The shape parameter is expected to remain unchanged through redesign. A zero-failure test plan is required to demonstrate, with 95% confidence, that the redesigned bracket has improved. Sixteen redesigned brackets are available for testing. What is the probability β of the test rejecting a redesign that has a 99% reliability at 2,000 cycles?

Test plan

$\lambda := 4.7 \qquad \sigma_o := 3200 \qquad n := 16 \qquad \alpha := 0.05$

(17.108): $T := \sigma_o \cdot \left(-\frac{1}{n} \cdot \ln(\alpha)\right)^{\frac{1}{\lambda}}$ $\qquad T = 2240$

Decision rule: If no units fail in T, the redesign has significantly improved the bracket fatigue life.

Producer's risk β

$R := 0.99 \qquad t := 2000$

required scale parameter (17.107): $\sigma_a := t \cdot \ln\left(\frac{1}{R}\right)^{-\frac{1}{\lambda}}$ $\qquad \sigma_a = 5322$

risk (17.110): $\beta := 1 - \exp\left[-n \cdot \left(\frac{T}{\sigma_a}\right)^{\lambda}\right]$ $\qquad \beta = 0.24$

Interpretation: The probability of rejecting a reliable redesign is 24%

EXAMPLE 17.15

Continuing with the preceding example, design a test plan with $\alpha = 0.05$, $\beta < 0.10$, and allowing for one or more failures during the test period T.

$\lambda := 4.7 \qquad \sigma_o := 3200 \qquad \alpha := 0.05 \qquad \sigma_a := 5322 \qquad n := 16$

$p_o(t) := 1 - \exp\left[-\left(\frac{t}{\sigma_o}\right)^{\lambda}\right] \qquad p_a(t) := 1 - \exp\left[-\left(\frac{t}{\sigma_a}\right)^{\lambda}\right]$

The simplest approach is to assume an increasing series of integer values for r_c and to solve equations (17.111) and (17.112) for the test duration T and the producer's risk β. The smallest value of r_c for which beta is just less than 0.10 is the solution.

Try: $r_c := 2 \qquad r := 0..(r_c - 1)$

(17.111) : $f(t) := \sum_r \frac{n!}{r! \cdot (n-r)!} \cdot p_o(t)^r \cdot (1 - p_o(t))^{n-r}$

$t := 2000 \qquad T := \text{root}(f(t) - \alpha, t)$ $\qquad T = 2488$

(17.112) : $\beta := 1 - \sum_r \left[\frac{n!}{r! \cdot (n-r)!} \cdot p_a(T)^r \cdot (1 - p_a(T))^{n-r}\right]$ $\qquad \beta = 0.071$

For $r_c = 1$ (ie. zero-failure test plan), we get $\beta = 0.24$. Hence, $r_c = 2$ is the solution, and the required test plan is a duration of 2,488 cycles on each of the 16 units. As soon as two units have failed, the test is stopped and it is concluded that fatigue life has not improved. If one or no units fail, it is concluded that the redesign was successful. The probability of rejecting a reliable redesign is less than required: 7%.

Index

Printed in the United States
By Bookmasters